Advanced Series in Agricultural Sciences 15

Elroy A. Curl
Bryan Truelove

The Rhizosphere

With 57 Figures

Springer-Verlag
Berlin Heidelberg New York Tokyo

Professor Dr. ELROY A. CURL
Professor Dr. BRYAN TRUELOVE
Alabama Agricultural Experiment Station
Department of Botany
Plant Pathology and Microbiology
Auburn University, Alabama 36849
USA

ISBN 3-540-15803-0 Springer-Verlag Berlin Heidelberg New York Tokyo
ISBN 0-387-15803-0 Springer-Verlag New York Heidelberg Berlin Tokyo

Library of Congress Cataloging in Publication Data. Curl, Elroy A. (Elroy Arvel), 1921–. The rhizosphere. (Advanced series in agricultural sciences ; 15) Bibliography: p. Includes index. 1. Rhizosphere. I. Truelove, Bryan. II. Title. III. Series. QK644.C87 1985 581.1'0428 85-17236

Typesetting, printing and bookbinding: Brühlsche Universitätsdruckerei, Giessen
2131/3130-543210

Preface

The Plant Root and the Rhizosphere was a major topical feature of the first International Symposium on Factors Determining the Behavior of Plant Pathogens in Soil held at the University of California, Berkeley in 1963. The symposium was edited by K. F. Baker and W. C. Snyder and published under the title *Ecology of Soil-Borne Plant Pathogens*. Since that time, several other international efforts, either on the root–soil interface specifically or on topics relating to the root environment, have provided a wealth of valuable information basic to promoting the culture of healthier, more productive plants.

For the writing of this book, inspiration has come, in large part, from 10 years of cooperative rhizosphere research in association with leading scientists participating in a regional effort within the southern United States. We have attempted to bring together in this work the major aspects of rhizosphere research and the principles of rhizosphere ecology for the benefit of developing young scientists and technologists, as well as for the established professional researcher and teacher. A prime objective and hope is that this volume might generate ideas that will bring forth new approaches and methodology leading to further advances in our understanding of rhizosphere interactions and their implications for agriculture.

Because of the enormous complexity of the chemical, physical, and microbiological environment of roots, the methods used by various workers are rarely standardized but must be devised or modified for each experiment. Consequently, conflicting results are often reported for apparently similar studies, thus making generalized statements of fact hazardous or in many cases impossible. This has necessitated very careful documentation, resulting in an extensive bibliography. However, a mere literature review is not intended, since much good work has not been cited. We have deliberately drawn upon some old research information on the rhizosphere and related areas, largely for the benefit of advanced students and young scientists, to show where rhizosphere research has come from and where it may be going. In doing this we believe we have revealed many of the gaps in our knowledge which are yet to be filled. For meaningful contributions to be made in the future the need for refined technology and a multidisciplinary pooling of expertise by soil microbiologists, phytopathologists, soil physicists and chemists, plant physiologists, and zoologists should be clearly evident.

Chapter 2, which deals with the structure and physiology of roots, may appear at first to dwell in unnecessary depth and detail on the structure and function of the aboveground plant parts as well as roots. However, this chapter is central to the primary purpose of rhizosphere study, i.e., to understand the relationship of rhizosphere ecology to the function and performance of the *whole* plant, not just the root system. Along with conventional plant breeding, the potential for genetically engineering specific modifications in shoot, root, and rhizosphere for desirable growth-enhancing characteristics would seem to emphasize further the essentiality of whole-plant involvement in rhizosphere investigations.

While we assume sole responsibility for any omissions or errors in the text, we are grateful to all those unselfish individuals who have contributed to the manuscript preparation over several years. We thank the many authors and journal editors or other officials who kindly granted permission for us to use data or to reproduce illustrations of figures, and we especially acknowledge the generous response of those who provided illustrative materials; proper credit for these contributions is given in the text. We are indebted to rhizosphere specialists at other institutions who gave of their time and expertise to examine specific chapters and offer constructive suggestions: Dr. G. J. Griffin, Virginia Polytechnic Institute and State University; Dr. L. F. Johnson, University of Tennessee; and Dr. N. C. Schenck, University of Florida. Further recognized are the valuable contributions of our own Auburn University colleagues, Dr. R. R. Dute, Dr. C. M. Peterson, Dr. J. D. Weete, and Dr. W. D. Kelley for examining various sections of the manuscript. For technical assistance in preparation of the manuscript, we express our sincere appreciation to Tammy Forbus, Susan Ledbetter, Barbara McFadyen, Robert Rush, Susan Scott, Jennifer Weete, and William Wiese.

Finally, we acknowledge the help of Dr. P. A. Lemke, Auburn University, who, through professional contacts in Germany, provided a most valuable service. In this regard, our special gratitude goes to Professor Dr. B. Hock, Lehrstuhl für Botanik der Techn. Universität München, and Professor Dr. Paul Tudzynski, Lehrstuhl für Allgemeine Botanik, Ruhr Universität, Federal Republic of Germany, for their successful efforts to locate a rare photograph of rhizosphere pioneer Professor Lorenz Hiltner; and we are equally indebted to Professor Dr. Karl Esser, Ruhr Universität, for providing the literal translation of Hiltner's lengthy address to the Deutsche Landwirtschaftsgesellschaft in 1904.

Auburn University, 1985

ELROY A. CURL
BRYAN TRUELOVE

Contents

Chapter 1 Introduction

1.1 Definitions and Historical

Speaking before an assembly of the Deutsche Landwirtschafts-Gesellschaft in 1904, Lorenz Hiltner, Soil Bacteriologist and Professor of Agronomy at the Technical College of Munich, emphasized the critical role of microbial activities in the "rhizosphere" in the nutrition and general health of plants. This initial use of the term was in reference to the zone of most intense bacterial activity around roots of the Leguminoseae. Hiltner stated, "The nutrition of plants in general certainly depends upon the composition of the soil flora in the rhizosphere." He further surmised, "If plants have the tendency to attract useful bacteria by their root ex-

Fig. 1.1. Lorenz Hiltner (1862–1923), Soil Bacteriologist and Professor of Agronomy at the Technical College of Munich, coined the term "rhizosphere" and emphasized the critical role of microbial activities in this root zone in relation to the nutrition and general health of plants. (Photograph from the archives of Bayerische Landesanstalt für Bodenkultur und Pflanzenbau, München)

cretions, it would not be surprising if they would also attract uninvited guests which, like the useful organisms, adapt to specific root excretions." Thus, Hiltner had already identified the two major influences of the rhizosphere on plants, and the topics which were to be most intensely researched during the next four decades. These topics are: (1) the relation of rhizosphere to plant nutrition, growth, and development, and (2) the influence of rhizosphere phenomena on pathogens and pathogenesis.

The term rhizosphere would seem to define itself, yet there is not total harmony among soil microbiologists and plant scientists regarding its precise meaning. Rhizo, or rhiza (from the Greek word for root), is straightforward enough, but sphere has many meanings, from a round body to one's social environment. Indeed, we are dealing with a unique environment inhabited by a "society" of microorganisms, a field of activity around a central point, which is another dictionary definition for sphere. The rhizosphere is that narrow zone of soil subject to the influence of living roots, as manifested by the leakage or exudation of substances that affect microbial activity. Early efforts also were made to define the boundary between the inner rhizosphere and root surface. Most researchers today accept the term rhizoplane suggested by F. E. Clark (1949) for the actual surface of plant roots together with any closely adhering particles of soil or debris. Though rhizosphere and rhizoplane are usually defined apart, it is difficult and often illogical to study the microbiology of one of these zones without including the other. The two zones together are sometimes referred to as the root–soil interface. In recent years, a distinction has been made between the rhizoplane and the mycorrhizosphere in plants that have ectomycorrhizae. In this case the sphere of influence is determined both by exudates from roots and the metabolites produced by the mycorrhizal fungi. Other regions of plant influence also have been recognized. The base of the plant at the soil–shoot interface, essentially representing the root–shoot transition zone, has been referred to as the laimosphere; then, continuing to the aerial plant parts, we encounter the phylloplane and phyllosphere (Preece and Dickinson 1971), which takes us out of the sphere of our principal topic.

The long series of events in agricultural research, which ultimately led to the "discovery" of the rhizosphere, probably began with the first compilation of Roman literature on agriculture by Petrus Crescentius of Bologna about the year 1240. E. W. Russell (1973) in his *Soil Conditions and Plant Growth* has discussed the landmark events that followed in an interesting short history of advances in agricultural research. These early events, however, were not directly concerned with the rhizosphere. Rather, they represent four major periods, beginning with the search for the "principle" of vegetation (1630–1750), which would establish the relationship between soil fertility and plant growth and dispel the convictions of Francis Bacon, Jan Baptista van Helmont, and others that water provided the essential ingredients for plant nourishment, while soil served only to hold the plant upright. Then followed what E. W. Russel has termed the phylogistic period of search for nutrients (1750–1800) and the modern period (1800–1860), which saw the foundation of plant physiology and the establishment of certain facts regarding nutritional requirements of plants; Justus von Liebig's killing of the humus theory and his assertion that plants obtain their carbon from carbonic acid in the air paved the way for these advances. The famous Rothamsted field experiments, started by Sir John B. Lawes and Dr. J. H. Gilbert in 1843 (see Hall 1917), and still continued with little change, have provided a wealth of data on soil fertility and plant growth.

The importance of nitrification and the utilization of nitrogenous compounds by plants became recognized and, with rapid advances in bacteriology, the mystery of nitrogen nutrition in leguminous plants was soon to be solved, largely through the experiments of Hellriegel and Wilfarth (1888); Beijerinck (1888) actually isolated the organism from legumes responsible for nitrogen fixation and named it *Bacillus radicicola,* now known as the genus *Rhizobium*. Recognition of the symbiotic association of a bacterium and plant roots attracted the attention of a number of soil scientists and bacteriologists, including Hiltner and his associates, who viewed the root-soil interface as a unique environment for other microorganisms as well.

The classical researches of R. L. Starkey of the New Jersey Agricultural Experiment Station between 1929 and 1940 demonstrated many of the principal effects of developing plants upon soil microorganisms. Clarification of the nature and role of root exudates in contributing to the rhizosphere effect is based substantially upon the investigations of A. D. Rovira and C. D. Bowen (1966) in Australia, though many others have contributed significantly. The investigations and philosophy of S. D. Garrett (1956) in England stimulated interest and activity in the ecology of soil-borne plant pathogens and subsequently the American Phytopathological Society in its *Plant Pathology, Problems and Progress 1908–1958* devoted a large section to soil microbiology and root-disease fungi, including a discussion of the rhizosphere relationship (Lochhead 1959). World attention was brought to bear on this subject at the International Symposium on Factors Determining the Behavior of Plant Pathogens in Soil held in California in 1963; these published proceedings (K. F. Baker and Snyder 1965) and many subsequent symposia (e.g., Schippers and Gams 1979; Harley and R. S. Russell 1979) and reviews have advanced our knowledge of the nature of the rhizosphere and firmly established its interdisciplinary significance.

Along with marked advances in rhizosphere research, improved techniques have been developed for the quantitative and qualitative assessment of microbial populations in soil, collection and analysis of root exudates, and study of microbial functions at the root surface. Some of the improved methodology in current use has emerged from international (Parkinson et al. 1971) and regional (L. F. Johnson and Curl 1972) efforts relating to the ecology of soil microorganisms and root diseases.

Numerous reviews of past investigations show clearly the direction that rhizosphere research has taken and where new emphasis has been applied (Katznelson et al. 1948; F. E. Clark 1949; Metz 1955; Starkey 1958; Schroth and Hildebrand 1964; D. A. Barber 1968; Bowen and Rovira 1969; Macura 1974; Rovira and Davey 1974; Rovira 1979; Bowen 1980). Investigations have dealt with many aspects of soil microbial ecology in both the rhizosphere and edaphosphere: growth, reproduction, and survival in soil; commensalism and symbiosis; antagonism, antibiotic production, and lysis; organic matter colonization, decomposition, and transformation; etc. Based on this foundation, concentration on the root environment continues to intensify with a new look toward practical and applied values in scientific agriculture.

1.2 The Root Environment

A study of the root environent cannot logically exclude consideration of the plant as a whole, i.e., root morphology, anatomy, physiology, and the root–shoot interrelationship (Scott 1965; Burstrom 1965; Aung 1974; Esau 1977). While focusing on the rhizosphere and root surface, one should maintain a constant awareness of the influences being exerted on the system by basic plant-growth phenomena. Most obvious amongst these is the relationship between the photosynthetic capacity of aerial plant parts and the carbohydrate content of the roots, along with the mutual interchange of materials between roots and foliage.

A well-established basis for the intensified microbial activity noted in the rhizosphere is the greater nutritional benefit derived by microorganisms from organic and inorganic components released from living roots, together with sloughed epidermal hairs and cortical cells. Other interrelated factors also are present at the root surface as oxygen tension decreases and carbon dioxide evolved by root cells and the microflora increases, along with bicarbonate formation and concomitant changes in pH of the soil solution.

Substances found in root exudates are sugars, amino acids, glycosides, organic acids, vitamins, enzymes, and a wide variety of miscellaneous compounds. The composition of the exudate varies with plant species, stage of growth, and the prevailing soil environment; consequently, the quantitative and qualitative nature of the rhizosphere flora also varies. The extent or bulk of the root system determines the total root surface area and, therefore, the quantity of exudates available to establish the rhizosphere effect. Grasses, with their enormous fibrous root systems and millions of root hairs, might be expected to exert a greater rhizosphere effect per unit area of soil than row crops such as cotton and soybean. In recent years some remarkable transmission and scanning electron microscope pictures have revealed the fine details of the root epidermis with its thin film of mucigel, adhering soil particles, and microflora components.

Nicholas (1965) made a conservative estimate of the numbers of microorganisms present in 1 g of fertile field soil: true bacteria 10^6–10^9, actinomycetes 10^5–10^6, fungi 10^4–10^5, and algae 10^1–10^3. These numbers are invariably higher for the rhizosphere-rhizoplane than for soil beyond the influence of living roots (the edaphosphere). Most estimates of the populations of the microflora have been based on soil-dilution and plate-count procedures which may reveal only a fraction of the true population as determined by the more time-consuming but more accurate direct counting procedures. Drawing upon some assumptions regarding the weight of soil per acre, and the numbers of organisms required to weigh 1 g, Nicholas (1965) estimated that bacteria might add as much as 700 lb dry weight of organic matter to the top 6 ins of an acre of arable land. Indeed, biomass is an important consideration when comparing the microbial content of soils. Population counts are only relative and vary considerably with the counting method used and with the change in physical and chemical properties of soils occasioned by climate and agricultural practices. The influence of individual plants is reflected in the rhizosphere as the R/S (rhizosphere to nonrhizosphere) ratio. For bacteria and fungi values commonly range from 5 to 20; actinomycetes, though somewhat less affected by the rhizosphere, may reveal R/S population ratios of from 2 to 12.

Nearly all of the animal phyla that are not exclusively marine forms are represented to some extent in the soil, where animal life is much more abundant and diverse than is generally recognized (Kevan 1965). We are concerned in the rhizosphere primarily with microscopic and submicroscopic forms inasmuch as they are the ones most likely to be affected by the immediate root environment. The term microfauna is used in a broad sense to include most members of the phyla Protozoa and Nematoda and some members of the Arthropoda, largely the mites (Arachnida, Acarina) and the springtails (Collembola). Though Charles Darwin, and perhaps others, made observations on earthworms as early as 1840, soil zoology did not emerge as a recognized area of research until Kühnelt (1950) summarized the known information on soil animals in his *Bodenbiologie,* and Franz (1950) in his *Bodenzoologie* emphasized the practical implications of studying the soil fauna. Kevan (1965) related these and other historical advances in a discussion of the nature and biology of the soil fauna. Though numerous books and research reports on soil zoology have been written, rhizosphere effects on soil animals (other than nematodes) remain largely uninvestigated. There is little doubt that the soil-dwelling microarthropods are important in the rhizosphere, but the actual root influence may be largely secondary, as members of the microfauna exploit the rhizosphere fungi and bacteria as food sources.

A rhizosphere effect even modifies the plant itself through microbial activities that provide plant-growth substances and increase the availability of nutrient elements at the root surface The rhizosphere effect upon plants may be a destructive one if the root environment should favor growth and reproduction (increase in inoculum density) of plant pathogens more than it favors the activities of competitive saprophytic microorganisms.

Apart from the rhizosphere, Raney (1965) and Chapman (1965) have discussed the physical and chemical features of soils affecting soil microorganisms. These same factors also apply in the rhizosphere, but they are modified under the influence of root growth and microbial activity. The physical movement of roots through soil initiates the first change as soil particles become appressed against the root surface, increasing bulk density and reducing pore size; this would be expected to affect the diffusion of exudates away from roots and nutrients toward roots. With intensive microbial activity, carbon dioxide accumulation and oxygen depletion are more likely to occur at the root surface than in a root-free soil of the same area.

1.3 Significance and Purpose of Rhizosphere Investigations

The rhizosphere is more than a zone of academic fascination. Plant growth and development are controlled largely by the soil environment in the root region, an environment which the plant itself helps to create and where microbial activity constitutes a major influential force. Understanding the relation of exudate-induced microbial activity in the rhizosphere to plant health and vigor is essential to the development of better crop production systems. To the plant physiologist the rhizosphere is important for processes relating to nutrient uptake, O_2 and CO_2 exchange, soil moisture gradients, and other factors affecting plant growth. Agronomists and soil scientists deal with the processes of mineralization, ammonification, nitrification, and symbiosis in the root zone. Plant pathologists have demonstrated that root exudates and microbial interactions at the root–soil interface can affect pathogen populations, propagule germination, and host susceptibility. Thus, a multidisciplinary approach is necessary for a complete understanding of rhizosphere influence on crop production.

In recent years, a number of international symposia conducted by soil scientists, microbiologists, and plant pathologists have featured the rhizosphere or some related aspects of the plant root and its environment. In the United States, a government-supported regional research project specifically dealing with rhizosphere ecology and plant health was initiated in 1972 with scientists from many states participating. These collective efforts have further advanced our understanding of root exudates and microbial interactions in the root ecosystem, and have shown a potential for practical application to agricultural systems.

Rhizosphere research can be divided into three broad, distinct but interrelated phases: (a) influence of roots on soil microorganisms, (b) influence of rhizosphere microorganisms on plant growth, and (c) rhizosphere influence on soil-borne pathogens and plant disease. The latter phase must necessarily consider interacting phenomena of the other two phases. The root influences the numbers, kinds, and activities of virtually all groups of the microflora that are common in soil near the root surface, as well as those in symbiotic association with roots. The symbionts are more root- or rhizoplane-dependent than other members of the rhizosphere flora, but they come under the influence of soil around roots before initiation of the symbiotic relationship.

The traditionally studied groups of bacteria involved in the nonsymbiotic fixation of nitrogen and the processes of ammonification, nitrification, denitrification, and the mineralization and transformation of various nutrient elements are subject to the influence of the rhizosphere. Following the discovery of the nitrogen-fixing *Azotobacter* group by Beijerinck (1901), and the insistence of Hiltner (1904) that they were stimulated by growing roots, volumes of largely German and Russian literature dealt with the subject of these and other bacteria in the rhizosphere. Contributions from scientists in other parts of the world followed and a wealth of information was amassed, yet we must generalize with caution regarding the rhizosphere's influence on these groups of bacteria. The *Azotobacter* group appears to be least affected by the rhizosphere, though seed inoculation with these organisms has been practiced for many years, particularly in the U.S.S.R (Allison 1947), with claims of significant benefit in terms of crop yield. Conflicting reports leave some question as to the influence of the rhizosphere on populations of nitrifying bacteria (*Nitrosomonas* and *Nitrobacter*), whereas R/S ratios of both the ammonifying and dentrifying bacteria are usually high. It would seem logical to assume a rhizosphere benefit to many bacteria, purely based on the nutrient value of root exudates. Thus, it is no surprise to find various nutritional groups present in high or low numbers depending on the chemical composition of root exudates of specific plants and to growth factors synthesized by associated bacteria. Among these organisms are those involved in the oxidation or reduction of inorganic compounds and the decomposition of organic compounds, thereby affecting the availability of phosphorus, potassium, sulfur, iron and other elements for plant utilization.

Significantly implicated in the competitive struggle for substrate and survival in the rhizosphere are soil-borne plant pathogens, particularly the fungi. The quantity and quality of root exudates, along with sloughed epidermal cells, directly or indirectly affect the growth and reproduction of these organisms, influencing inoculum density and disease potential. Resting spores and sclerotia lying in the grip of soil fungistasis, which naturally prevents germination, may be induced to germinate when root exudation provides the nutrients necessary to overcome the inhibitory factor. Even exudates of nonsusceptible weeds or other plants contribute to propagule germination, in which case new propagules are produced or the germlings are destroyed by lysis.

Motile agents of plant diseases exhibit a direct sensitivity to root influence, as seen by the chemotactic response by Phycomycete zoospores to the exudation sphere of roots, most prominently in the region of elongation. This phenomenon has been well established for species of *Phytophthora* (Zentmyer 1970). Organic components of root exudates, as well as some inorganic compounds, induce hatching of eggs in the cyst-forming nematodes (*Globodera* and *Heterodera* spp.); these and other plant parasitic nematodes are commonly observed to group around their feeding sites on roots.

The effects of living plant roots on microbial behavior reveal true significance only when we turn the topic around and view the effects of microorganisms on plants. In the broad view, plants are affected by microbially related phenomena primarily in three interrelated categories: (a) microbial activities that either provide or deplete nutritional materials needed for plant growth, (b) toxins and gases of microbial origin that suppress plant growth, and (c) microbial actions that affect soil-borne pathogens and host susceptibility to diseases. Among the organisms which are especially important to plant nutrition are those living in symbiotic alliance with roots, the rhizobia and the endo- and ectotrophic mycorrhizal fungi. Common bacteria of the rhizosphere-rhizoplane and those embedded in the root epidermal mucigel synthesize both vitamins and toxins available for uptake by roots; add to these ethylene, CO_2, and other gases of microbial origin capable of affecting plant growth. Microbial competition for nutrients at the root surface may reduce the levels of essential elements for root absorption.

The relation of all the foregoing factors to plant disease, or the potential for disease, is well established; indeed plant pathology textbooks have traditionally

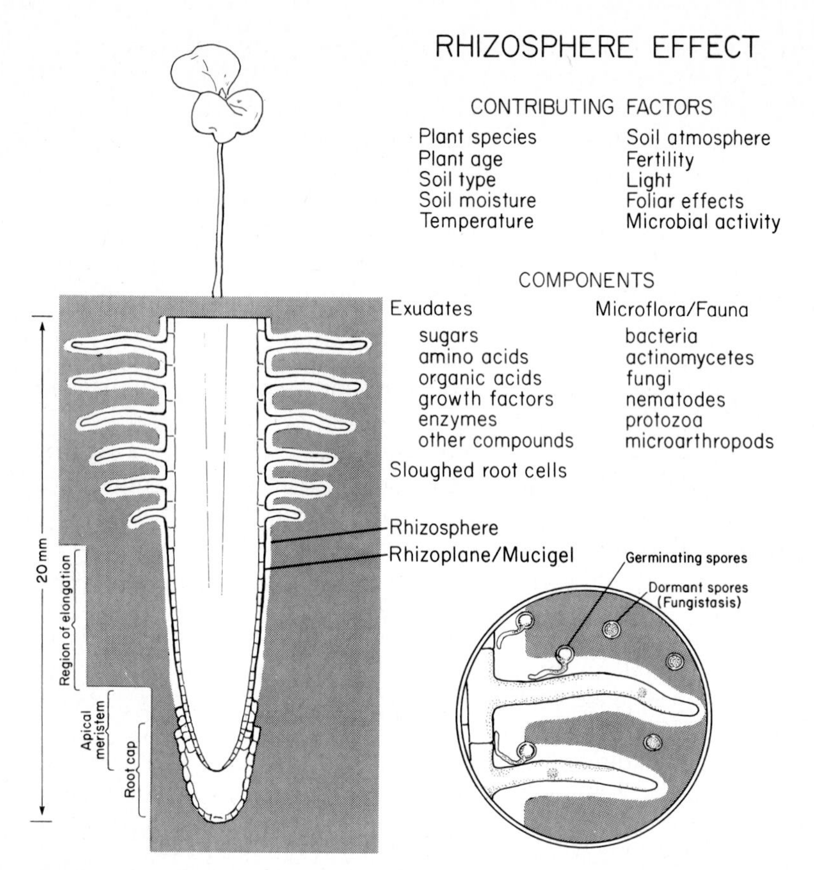

Fig. 1.2. Diagram of a young root featuring the rhizosphere and rhizoplane. Major organic materials released by the root, and groups of the microbiota affected are indicated along with factors governing the extent of root influence. *Inset* shows fungal spores germinating in the rhizosphere but not outside this nutrient zone

treated nutrient deficiency and mineral toxicity as nonparasitic diseases and as predisposing factors to pathogenesis. Wheeler (1975), in his excellent account of plant pathogenesis, made reference to the "environmental referees", any one of which may determine the outcome in the plant versus pathogen struggle. Along with climatic conditions and the physical and chemical properties of soils, the activities of rhizoplane and rhizosphere microorganisms profoundly influence inoculum availability and survival of pathogenic root-infecting fungi.

Thus, the effects of plants on soil microorganisms and the effects of microorganisms on plants are interacting, interdependent, and enormously complex. Plants provide in root exudates the raw materials that feed the microbial machinery of the rhizosphere, which possesses the power to promote good health and growth of a plant or to create the monster that destroys it.

In the ensuing chapters of this book the unique zone of root influence on soil microorganisms, as diagramed in Fig. 1.2, will be discussed in relation to its significance in plant growth and development, root diseases, and crop production.

Chapter 2 The Structure and Function of Roots

2.1 Introduction

In the latter part of the 17th century, Robert Hooke was one of the most influential scientists in the world. This versatile genius, who was adept in all areas of science known at that time, was Curator of Experiments, and also for a time served as Secretary, to the Royal Society in London. Among his many duties he was required, on occasion, to comment on the scientific papers presented by members at the weekly meetings of the Society. One such paper described studies of the translocation of substances within trees. It contained some interesting experiments and observations, but it was not a particularly distinguished investigation. Hooke's comments following the presentation, however, are, when seen with the perspective of time, truly remarkable. He said:

> "To me it seems very probable, that the bodies of plants, as well as those of moving animals, are nourished and increased by a double food, the one an impregnated water, and the other an impregnated air, and that without a convenient supply of these two, the vegetable cannot subsist, at least not increase; these do mutually mix, and coalesce, and parts of the air convert to water, and parts of water to air, as some of these latter are rarify'd and freed from their chains, and become spiritual and airy, so others of the forementioned are clogged, and fettered and become debased. To this purpose all plants, as well as animals, have a two-fold kind of root, one that branches, and spreads into the earth, and other that spreads and shoots into the air, both kinds of roots serve to receive and carry their proper nourishment to the body of the plant, and both serve also to convey and carry off the useless recrements; useless I mean any farther, within the body of the plant, though useful to it when they are separated, and without it, the one for seasoning the earth and water, wherein it is planted, and the other for seasoning the air."

This statement contained two powerful, and at that time, new, concepts. Firstly, it presented the idea that plant growth depends, in part, on some factor (CO_2) obtained from the air. The idea of "air" contributing to the growth of plants is generally ascribed to Stephen Hales, who discusses the possibility in his book *Vegetable Staticks* which was published in 1724; but Hooke's comments, which were made on May 18th, 1687, presaged Hales' publication by 40 years. Secondly, and more pertinent to our present topic, it contained the idea that roots not only absorb nutrient materials from the soil, but also return to the soil what Hooke described as *"useless recrements"* (root exudates) which *"season the earth and water wherein it is planted."* At a time when so little was known about plant nutrition, and the relationship between plant growth and the soil environment was not even a subject of conjecture, only an intellect as remarkable as Hooke's could have envisioned such a possibility. Almost 300 years were to pass before scientists began seriously to consider the effects of root exudates and associated microbial activity on soil properties and plant performance.

The two universal properties of all roots are anchorage and the absorption of water and mineral nutrients. The anchorage ability of roots is apparent to anyone who has ever tried to pull a well-established plant out of the ground, and the function of roots as absorptive organs providing nutrition to the plant was being taught by Aristotle in the 4th century B.C., and undoubtedly by other scholars before his time.

While these two primary functions of all roots were established empirically a considerable time ago, our understanding of how these functions are accomplished did not really begin until the 19th century, when the systematic study of the anatomy, morphology, and physiology of roots commenced.

In addition to their absorptive and anchorage functions, roots may function for specific purposes such as the storage of reserves, as in carrot (*Daucus carota* L.) and sweet potato (*Ipomoea batatas* Lam.), as prop roots and aerating organs in mangroves, and as contractile roots which draw the shoot system deeper into the soil in a number of bulbous species. However, such modifications of roots are outside the scope of the present discussion.

Only a relatively brief description of root structure and organization can be presented here. For more detailed information, the reader should consult one of the standard plant-anatomy texts such as the experimental approach to anatomy by Cutter (1971, 1978), or the more classical, descriptive plant anatomy text of Esau (1977).

2.2 Root Morphology

Under conditions conducive to germination, water is imbibed by a seed and a series of physiological and developmental changes are initiated leading to the production of a new plant. One of the first observable changes is the onset of division of the apical meristematic cells located at the root end of the embryo. These cell divisions, with subsequent enlargement of the daughter cells produced, lead to extension of the radicle (embryonic root), which forces its way through the seed coat and enters the soil.

The primary root develops from the radicle. In most dicotyledons and gymnosperms this primary root (taproot), bearing many lateral branches which in turn are branched, comprises the root system which persists throughout the life of the plant and undergoes secondary growth. In the majority of monocotyledons, however, the primary root frequently is of limited life span and is rapidly replaced by a series of adventitious roots usually originating from the base of the stem. The adventitious roots, which also show a hierarchy of lateral branching, form a somewhat homogeneous root system termed a fibrous root system. The adventitious roots of a fibrous root system may also show secondary growth with increasing age.

While the morphology of the root system within a species and variety may be relatively uniform under controlled laboratory growth conditions, it can show considerable variation due to factors such as nutrient deficiency (Hackett 1968) or changing temperature (Crossett et al. 1975). The soil environment is heteroge-

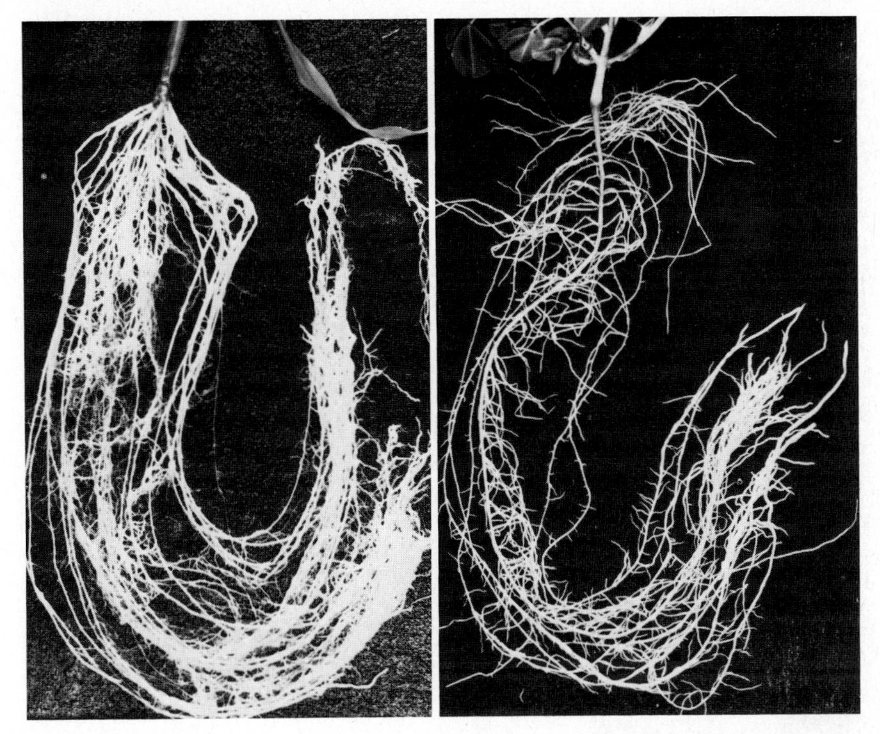

Fig. 2.1. Monocotyledon and dicotyledon root systems showing first- and second-order lateral roots. (*Left*) Fibrous root system of corn (*Zea mays* L.); (*right*) taproot system of peanut (*Arachis hypogaea* L.)

neous, intrinsically variable, and subject to continuous change due to climatic changes etc. Hence, one has to be cautious in extrapolating data obtained with root systems growing under favorable laboratory conditions to the field situation. For a fuller description of the effects of the soil environment on root growth and form see R. S. Russell (1977).

Because they lack a taproot, which is of indeterminate growth, species having fibrous root systems (Fig. 2.1) tend to be more shallowly rooted than plants with a persistent taproot. Plant size offers little indication of rooting depth, which to a large extent is characteristic of the species. However, rooting depth, as well as the overall root morphology, can be considerably modified by the environment. The rooting depth of a tree species is rarely as great as might be imagined; most of the roots in a well-drained soil do not extend more than 2 m below the soil surface, with a few roots reaching a depth of possibly 4 m. The majority of the roots of most herbaceous crop species are also concentrated in the upper 2 m of soil, with occasional roots reaching a depth of 3 m. Clearly, such factors as soil structure, water availability, and nutrient content have the potential for affecting root form and distribution, but perhaps not so obvious is the fact that the form of the root system also can be modified considerably by environmental factors to which the aerial portion of the plant is subject. Working with bermudagrass (*Cynodon* sp.) cuttings in a controlled environment, R. E. Burns (1972) showed that total

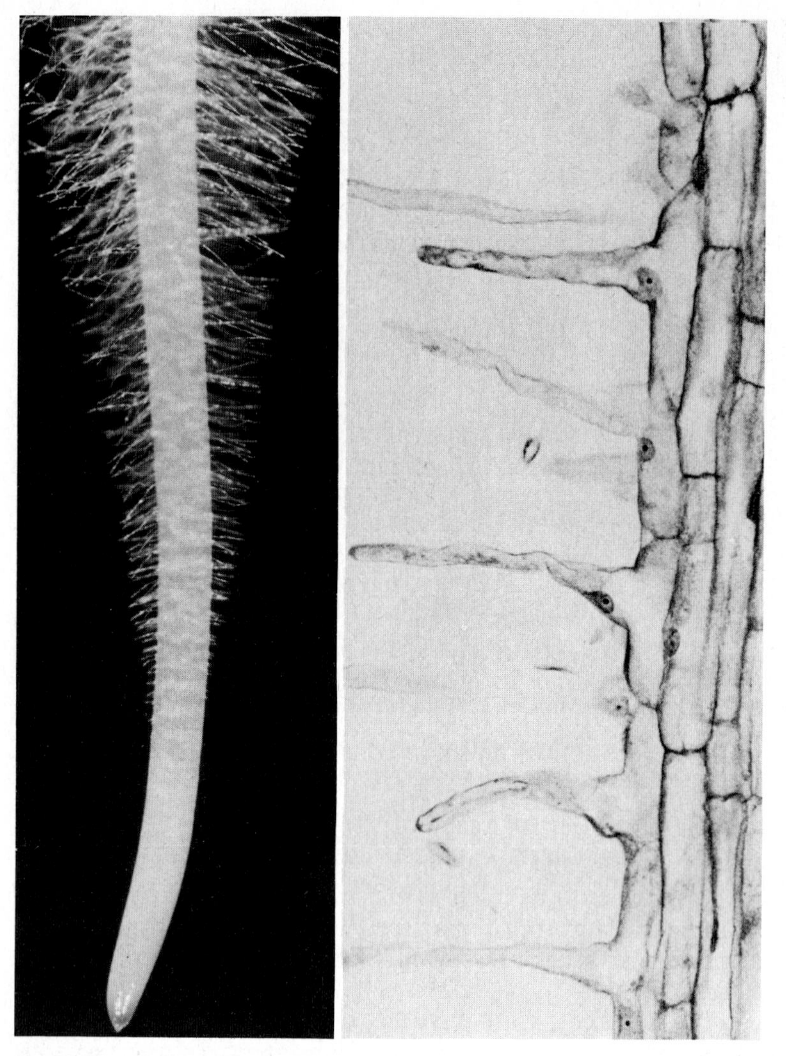

Fig. 2.2. Root hairs. (*Left*) Seedling of cress (*Lepidium sativum* L.) (courtesy of A. W. Charlton, Manchester University); (*right*) epidermis of young corn (*Zea mays* L.) root showing developing root hairs

root length, average length of roots, total root number, and the number of roots produced per node were all reduced when the cuttings were subjected to the lower of two temperature, light intensity, and daylength regimes. Root branching also was reduced by the lower light intensity and lower temperature. These changes could not be related entirely to the levels of stored carbohydrates because, while carbohydrate reserves were reduced with a shorter daylength and lower light intensity, they were increased by the lower temperature used in the experiments.

Like rooting depth, the total volume of a plant's root system is not necessarily reflected in the size of the shoot system. Because of the repeatedly branching nature of the root, a mature root system may have literally millions of actively growing root tips which, under favorable conditions, may continually invade and exploit fresh regions of the soil. Even though a single plant may have millions of root axes exploiting the surrouding soil, best estimates would indicate that the volume of the root system represents less than 5% of the volume of the soil in which it is growing (see, e.g., Wiersum 1961).

In most species, at a short distance behind each root tip there is a section of the root which bears root hairs (Fig. 2.2). Root hairs are unicellular extensions of the epidermal cells. They are usually short-lived but, because of the great numbers present, they increase enormously the total surface area of the root in contact with soil solution.

In many, perhaps the vast majority, of species the absorptive capacity of the root system is increased even further through a symbiotic association with certain fungi. These fungus–root associations, which are termed mycorrhizae, may be of two types. In ectomycorrhizal associations, the hyphae of many Basidiomycetes form an external weft or mantle which envelops the root surface and penetrates between the cortical cells but does not invade the cells. In endomycorrhizal associations, a mantle around the roots is absent, but fungal hyphae, largely of the Zygomycetes, actually penetrate the cortical cells of the root. Both types of association have been shown to increase the rate of entry of mineral nutrients and reduce resistance to the entry of water into the root system, which in turn releases organic compounds used by the fungus as a nutritive source. Mycorrhizae are discussed more extensively in Chaps. 6 and 8.

2.3 Root Structure

2.3.1 Organization of the Apical Region

The Apical Meristem. All of the cells of the root have their origin from the divisions of a relatively small number of cells termed initials which are part of the root apical meristem. The term apical meristem is not very appropriate for this region of active division in roots because, unlike the situation in the stem, this group of dividing cells is not truly apical, being covered by a protective tissue called the root cap.

The cells of the meristem are thin-walled, small, densely cytoplasmic and, relative to the cell size, their nuclei are large. The origins of specific root tissues can be traced in some species back to particular groups of initials in the meristem, termed histogens. In other species, however, several different tissues may arise from a common group of initials. In the roots of many vascular nonflowering plants (e.g., certain ferns) the apical meristem is represented by a single, tetrahedral cell (the apical cell) from which all of the root tissues arise (Fig. 2.3.). Collectively, the initials of the histogens are termed the promeristem (Clowes 1961). In many roots the promeristem is an inverted-cup-shaped group of cells bounding a region termed the quiescent center, which consists of cells showing low metabolic activity and a much slower rate of cell division. The quiescent center (see, e.g., Clowes 1956 a, b, 1958, 1959) is the subject of considerable interest. It ap-

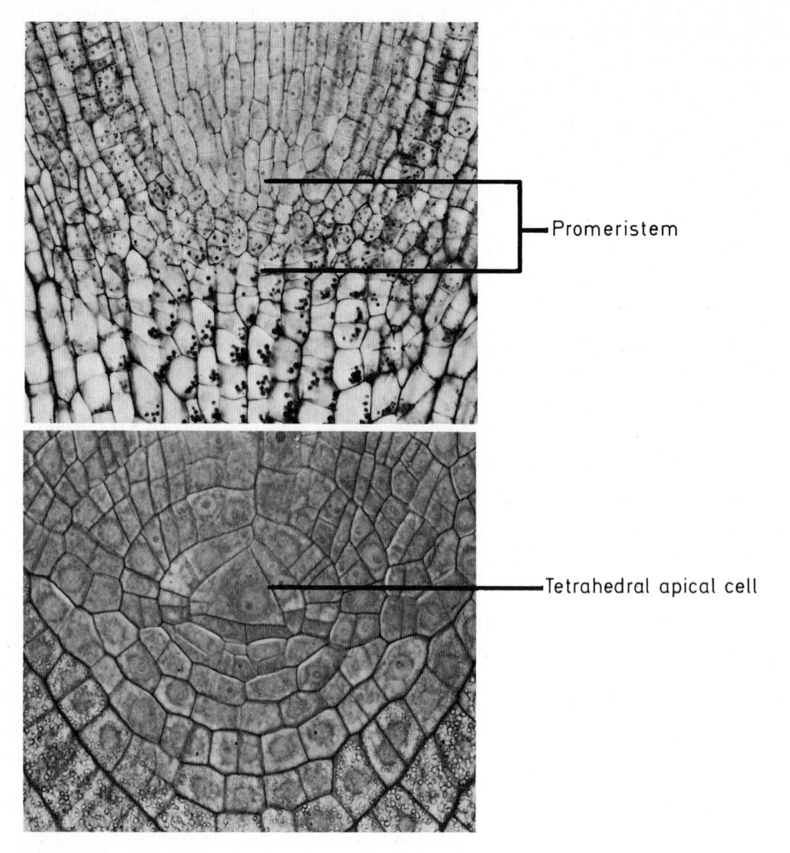

Fig. 2.3. Forms of apical meristem. (*Above*) Apical region of soybean [*Glycine max* (L.) Merr.] with promeristem of histogen initials; (*below*) apical region of fern (*Botrychium* sp.) root with a single cell constituting the apical meristem

pears during root development, being absent from the embryonic primary root and young lateral roots. It can be shown that nuclei in the cells of the quiescent center do not synthesize DNA at a rate comparable to that of nuclei in the surrounding cells of the meristem. Consequently, the rate of cell division in the quiescent center is, on average, at least ten times slower than that of the meristem initials. However, cells of the quiescent center can begin dividing under certain circumstances (e.g., following injury to the meristematic initials). It has been suggested that the quiescent center may serve as a reservoir of cells less prone to environmental stress damage which, when necessary, can take over the role of injured promeristem cells.

The Root Cap. Immediately distal to the apical meristem, and the most apical structure of the root, is the root cap. It is composed of living parenchyma cells produced by division of the most distal cells of the apical meristem. When the root-cap-forming promeristem cells appear as a distinct group of initials, as in many monocotyledons such as corn (*Zea mays* L.), they are referred to as the calyptrogen (Fig. 2.5).

The root cap appears to have two primary functions. Firstly, it offers protection to the delicate, thin-walled cells of the apical meristem as the root elongates and pushes through the soil. During the process of root elongation, the outermost, loosely connected cells of the root cap are being constantly disrupted and abraded by the soil mineral particles. These damaged and aging root-cap cells are sloughed off from the outside of the root cap, decay, and add organic nutrient materials to the rhizosphere around the advancing root tips. In spite of this constant loss of the outermost cells of the root cap through abrasion and death, the total volume and integrity of the root cap is maintained through divisions of the apical meristem initials. As might be inferred, the life span of root-cap cells is relatively short. Harkes (1973) estimated for oat (*Avena sativa* L.) that the time between the formation of a root-cap cell by the apical meristem and its loss from the root surface was not more than 5 to 6 days.

A second role of the root cap is a physiological function of considerable interest. Roots respond to the stimulus of gravity by growing downward (positive gravitropism). This active bending of roots, directing the root tips toward the center of gravity, is due to a differentially greater rate of expansion of cells on the upper side (the side farthest from the center of gravity) of the root in the region of cell expansion, which lies a short distance behind the root tip (Shen-Miller et al. 1978). While the bending occurs at some point behind the root tip, the gravity stimulus is perceived by the root cap. This has been convincingly demonstrated with the seedling roots of certain cereal grasses which have a large root cap that can be surgically removed with minimal damage to the root apex (Juniper et al. 1966; Pilet 1971). Roots from which root caps have been removed continue to grow but no longer respond to the gravity stimulus; the response returns, however, following the regeneration of new root-cap tissue. How the root cap perceives gravity and then regulates the activity of enlarging cells some distance back in the root is not fully understood. However, most scientists believe that gravity perception is through the movement of starch-containing plastids (amyloplasts) found within certain cells (statocytes) of the root cap. Amyloplasts are relatively large, dense bodies which tend to lie in groups pressed against the endoplasmic reticulum and cell membranes adjacent to the lowermost cell walls. When a root is turned from its normal downward orientation to a more horizontal position, the amyloplasts can be seen to fall through the cytoplasm under the influence of gravity and take up positions pressing against the cell membranes of the new lowermost cell walls (cell walls which had been in a lateral position when the root was vertical). It is believed that the pressure exerted by the amyloplasts on the membranes lining the lowermost cell walls represents a recognition signal to the plant defining the direction of gravity. Recently, calcium has been found to be associated with the amyloplasts of several species (Chandra et al. 1982) and, from this and other evidence (Lee et al. 1983), it is now thought that calcium gradients in the root tip may play an important role in linking the perception of gravity to gravitropic root curvature. Once the signal for gravity direction has been perceived, it is translated in the region of cell expansion by the unequal growth of the cells of the upper and lower sides of the root.

Cell expansion in most plant tissues is intimately related to the concentration within the tissues of the plant hormone indole-3-acetic acid (IAA). At extremely low concentration IAA is stimulatory to root growth, but it becomes inhibitory when the concentration exceeds some threshold level. For a long time it was believed that an asymmetrical distribution of IAA was responsible for the gravitropic response, a growth-inhibiting concentration of the hormone accumulating on the lower side of a horizontally positioned root (Went and Thimann 1937). For a number of reasons this is no longer an acceptable explanation. Firstly, the IAA content of a root is extremely low and is essentially restricted to the vascular tissue, while it is the cortical cells which enlarge. Secondly, any explanation based on an asymmetrical distribution of IAA requires that the hormone move from the root apex backward along the root, and also laterally across the root in a downward direction. Movement of IAA in the root, however, is within the stelar tissue and it shows strictly acropetal (toward the root apex) movement. Also, experiments have generally failed to show any lateral, downward, movement of IAA in gravitropically stimulated roots (Wilkins 1984). Today, it is believed that growth inhibition of the lower side of horizontally positioned roots is due not to a supra-optimal level of a growth-stimulating hormone such as IAA, but to the accumulation on the lower side of the root of root-cap-synthesized growth inhibitor(s) (Gibbons and Wilkins 1970; Shaw and Wilkins 1973; Wareing and Phillips 1981; Wilkins 1984).

The precise growth inhibitor(s) involved has not yet been determined. Abscisic acid (ABA) is present in root-cap cells, but other, as yet unidentified, growth inhibitors are also present (Suzuki et al. 1979; Wilkins 1984). It is suggested that the growth inhibitor responsible shows downward, lateral transport in the root cap of horizontally placed roots and is then translocated basipetally along the root to inhibit cell enlargement in the zone of elongation. Several investigators have tried without success to demonstrate an asymmetrical distribution of ABA in horizontally positioned roots, but such a distribution has been found for a presently unidentified growth inhibitor in the root cap of corn (*Zea mays* L.) (Suzuki et al. 1979).

The problem may turn out to be even more complex than presently envisioned, involving a balance between growth inhibitors and growth stimulators. Indeed, the involvement of IAA in the system, albeit indirectly, has been demonstrated in the work of Feldman (1981), who showed that growth-inhibitor production by corn root-cap cells involved protein synthesis and required a low concentration of IAA.

The Mucigel. Root tips as far back as the root-hair zone frequently are covered by a partially granular, partially fibrillar mucilaginous sheath termed the mucigel (see, e.g., Jenny and Grossenbacher 1963; Leiser 1968; Dart 1971; Floyd and Ohlrogge 1971; Greaves and Darbyshire 1972; Guckert et al. 1975). The mucigel is probably produced by apical tissues in general, but the root cap appears to be a primary source (Greaves and Darbyshire 1972). It is composed of highly hydrated, complex polysaccharides, probably mostly pectin and hemicellulose (Samstevich 1968; Floyd and Ohlrogge 1971; Miki et al. 1980). The mucilage appears to be synthesized from simpler monosaccharides within the cisternae of cell dictyosomes; it is transported from there to the cell membrane in vesicles and then moves across the membrane and through the cell wall to the outside (Juniper and Roberts 1966; Northcote and Pickett-Heaps 1966; Morré et al. 1967; Wright and Northcote 1974; Paull and Jones 1975). Some observations suggest that the mucigel is bounded by a thin membrane.

A variety of functions have been ascribed to the mucigel including: (a) a role in nutrient absorption, (b) the protection of root tips from injurious soil products, and (c) protection of the delicate apical region of the root against desiccation. However, the precise functions still remains to be resolved.

Mucigel can fill the spaces between the root surface and adjacent soil particles. Soil mineral particles are frequently embedded in the mucigel particularly in the root-hair zone (Jenny and Grossenbacher 1963; Greaves and Darbyshire 1972; Sprent 1975; Vermeer and McCully 1982). The total complex of root hairs, mucilage and other root-derived materials, mineral particles, and the associated microorganisms has been termed the rhizosheath (Fig. 2.4; Wullstein and Pratt 1981).

Plants grown under sterile (axenic) conditions generally have less mucigel than those grown under nonsterile conditions. Jenny and Grossenbacher (1963) have suggested that microorganisms may be contributing directly to the mucigel and, under such nonaxenic conditions of growth, a more distinct external boundary to the mucigel layer has been observed. Wullstein et al. (1979) isolated from the rhizosheaths of certain grasses a *Bacillus polymxa*-like bacterium which produced mucilage and contributed material to the rhizosheath complex.

The role of the mucigel in relation to rhizosphere microorganisms is unclear. Dart and Mercer (1964) suggest that in legume roots the mucigel may be a region of high root exudate accumulation which provides an ecological niche for the rapid proliferation of root-nodule bacteria. Greaves and Darbyshire (1972) proposed that the mucigel may aid in the establishment of pioneer microbial species, thus preventing later colonization of the root surface by other, competing microorganisms. Whatever the precise functions may be, the close association of roots with microorganisms within the mucigel must affect root physiology. Microorganisms within the mucigel will compete with roots for certain external nutrients, but they will also metabolize a number of soil components which might otherwise be unavailable to higher plants to compounds which can be absorbed by roots and function as nutritional sources. Similarly, root exudates, and the polysaccharide components of the mucigel itself, may be utilized directly by rhizosphere organisms within the mucigel, or may be metabolized by those microorganisms to some other chemical form affecting rhizosphere microbial activity and thereby indirectly affecting root-soil interrelationships.

2.3.2 The Subapical Region of the Root

The growth and development of a root involves cell division, cell elongation expansion, and cell differentiation. The apical meristem described earlier is the primary region of root tissue initiation and it is there that the pattern of the three fundamental tissue systems (dermal, ground tissue, and vascular tissue) of the

Fig. 2.4 A, B

root is established. It is not, however, the region of the root showing the highest rate of cell division. The most rapid rate of division generally occurs in the cells of a short section of the root immediately behind the apical meristem (see, e.g., Jensen and Kavaljian 1958). The apical meristem together with this subapical, actively dividing zone make up the root apex.

The precise distance from the root apex at which the maximum rate of cell division occurs depends on a variety of factors such as species, root age, etc., but it is usually within the first millimeter immediately proximal to the apical mersitem. Few divisions occur in cells lying at a distance greater than about 2 mm from the apical meristem. As the root cap and apical meristem are usually confined to the terminal 0.5 mm section of the root, this means that, in the vast majority of species, essentially all divisions will have been completed in the apical 2 mm of the root and, from that point backward, subsequent development represents mostly cell elongation and differentiation.

For convenience, the various regions of cell activity in the root tip are described as if they were discrete, functional units spatially separated from the adjacent zones. A longitudinal section of a young root tip, however, shows that the regions of cell division, elongation, and differentiation are not distinct (Fig. 2.5). There is no clearly delineated point of transition between the three zones as one moves basipetally (from apex to base) along the root.

Within the zone of most active cell division one can recongize, particularly toward the proximal end, an increase in the cross-sectional area and length of the cells. As one progresses basipetally along the root, cell division ceases and the meristematic zone gradually merges into the zone of cell elongation over most of which only the enlargement, with accompanying vacuolation, of pre-existing cells is occurring.

Proximal to the zone of elongation is the region of differentiation where most of the cells show the final transformations and modifications which produce the features characteristic of a specific tissue. Again, however, the boundaries of this zone are indistinct, the onset, or even completion, of differentiation of some tissues being evident in the zone of elongation. In onion roots, for example, cells of the phloem are functionally mature at a distance of about 1 mm behind the apical meristem which, for that species, is still within the zone primarily devoted to cell elongation.

Commencing in the zone of elongation and extending basipetally is the zone of differentiation in which cells of the epidermis produce papillae which then extend outwards to form the delicate lateral extensions called root hairs. These epidermal extensions increase the total root surface area many-fold.

Fig. 2.4 A, B. Rhizosheath covering root segment of Indian ricegrass [*Oryzopsis hymenoides* (Roem. and Schult.) Ricker]. (A) Fine sand grains cemented together and attached to the root hairs can be seen ensheathing the root; *arrows* indicate positions where the tips of root hairs protrude through the rhizosheath. (B) A short section of root from which the sand grains have been removed to reveal a dense mass of root hairs on both primary and lateral roots. (Wullstein and Pratt 1981)

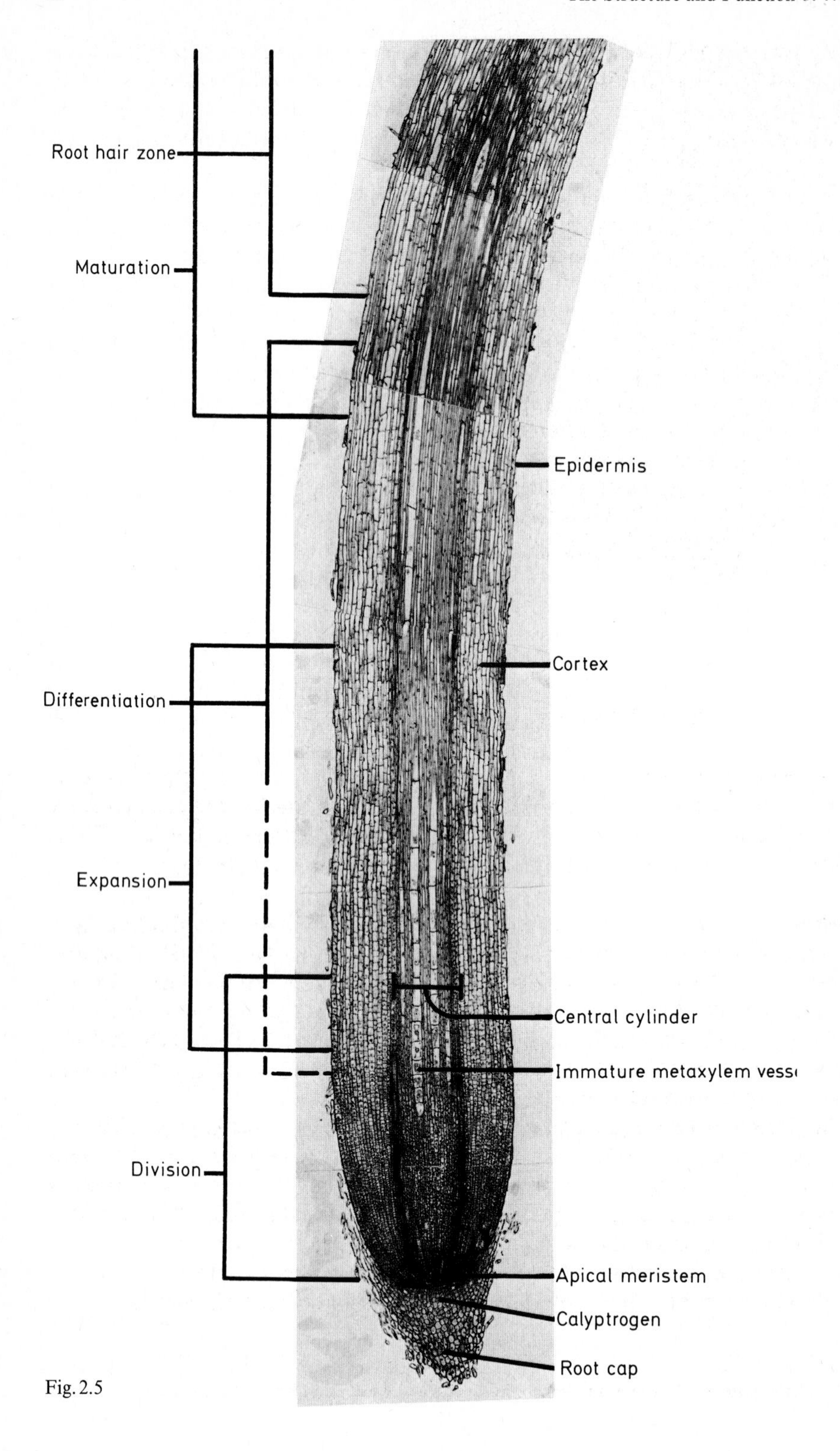
Root hair zone
Maturation
Differentiation
Expansion
Division
Epidermis
Cortex
Central cylinder
Immature metaxylem vess
Apical meristem
Calyptrogen
Root cap

Fig. 2.5

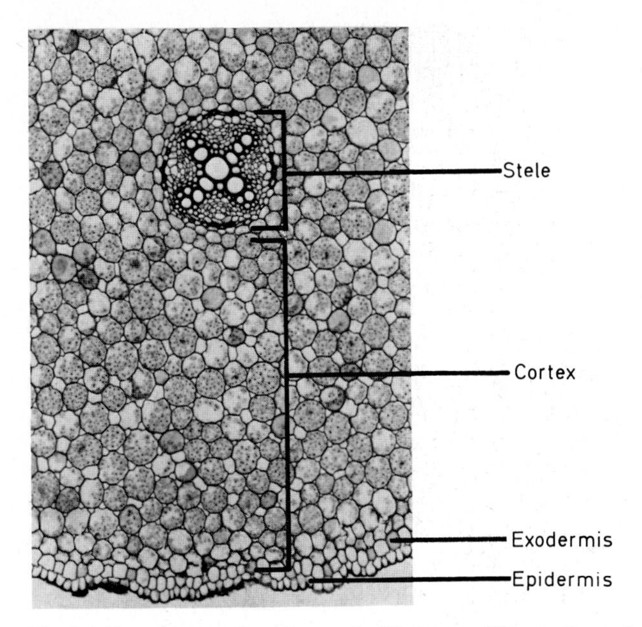

Fig. 2.6. Transverse section of root of tall buttercup (*Ranunculus acris* L.)

2.3.3 The Tissues of the Root

The three primary root tissues are the epidermis, the outermost layer of cells covering the root surface, the cortex or ground tissue which surrounds the stele, and the vascular tissue or stele, which occupies a central position (Fig. 2.6). These tissues will be described briefly starting from the outside of the root.

Epidermis. The root epidermis is usually a one-cell-thick layer, though a multiseriate epidermis (velamen) is found in the roots of many tropical Orchidaceae, epiphytic Araceae, and some terrestrial monocotyledons. With increasing root age and the onset of secondary thickening, the epidermis is frequently lost and is replaced by a periderm (cork tissue) which originates in the pericycle or phloem. The epidermis may, however, persist for many years and is always present over all parts of the young root except for the root cap.

The epidermis is frequently covered by a thin cuticle (waxy layer) which may even extend over the root hairs. Where the epidermis is persistent, the cell walls may become markedly cutinized; this process involves the deposition within the cell walls of polymerized, waxy materials laid down between the cellulose microfibrils which constitute the bulk, and primary component, of cell walls.

In some species, any of the epidermal cells in the region of the root immediately proximal to the zone of most active cell division have the potential to give

◄———

Fig. 2.5. Median longitudinal section of the apical region of a young corn (*Zea mays* L.) root

rise to a root hair. In other species, however, only certain, usually shorter, epidermal cells (trichoblasts) are capable of giving rise to root hairs; this is particularly evident in some grasses where trichoblasts may be arranged in a regular pattern along the root producing a symmetrical, equidistantly spaced root hair arrangement (see Cutter 1978).

The absorption of water through root hairs is probably no greater than through the epidermal surface without hairs but, because root hairs are so numerous and ramify radially from the root into the surrounding soil, it is thought by some that they may serve to exploit what might otherwise be untapped pockets of soil water.

Ground Tissue. The cortex may consist of a variety of cell types but it is primarily composed of thin-walled parenchyma cells which, as seen in transverse section, are frequently arranged in radial rows or concentric circles. In perennial dicotyledonous plants where there is secondary thickening of the root, the cortex of older roots is frequently shed. Many monocotyledons, on the other hand, retain their cortex, and these roots may have considerable amounts of sclerenchyma, which is usually found in a cylindrical arrangement immediately beneath the epidermis or adjacent to the endodermis.

The cells of the outermost cortical layers are frequently elongated and differentiated as an exodermis (hypodermis) (Fig. 2.6). The cells of the exodermis resemble in many ways endodermal cells with thick, suberized, and frequently lignified cell walls. In some species, short nonsuberized cells are interspersed with the elongate, suberized cells of the exodermis.

Conspicuous intercellular spaces may be present within the cortex, forming an interconnected system which may permit the longitudinal diffusion of gases through the root. Such an aerenchyma is particularly evident in the roots of many aquatic species.

The root cortex frequently represents a major storage region, the parenchyma cells being packed with starch grains or other stored, reserve compounds.

The innermost layer of the cortex is differentiated as an endodermis. The cells of the endodermis differ both structurally and physiologically from the cells bounding them on either side.

Endodermis. The endodermis is a one-cell-thick layer enclosing the vascular cylinder. The cells are somewhat elongated and characterized by the presence of a Casparian strip which is a band of lignin and suberin impregnating the radial and transverse walls and completely encircling the cell (Fig. 2.7). The Casparian strip varies considerably in width in different species; often it is much narrower than the cell wall, but it can cover one third of the total radial wall-surface area. The cell membrane is firmly anchored to the Casparian strip region of the cell wall. Because of its water-repellent properties, the Casparian strip constitutes a barrier to the free diffusion of water and dissolved salts through the cell wall-intercellular space route from the cortex to the vascular tissue.

The Casparian strip can be considered as the primary developmental modification of the endodermis and in dicotyledons showing secondary growth it is frequently the only endodermal state noted. Where secondary growth involves the

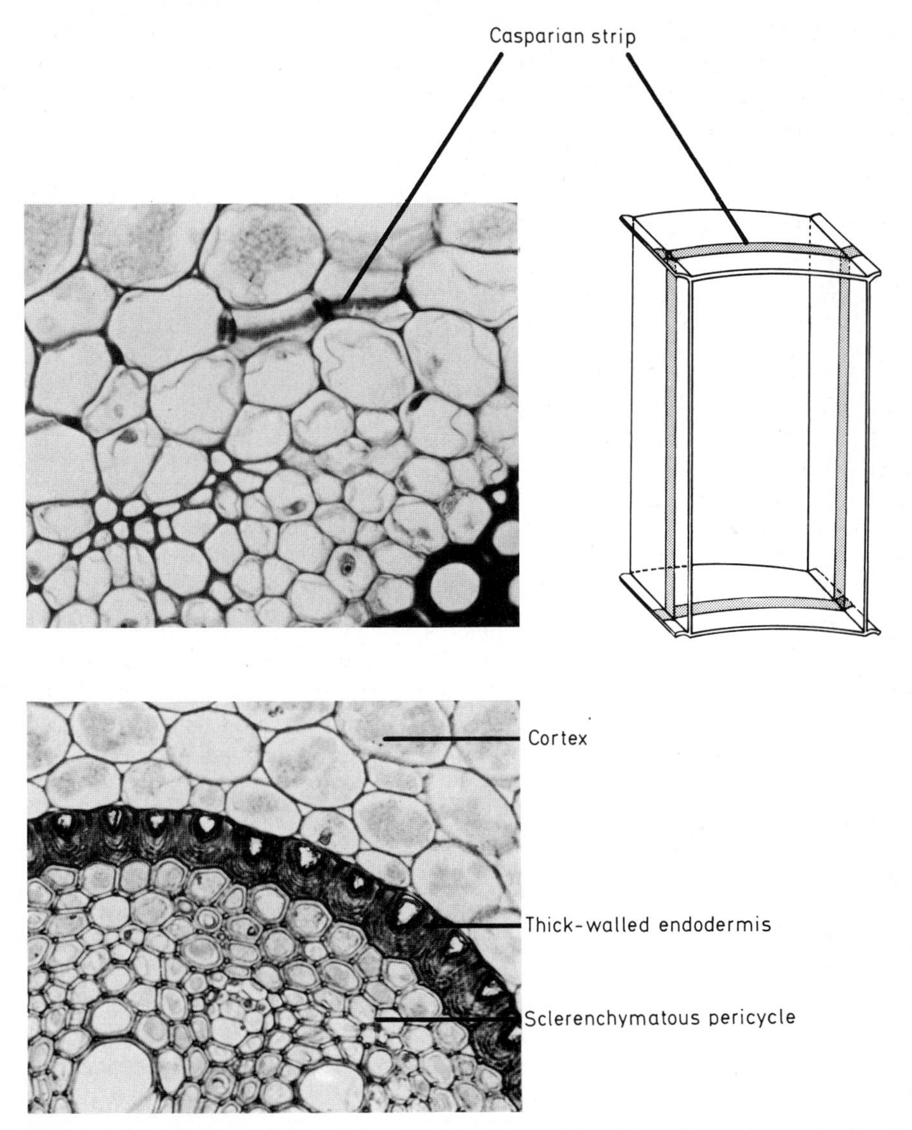

Fig. 2.7. Root endodermis. (*Above, left*) Transverse section of fern (*Botrychium* sp.) root showing endodermal cells with Casparian strip visible in their transverse walls; (*right*) a diagrammatic representation of an endodermal cell to show the axis of orientation of the Casparian strip; (*below*) transverse section of root of *Smilax* sp. showing extensive thickening of radial and inner-tangential walls of endodermal cells

development of a periderm through meristematic activity originating in the pericycle, the endodermis may be shed along with the rest of the cortical tissue. In roots which do not show secondary growth (the majority of monocotyledons and some dicotyledons), however, the endodermis frequently shows additional, secondary modifications as one moves basipetally (from apex to base) along the

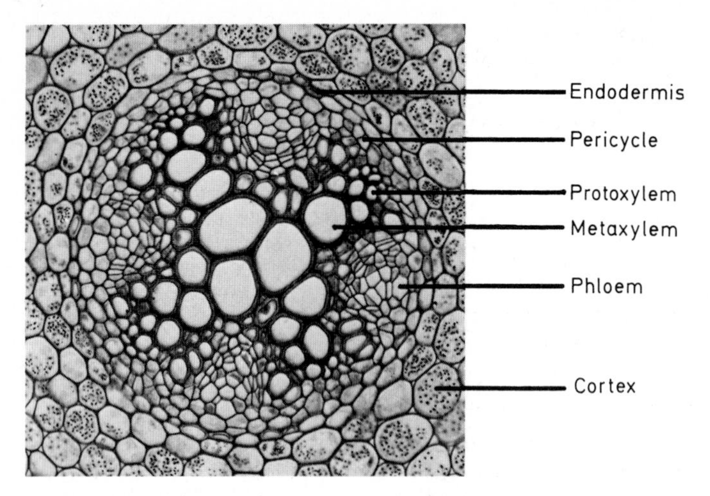

Fig. 2.8. Transverse section of tall buttercup (*Ranunculus acris* L.) root showing stele with tetrarch arrangement of xylem with phloem located between the xylem arcs

root. Firstly, suberin lamellae are deposited over the entire inner surface of the cell walls and, secondly, massive quantities of cellulose are laid down on the inner tangential walls and the adjacent areas of the radial longitudinal walls. These thick, secondary cell walls have thin areas (pits), and plasmodesmata have been seen to penetrate from the base of these pits into the adjacent pericycle cells (Robards et al. 1973).

The stimulus for formation of the Casparian strip, and the subsequent thickening of the cell walls, occurs first in those endodermal cells adjacent to the phloem and then spreads laterally to the cells facing the xylem. This developmental pattern can result in thick-walled endodermal cells adjacent to the phloem and endodermal cells with only a Casparian strip (passage cells) facing the xylem. The passage cells were so named because it was believed that they represented a pathway for transfer of materials between the cortex and the vascular cylinder. Some species such as barley (*Hordeum vulgare* L.), however, rarely show unthickened passage cells and their importance to radial transport of water and solutes has not been established.

Vascular Tissue. The central region of the root consists of the xylem and phloem together with associated parenchymatous tissues (Fig. 2.8). Immediately inside the endodermis there are one or more layers of parenchymatous cells which constitute the pericycle. In many monocotyledons the pericycle cells of older roots may show extensive sclerification (Fig. 2.7). The pericycle is the site of extensive meristematic activity. It is here that lateral root primordia originate, and the meristem (phellogen) giving rise to cork is formed; it also gives rise to a part of the vascular cambium producing secondary xylem and phloem.

Xylem. The xylem may occupy the center of the root as in most dicotyledons or there may be a central parenchymatous or sclerenchymatous pith (as in many monocotyledons). When the xylem occupies the center of the root it has a variable number of extensions projecting outward toward the endodermis, the thin-walled phloem tissue lying between these radiating arcs of xylem (Fig. 2.8). Where a pith is present, the vascular tissue takes the form of discrete strands of xylem with alternating strands of phloem.

Differentiation of xylem proceeds in a centripetal direction with the first-formed elements, protoxylem, to the outside and the later-formed metaxylem elements toward the center. Depending on the number of protoxylem loci or poles as seen in transverse sections, roots are classified as monarch, diarch, triarch, etc., to polyarch where there are many protoxylem poles. The suffix "arch" is derived from a Greek word meaning beginning (the same root as is used in such words as archaeology and archaeopteryx). The majority of dicotyledons have relatively few protoxylem poles and are usually di-, tri-, or tetrarch, but this is not a rigidly fixed character and the basal region of a taproot may show more protoxylem poles than the more apical regions. In monocotyledons there is a great deal more variation in the arrangement of the primary xylem than is usually found in the dicotyledons.

Xylem is composed of tracheary elements, fibers, and parenchyma. Fibers are elongated cells with pointed ends, a narrow lumen, and extensive lignification of the cell walls. The parenchyma cells found throughout the xylem frequently store starch or other reserve materials in young roots, but in older roots they may become lignified. The tracheary elements, vessel elements and tracheids, are the cells through which water absorbed by the roots is moved to the aerial parts of the plant. Both vessel elements and tracheids occur in the angiosperms, whereas the vast majority of gymnosperms have only tracheids. In both cell types the cell walls show extensive thickening and lignification and at maturity they are devoid of living contents. Metaxylem and secondary xylem elements show considerably more wall thickening than is found in the protoxylem elements of the primary xylem. For a thorough discussion of tracheary element structure and development see Cutter (1978).

Phloem. Unlike the situation in stem tissue, root phloems lie adjacent to and not on the same radii as the xylem elements. Development, like that of the xylem, is centripetal with the early-maturing protophloem to the outside and the later-maturing elements of the metaphloem toward the center. Phloem is the tissue through which photosynthate manufactured by the leaves and other green parts of a plant is translocated to other regions of the plant. At times of reserve mobilization, the phloem is also the tissue through which organic materials are transported from storage regions such as roots and tubers to support new or continued growth of other parts of the plant. It is generally agreed that the movement of materials within the phloem follows a source-to-sink pathway but, despite many years of research, there is still no consensus of agreement on the mechanism of phloem transport (see, e.g., Aronoff et al. 1975; Evert 1982).

Protophloem matures at a position nearer to the root apex than the protoxylem. Full development of the protoxylem may not be observable in tissue closer than 2 to 3 mm from the root apical meristem, a distance well beyond the zone of cell expansion in most species.

Phloem consists of sieve elements, fibers, parenchyma cells, and occasionally sclereids. Fibers, like those of the xylem, are elongated cells with heavily lignified walls. In angiosperms, fibers occur in both the primary and secondary phloem, whereas in gymnosperms the primary phloem is usually devoid of fibers.

The conducting elements of the phloem are the sieve elements. Sieve elements are of two types: sieve cells, found in gymnosperms, and sieve-tube members, which usually occur only in angiosperms. Both these cell types have in common that they contain living cytoplasm, and transport solutions of inorganic and organic solutes under a positive pressure. The study of phloem is more difficult than that of xylem because of this positive internal pressure, any damage tending to cause extensive cytoplasmic disruption with the production of artifacts. During their development, sieve elements lose their nuclei. In many species, much of the mature sieve-element cytoplasm is in the form of P-protein, a fibrillar protein occurring as a loose network of filaments filling the cell lumen or arranged in a parietal position. The role and function of P-protein in phloem transport is still unresolved. In addition to the P-protein, sieve element cytoplasm also contains endoplasmic reticulum, plastids, and mitochondria.

Sieve cells and sieve-tube members differ principally in the degree of development of clearly defined, depressed areas of the walls with groups of perforations (sieve areas or sieve plates) through which protoplasmic connection between contiguous cells is maintained.

Sieve cells are usually long and narrow with fairly steeply inclined end walls. Sieve areas of a relatively unspecialized type occur over both the side and end walls, though they may be more numerous on the end walls. The cells are arranged in files with the tapered, end walls overlapping.

Sieve-tube members are also elongated cells with end walls which may be inclined to transverse. The sieve areas of their end walls (sieve plates) show considerable specialization with much larger pores, whereas their side walls generally show simpler, unspecialized sieve areas. Sieve-tube members are cemented together to form long sieve tubes, the common walls of adjacent cells bearing sieve plates.

The parenchyma cells of phloem can exist in several forms. Some parenchyma cells are unspecialized, elongated cells which may be storage zones for carbohydrates, fats, or other nutritional reserves. Of more interest, however, are certain parenchyma cells (companion cells) closely associated with the sieve tube members.

Companion cells are highly specialized, elongated cells which are both structurally and functionally associated with adjacent sieve-tube members of the metaphloem and secondary phloem, but not the protophloem, of roots. The companion cell and its adjacent sieve-tube member arise from the same meristematic cell. There may be one or several companion cells associated with a sieve-tube member cell, and there are specialized cytoplasmic connections between the two cell types. Companion cells vary in length but they can be as long as the accompanying sieve-tube member cell. The role of the companion cell in sieve-tube functioning is unclear, but a considerable body of evidence indicates that they may function in the loading and unloading of sieve tube members.

The sieve elements of the phloem of gymnosperms are not associated with companion cells but they do show a relationship to certain of the phloem parenchyma cells which, based on their staining, have been termed albuminous cells. The albuminous cells differ little from the other phloem parenchyma cells, but there appears to be cytoplasmic connection between albuminous cells and adjacent sieve cells which show conspicuous sieve areas on those walls abutting onto the albuminous cells. The relationship between these two cell types would appear to be similar to that between sieve-tube members and companion cells.

For a more extensive discussion of phloem structure the reader is referred to Esau (1969) and Evert (1977).

2.3.4 Lateral Root Formation

Except for a few monocotyledons, primary root axes give rise to first-order lateral roots which, in turn, can produce second-order laterals, and so on.

Lateral roots have their origin deep within the root tissue, and longitudinal sections of roots may show lateral root primordia to be present quite close to the root apex, though these may not enlarge and appear as lateral roots until the axis has increased considerably in length.

The primary tissue of lateral root origin is the pericycle, though in some species the endodermis may also participate in lateral-root primordium formation. At the point of lateral-root origin, pericycle cells show meristematic activity and there is a series of cell divisions with new cell walls being laid down parallel to the long axis of the root. These divisions are followed by further divisions, this time with new cell walls laid down both parallel and at right angles to the root axis. Through this activity, which may also involve division of cells of the adjacent endodermis, an apical meristem becomes organized which is similar in structure to the apical meristem of the parent axis. Through continued cell division and expansion this new apical meristem forces its way through the surrounding cortex (Fig. 2.9). The disruption produced in the cortical cells due to the growth of the lateral primordium is partly the result of mechanical pressures exerted by the growing tissue, but it also appears to be due in part to the lysis and digestion of cortical cells by enzymes secreted by the emerging lateral root. As the primordium expands and pushes its way through the cortex, vascular tissue is differentiated in the primordium which joins up with the xylem and phloem elements of the main axis.

The lateral root finally emerges by bursting through the outer cortical cells and epidermis of the root axis. This process releases to the soil damaged epidermal cells and the contents of the crushed cortical tissue thus enriching the rhizosphere with a wide range of organic materials.

2.3.5 Secondary Growth of Roots

Secondary growth resulting in the deposition of cylinders of secondary xylem and phloem is rare in monocotyledons and herbaceous dicotyledons but usual in the woody dicotyledons.

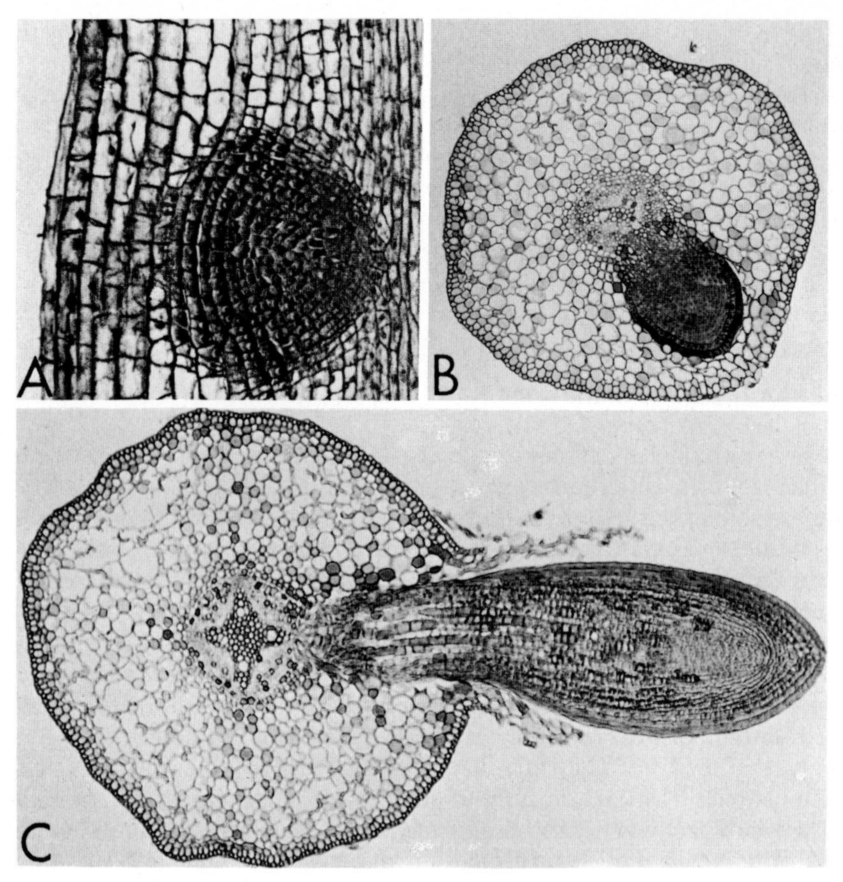

Fig. 2.9 A–C. Lateral root development. (A) Longitudinal section of corn (*Zea mays* L.) root showing a lateral root primordium produced from division of cells in the pericycle of the stele; (B, C) successive stages in the development of a lateral root in willow (*Salix* sp.); (C) shows extensive damage to the cortical tissue occurring as the lateral root emerges

Secondary vascular tissue is produced through the activity of a vascular cambium which arises first from undifferentiated procambial cells lying between adjacent strands of primary phloem and xylem. These areas of cambial activity are at first discrete, but they are soon connected by bands of cambium which arise in the pericycle on the outer face of the protoxylem poles. Because the pattern of primary xylem varies, usually with a central core of metaxylem and radiating arcs of protoxylem, this means that the newly formed vascular cambium frequently has, initially at least, a wavy outline. Very soon, however, the cambial cells lying between the protophloem and the xylem arcs begins to divide by periclinal divisions (cell walls laid down parallel to the long axis of the root) cutting off cells to the inside which differentiate into secondary xylem. Because activity of this region of the cambium precedes that of the cambial cells on the outer face of the protoxylem, the cambial layer very quickly becomes circular, as seen in cross-sec-

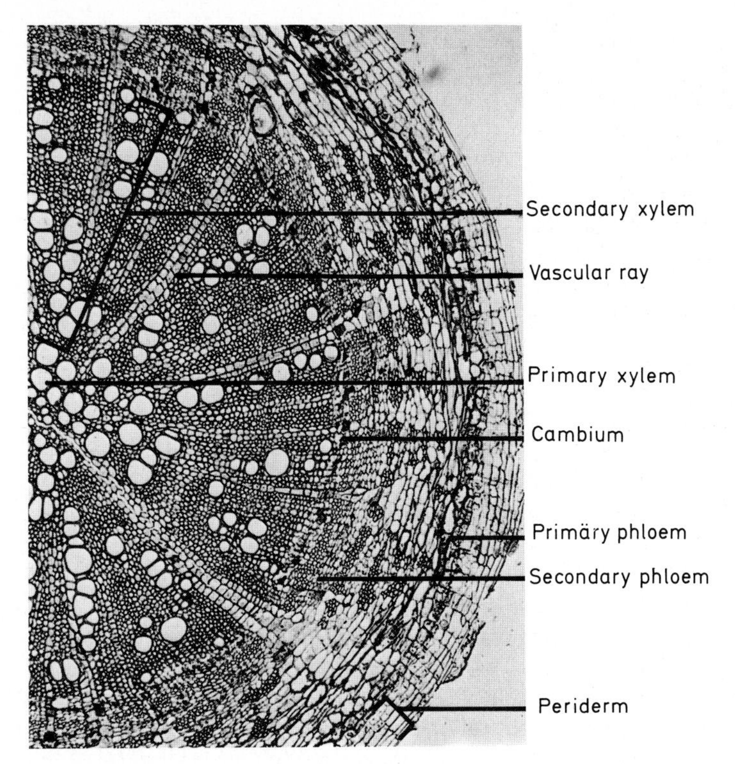

Fig. 2.10. Transverse section of mature cotton (*Gossypium hirsutum* L.) root showing extensive secondary thickening. The most abundant tissue present is secondary xylem. The primary phloem has been crushed and is represented only by groups of phloem fibers. A well-developed periderm is present

tion. The total cambium continues to divide periclinally, cutting off secondary phloem toward the outside of the root and secondary xylem toward the center. With increased deposition of secondary vascular tissue the primary phloem may be obliterated (Fig. 2.10).

Concurrent with the activity of a vascular cambium, other cells in the pericycle also begin to divide periclinally. A phellogen or cork cambium arises in the outer layers of this proliferated pericycle tissue which cuts off cork cells (phellem) to the outside and what might be considered as a secondary cortex, the phelloderm, toward the inside. The cells of the cork tissue are generally heavily suberized (Schmidt and Schonherr, 1982).

The increased diameter of the root occasioned by the deposition of secondary vascular tissue and cork cells causes the disruption of the outer primary tissues of the root (endodermis, cortex, and epidermis), and these tissues are frequently ruptured and sloughed off, adding rich organic material to the rhizosphere.

2.4 Root Physiology

The physiology of roots is an extremely complex area which has been the subject of a great deal of speculation and scientific investigation. A complete treatment of the subject would require several volumes of this size. We must, therefore, confine the present discussion to a brief review of plant roots in relation to water, and the phenomenon of exudation from roots. For a more comprehensive treatment of plant roots in relation to water, advanced treatises are available, e.g., Kramer (1983).

2.4.1 Plants and Water

It is generally agreed that life most probably originated in the vast areas of water which we believe covered the primitive earth billions of years ago. It was in this evolutionary womb that the first primitive plant forms evolved. In an aquatic environment, minerals and gases are present in solution and can diffuse into an organism over its entire surface. We can see this in many present-day algae and other plant aquatic organisms. Within these oceans more complex forms of plant life evolved; among these were forms which, when the time was propitious, were capable of colonizing the land surface. Land, however, represented an essentially hostile environment for plants. The raw materials for growth – water, essential elements, and CO_2 – were now spatially separated. The water and mineral nutrients were, for the most part, in the soil and had to be extracted by a subterranean part of the plant (the root system) and translocated to the aerial parts of the plant (the shoot system) which extracted from the air the CO_2 necessary for photosynthesis. This entry of CO_2 into the shoot system, and the release from the shoot system of O_2, which is a by-product of photosynthesis, required openings in the surfaces of the shoot system (stomata) through which these gases could move freely.

Evolution of our present-day land flora did not take place through giant strides, but gradually through small steps. In a recent publication, Mason and Marshall (1983) expressed the measured pace of evolution very nicely when they wrote:

> "A gradual accumulation of little developments is far more plausible than a single, giant leap. The chance of being dealt the ace, king, queen, jack, and ten of spades as a five-card poker hand is only about one in three million. In 45 years of continuous around-the-clock poker playing, we could expect that to happen only once. Still, the chance of being dealt each of those five cards separately during an evening's play is not at all remote. If there is a way of accumulating the individual items one at a time and holding onto each one until the rest can be picked up, a royal flush becomes a reasonable likelihood. In poker this is called cheating. In biology it is called evolution."

The accumulation with time of those small changes which constitute the "royal flush" of our present-day flora has brought with it some interesting physiological consequences in relation to plants and water. For example, the stomatal pores of leaves and stems, which are essential for the entry of CO_2 in amounts sufficient for a high rate of photosynthesis, expose the water-saturated cell walls of the plant interior to the much drier air of the atmosphere, leading to the loss of a very large part of the water previously absorbed by the root system. Similarly, separation of the aerial shoot system from direct contact with soil mineral-nutrients means that nutrients have to be transported to the top of the plant in solution. The soil solution represents an extremely dilute solution of mineral salts; therefore, supplying a large plant with the required amounts of such nutrients as nitrogen and phosphorus necessitates considerably more water being absorbed and moved than would be needed simply to satisfy the plant's metabolic need for water.

2.4.2 Plant Cell-Water Relations

The Concept of Water Potential. Movement of water in a plant-soil system, be it transfer of water from soil to root, root to shoot, or leaf to air, depends on gradients in the free energy of water at different points throughout the system. These differences in the free energy of water are usually described, and quantitatively expressed, in terms of *water potential*, which is represented by the Greek letter psi (ψ).

Water potential is really an expression of the differences between the chemical potential, i.e., free energy per mole (μ_w), of water at some specified point in the system and the chemical potential of pure water under conditions of standard temperature and pressure (μ_w^0). While precise values for the chemical potential of water are difficult to determine, the value for water potential can be determined since:

$$\psi = \mu_w - \mu_w^0 = RT \ln (e/e^0),$$

where R = the ideal gas constant, T = absolute temperature, e = vapor presure of the water under study at temperature T, and e^0 = vapor pressure of pure water at temperature T. Solving this equation gives the value of ψ in energy units (joules mol^{-1}) but, for convenience, these values are usually converted and expressed in terms of pressure units, bars (1 bar = 10^5 pascals, or 0.987 atmospheres).

In considering water movement, the absolute values for water potential are much less important than the difference in water potential ($\Delta\psi$) between any two points in the system. Water potential is increased with an increase in either temperature or pressure, and is reduced by dissolved solutes. For convenience, and as a reference point, pure water at standard temperature and pressure has been arbitrarily assigned a water potential value of 0. Because solute particles lower water potential, this means that all solutions at standard pressure and temperature have water potential values of less than 0 bar (i.e., a negative value). Precisely how negative the value is will depend on the concentration and nature of the solute. The greater the number of solute particles per unit volume of solution the lower (more negative) the water potential of the solution will be. For example, at 20 °C the water potential of a 0.1 M sucrose solution is −2.6 bar, while that of a 0.2 M sucrose solution is −5.4 bar.

Water diffuses from regions of higher water potential toward regions of lower (more negative) water potential or, in other words, from regions with water at a high free energy to regions with water at a lower free energy. To illustrate this point, imagine two sucrose solutions, one at a concentration of 0.1 M and the

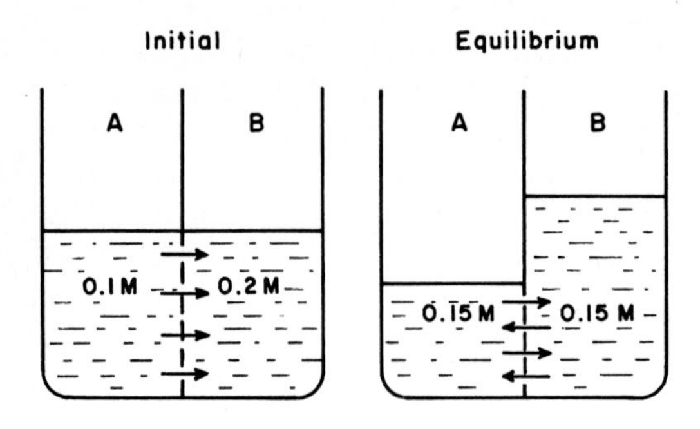

Fig. 2.11. Diffusion of water across a semi-permeable membrane from a region of higher water potential (0.1 M sucrose solution, compartment *A*) to a region of lower water potential (0.2 M sucrose solution, compartment *B*). When equilibrium has been attained, the solutions in the two compartments will be at the same water potential and no further net movement of water will occur in either direction

other at a concentration of 0.2 M, which are separated by a membrane which is permeable to water molecules but which is impermeable to sucrose molecules (a semi-permeable membrane), Fig. 2.11. The water potential of the 0.1 M sucrose solution (compartment A) = −2.6 bar, that of the 0.2 M sucrose solution (compartment B) = −5.4 bar. The difference in water potential ($\Delta\psi$) between the compartments is therefore:

$$\Delta\psi = -2.6 - (-5.4) = 2.8 \text{ bar.}$$

Under these conditions, water molecules will leave compartment A, pass through the membrane, and enter compartment B (movement towards the lower, more negative, ψ). This process is termed *osmosis*. Water will continue to move from the dilute to the more concentrated solution until the concentration of the solution initially at 0.2 M has been diluted to 0.15 M, and the solution initially at 0.1 M has been concentrated to 0.15 M. At that point, the water potentials will be the same on both sides of the membrane (ψ 0.15 M sucrose at 20 °C = −4.1 bar), the $\Delta\psi$ value will be 0, and net water movement across the membrane will cease. Such solutions are said to be at equilibrium.

Water potential is made up of different component forces (potentials) and three of these, *osmotic potential* (ψ_s), *pressure potential* (ψ_p), and *matric potential* (ψ_m) are extremely important for an understanding of plant water relationships. Water potential is the sum of these components, i.e.:

$$\psi = \psi_s + \psi_p + \psi_m.$$

Osmotic and Pressure Potentials. A system in which two solutions with different free energy (water potential) values were separated by a semi-permeable membrane was illustrated in Fig. 2.11. In this example, as water moved by osmosis from compartment A to compartment B, the energy released was available to perform work. The work it performed was to raise the level of the solution in compartment B. The magnitude of the force involved could have been measured if we had placed a piston over the surface of the 0.2 M solution in compartment B and measured the pressure which had to be applied to the piston in order to prevent the movement of water across the membrane from the 0.1 M solution of compartment A into the 0.2 M solution. It would have taken a pressure equivalent to $\Delta\psi$ (2.8 bar) to prevent such movement. Had compartment A held water ($\psi = 0$) in-

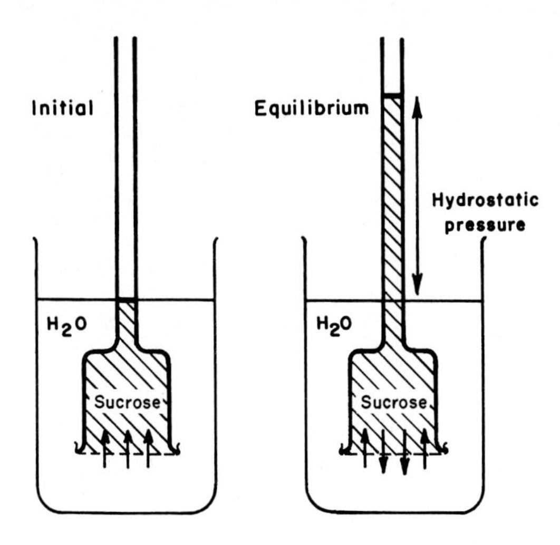

Fig. 2.12. A simple form of osmometer. In this apparatus, a sucrose solution is separated from water by a semi-permeable membrane. Under such conditions there is a net influx of water across the membrane to enter the sucrose solution. A hydrostatic pressure develops, the magnitude of which depends on the osmotic pressure of the enclosed sucrose solution

stead of a 0.1 M sucrose solution, it would have required a pressure of 5.4 bar to prevent water moving into compartment B (because in that case the difference between A and B would have been 5.4 bar). Note that this pressure (5.4 bar) is equal in magnitude, but opposite in sign, to the osmotic potential of the 0.2 M sucrose solution. This value is the *osmotic pressure* of the solution.

Osmotic pressure can be measured, somewhat less accurately, but more conveniently, with a simple form of apparatus called an osmometer (Fig. 2.12). In an osmometer, a solution (e.g., 0.2 M sucrose) is enclosed in a chamber, one end of which is attached to a long tube with a narrow bore, and the other end is covered with a semi-permeable membrane. If this apparatus is lowered into a beaker of water such that the levels of the solution and the water are initially equal, water will pass across the membrane into the solution [$\Delta\psi = 0 - (-5.4) = 5.4$ bar] and the solution will rise up the bore of the exit tube. The solution within the osmometer will increase in volume and a hydrostatic pressure will develop. The magnitude of the hydrostatic pressure can be measured by the solution-column height within the exit tube. At equilibrium (i.e., when water ceases to enter) the hydrostatic pressure will be equal to the osmotic pressure of the solution. In this particular example the hydrostatic pressure developed would be 5.4 bar.

As we saw earlier, at the equilibrium position there is no net movement of water because $\Delta\psi = 0$. In the osmometer illustration just presented, equilibrium in also attained when $\Delta\psi = 0$. Because the osmometer was placed in water ($\psi =$ 0), this must mean that the water potential value of the sucrose solution in the osmometer must also have achieved a water potential value of 0. Initially, the enclosed sucrose solution had an osmotic potential $= -5.4$ bar; therefore, because $\psi = \psi_s + \psi_p$, the water potential of the solution was also -5.4 bar (atmospheric pressure by definition is assigned a value of 0, therefore initially $\psi_p = 0$). As the solution column increased in height, however, a pressure was applied to the sucrose solution raising its water potential value. When the hydrostatic pressure reached a value of $+5.4$ bar the water potential of the solution became 0

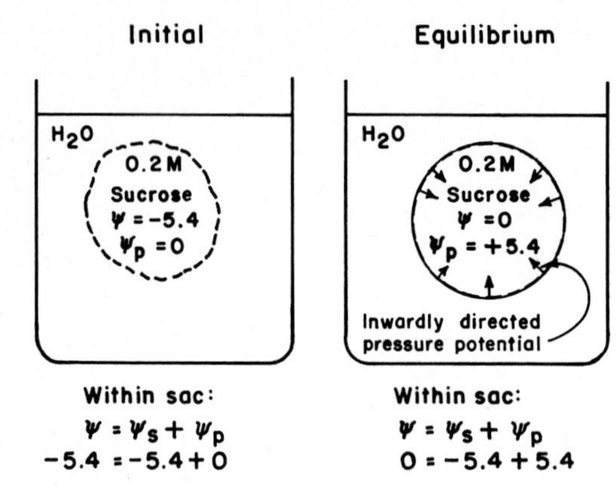

Fig. 2.13. A 0.2 M sucrose solution is enclosed within a semi-permeable membrane sac and placed in water. Net movement of water occurs across the membrane into the sucrose solution and a positive pressure (pressure potential, ψ_p) develops raising the water potential of the solution. Equilibrium is attained when the pressure potential value reaches 5.4 bar; at that time, water potential will be the same on both sides of the membrane and there will be no further net movement of water

because:

$$\psi = \psi^s + \psi_p$$

or

$$0 = -5.4 + (+5.4).$$

(Note that these values are not precisely right because water entering the osmometer would dilute the sucrose solution, lowering its concentration and, hence, its ψ_s value. Equilibrium would, in fact, be reached with a hydrostatic pressure somewhat lower than $+5.4$ bar but still numerically equivalent to the ψ_s value of the diluted sucrose solution).

The take-home message of the foregoing is that the osmotic pressure and osmotic potential values of a solution are numerically the same but opposite in sign (ψ is always a negative value at standard temperature and pressure), and that water potential is the sum of ψ_s plus any applied pressure (ψ_p). For example, if we apply a positive pressure of 2 bar to a sucrose solution with $\psi_s = -5.4$ bar the ψ of that solution $= -5.4 + (+2)$, or -3.4 bar. If the pressure applied to the solution was $+5.4$ bar, then the water potential of the solution would be zero. At standard temperature and pressure, of course, the water potential and osmotic potential of a solution are the same because in that case $\psi_p = 0$.

At this point the reader may feel, somewhat justifiably, that a discussion of the component forces of water potential has strayed a long way from the subject of plants in relation to water, but this is not so. Movement of water between any two phases in the soil-plant-atmosphere continuum is intimately related to differences in the water potential of the various phases and, indeed, a modified version of the osmometer depicted in Fig. 2.12 can serve as a model for understanding plant cell-water relationships. Imagine now that a 0.2 M sucrose solution ($\psi = -5.4$ bar) instead of being enclosed inside an osmometer is contained within a sac made from a semi-permeable membrane and that this sucrose package is dropped into a beaker of water (Fig. 2.13). Because the membrane sac is not initially overfilled with solution, the sucrose solution will be at atmospheric pressure and, therefore, the ψ of the solution will be -5.4 bar ($\psi = \psi_s + \psi_p$, or $\psi = -5.4 + 0$). Immediately upon entry into the water, water molecules will pass across the membrane and enter the sucrose solution. Unlike the osmometer apparatus described earlier, this sac has no exit tube permitting expansion of the enclosed liquid, hence, the bag will begin to swell and the walls of the sac will become stretched. In other words, a hydrostatic pressure will build up within the sac. This is the *turgor pressure* or pressure potential (ψ_p) of the system. When sufficient water has entered to produce a ψ_p of $+5.4$ bar, further entry of water will cease, because the ψ of the solution will then be zero [$\psi = \psi_s + \psi_p$, or $0 = -5.4 + (+5.4)$], and the system will have attained equilibrium. Such a solution-filled sac is described as being fully turgid.

The foregoing model approximates the condition of a living, mature plant cell. Surrounding every cell there is a cell wall capable of being stretched to a degree but offering resistance to continued expansion. The presence of this cell wall, and the resistance to expansion presented by the packing of the cells in a tissue, means that as water enters a cell a hydrostatic pressure (ψ_p) quickly develops. Inside the cell wall the cytoplasm, with its outer and inner bounding membranes (the plasmalemma and tonoplast, respectively), acts as a differentially permeable membrane (a differentially permeable membrane is like a semi-permeable membrane but not so perfect in its ability to exclude solutes) enclosing one or more fluid-filled vacuoles. The vacuolar fluid (cell sap) is a solution of many different inorganic and organic compounds including sugars, organic acids, amino acids, proteins, gums, tannins, phenolics, alkaloids, etc. The concentration of these substances varies with the species, tissue, cultural conditions, etc., but measurements have shown that the osmotic potential of this solution usually lies within the range of -5 to -30 bar (Sutcliffe, 1968). When a plant cell which is not already fully turgid is exposed to water, the water crosses the cell wall and cytoplasm and enters the vacuole because of the difference in water potential between the water outside and the vacuolar sap inside the cell. As a consequence, the vacuolar volume increases and the membrane-bounded cytoplasm pushes against the cell wall. Because the cell wall can only be streched to a limited extent, a hydrostatic pressure (ψ_p) rapidly builds up. When this pressure potential has reached a level which is numerically equivalent to the ψ_s value of the cell sap, the ψ value of the cell sap will be increased to zero and net movement of water into the cell will cease. The cell will then have reached equilibrium with the surrounding water (Fig. 2.14). Figure 2.15 shows the changes in ψ, ψ_s, ψ_p, and cell volume that occur when a plant cell that is not already fully turgid is exposed to water.

Matric Potential. Matric potential (ψ_m) is that component of water potential which results from the interaction of water molecules with dry, hydrophilic, porous surfaces. Seeds intended for planting are usually stored with a water content of about 4 to 6%. Usually, when such essentially dry seeds are placed in water, there begins an immediate, rapid, water-uptake phase termed imbibition. During this process, the previously dehydrated cellular components, such as protein molecules and cell-wall materials, are rehydrated. The binding of water to these cell marcromolecules during the hydration process leads to a massive reduction in the free energy of the water molecules and, therefore, to a corresponding reduction

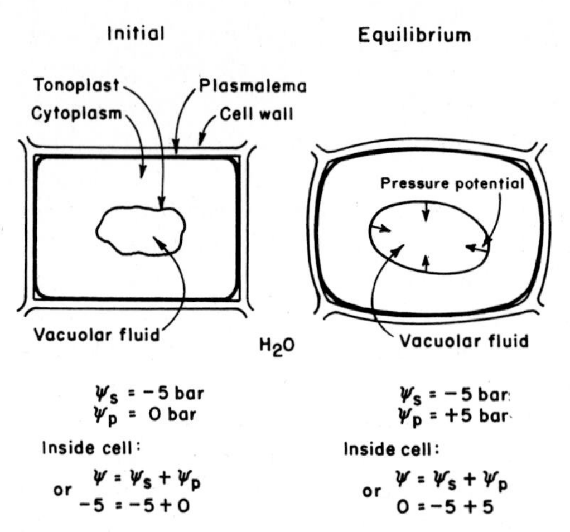

Fig. 2.14. A mature plant cell is placed in water ($\psi = 0$). The cytoplasm, bounded to the inside by the tonoplast and to the outside by the plasmalemma, behaves like a semi-permeable permeable membrane enclosing the vacuolar fluid which is a dilute solution of salts ($\psi_s = -5$ bar). Under such conditions, water will enter the cell vacuole which will swell. When the pressure potential (ψ_p) developing within the vacuole becomes numerically equal to the osmotic potential of the vacuolar fluid, the water potential of the vacuolar fluid will be zero. No further net movement of water occurs, and the cell is at equilibrium with the surrounding water

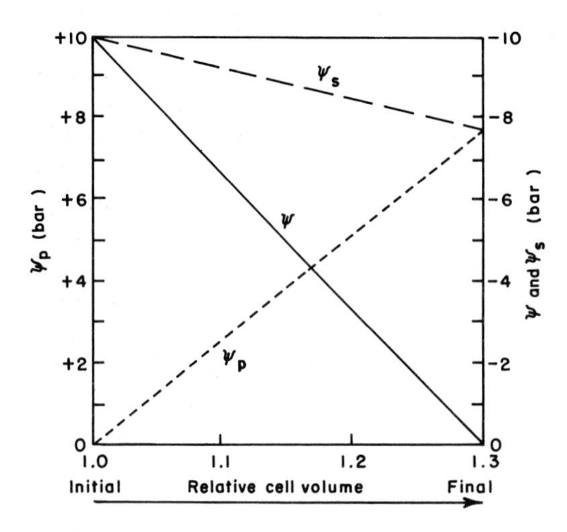

Fig. 2.15. Diagram to illustrate the changes in the values of ψ, ψ_s, and ψ_p which would be anticipated if a cell placed in water had a volume at equilibrium of 1.3 x the original volume. For illustration, it is assumed that initially the cell had a ψ_s value of −10 bar and a ψ_p value of 0 bar

in water potential. The excess energy is liberated as heat. Water imbibition, of course, results in seed swelling and if dry seeds are confined in an enclosed space but have access to water considerable pressures can be developed. It has been estimated that imbibing pea (*Pisum sativum* L.) seeds, for example, can develop pressures of about 1,000 bar (approximately 15,000 p.s.i.). It is not surprising, therefore, that germinating seeds can develop sufficient pressure to fracture asphalt or even cement paving. These imbibitional forces have been exploited in quarrying since ancient times, when it was discovered that marble or granite could be split by driving dry, wooden wedges into a rock face and then soaking them with water.

The force with which water molecules are bound to an absorbent surface is a function of the distance between the water molecules and the surface. Those water molecules actually interacting with the surface are held very much more tenaciously than those some distance away. If you think of an absorbent material as a particle, then you can visualize it as being surrounded by concentric shells of water molecules, each shell, as one moves outward, being held much more loosely than the adjacent one toward the center. At a certain distance from the particle there will be no further effect on the water molecules and water at this distance will have a free energy value of zero, like that of pure water. Matric potential is, in effect, a measure of the tenacity with which the least tightly held water molecules are retained. Hence, when ψ_m is measured, the value found will depend not only on the type of absorbent involved, but also on the relative proportions of absorbent and water. A dry colloidal material (such as gelatin) may have a ψ_m value that is extremely low (ca. $-3{,}000$ bar). If such a colloid interacts with an amount of water which is insufficient to saturate all of the absorbent surfaces and satisfy this very low ψ_m value, then the overall ψ of this system will also be very low. If, however, this absorbent, colloidal material is placed into a large volume of water, the absorbent's surfaces will become fully water-saturated, the matric forces will be satisfied, and the overall ψ value of the system will be zero.

From the foregoing, it should be evident that in a growing plant, as opposed to seeds and a few other specific situations, ψ_m is usually not a factor affecting the value of ψ to any great extent because, for the most part, molecular surfaces are already water-saturated. Unlike the growing plant, however, soils are subject to very wide fluctuations in their water content and the value of ψ_m is frequently the major component force affecting soil water potential. The liquid phase of soils is a very dilute solution of mineral salts, etc., and, except in soils of high salinity, the ψ_s value of the soil solution rarely exceeds -1 bar. The clay and organic particle components of soils, on the other hand, exhibit a very high ψ_m, which is an expression of the various physical and chemical forces binding water molecules to these materials. The ψ_m value of a dry clay, like organic colloidal material, may be as low as $-3{,}000$ bar. This means that the films of water molecules immediately adjacent to clay particles are held there by enormous binding forces, but films or shells of water progressively farther away from the particles are held much less tenaciously. In a well-watered soil, the ψ value will be the same as that of the soil solution (i.e., <-1 bar). When a soil begins to dry out, however, the forces binding the remaining water to the clay and organic matter become increasingly greater and eventually, the value of ψ_s becomes insignificant relative to that of ψ_m in establishing the ψ value of the soil because:

$$\psi = \psi_s + \psi_p + \psi_m.$$

Clearly, if ψ_m falls to $-1{,}000$ bar, then the small contribution made by ψ_s will not materially affect the final value of ψ. Because water will only move toward regions of lower (more negative) water potential, it should be clear that when a soil begins to dry out the remaining water will no longer be able to move into a plant. The literature will frequently state that when a soil dries to the point when the ψ value of the soil is approximately -15 bar plants are unable to obtain sufficient water from the soil to replace that lost from the shoot system, and wilting occurs. In soils having a high percentage of clay, a value of -15 bar may be attained when a soil still contains about 20% by weight of water. In a soil primarily composed of sand, however, a ψ value of -15 bar may not be achieved until the water content of the soil has been reduced to as little as 2 to 3% of its weight.

The percentage of water remaining in a soil at the time when a plant is no longer able to extract any more water is known as the *permanent wilting percentage*. Because a plant growing in such a soil continues to lose water from the shoot system to the atmosphere, cell turgidity is quickly lost under such conditions and a potentially very damaging wilting occurs. This wilting will eventually lead to plant death unless the situation is reversed by the addition of water to the soil to overcome the extremely negative soil water potential value developed during the drying period. Such wilting should not be confused with the temporary wilting noted in many mesophytes during the warmest parts of the day. Here, wilting is not due to a soil water deficit but is the result of water being lost from the shoots at a rate greater than it is being absorbed by the roots. In this situation, cell turgidity is generally fully restored at night when plant water loss to the atmosphere is considerably reduced but water uptake by the roots continues.

2.4.3 Movement of Water Through the Plant

All of the water within a plant can be considered as connected through the cell cytoplasm, wet cell walls, and the continuous water columns within the lumina of the xylem elements. Thus, in an actively growing plant there is a continuous liquid water phase extending all the way from the root epidermis to the cell walls of the leaf parenchyma.

It is generally accepted that the movement of water from soil to air through a plant can be explained on the basis of gradients of water potential along the pathway. Indeed, during the past 30 years, many attempts have been made to describe and define water movement within plants strictly in mathematical terms (see Molz and Ferrier 1982). Nevertheless, over the years, investigators have continued to present evidence for non-osmotic, active water transport in roots which is directly powered by metabolically derived energy (see W. P. Anderson 1976). While such reports cannot be discounted, it has not been found necessary to invoke the concept of active water transport in order to explain the vast majority of these reports, and this aspect of water movement will not be considered further here.

The growth of green plants depends on photosynthesis, which necessitates the uptake of CO_2 with an accompanying release of O_2 to the atmosphere. This gas exchange occurs primarily through the stomata of the leaf. While most of the leaf surface is covered with a waxy cuticle which is relatively impermeable to water, the stomatal apertures represent points where the wet cell walls of the leaf interior are in direct contact with the outside air. Because the atmosphere within the intercellular spaces of the leaf spongy mesophyll is essentially at saturation vapor pressure while the outside air is rarely water-saturated, water vapor moves from the leaf interior down a water potential gradient to the outside atmosphere. This process, which is termed transpiration, is the major driving force for water movement through the plant. While transpiration may be regarded as an inevitable and, in times of low soil water content, undesirable, consequence of the structural organization of plants, it probably fulfills at least two very important functions. Firstly, a significant part of the solar energy absorbed by leaves is used in the evaporation of water. This serves to cool the leaf and maintain a leaf temperature favorable to the orderly progression of biochemical reactions. Secondly, many of the essential mineral nutrients required by plants are present only at very low concentration in the soil solution. The transpiration stream serves to transport these nutrients from the soil to the tissues of the shoot system.

The volume of water transpired by plants can be relatively enormous. For example, on a warm day under favorable growing conditions a single corn (*Zea mays* L.) plant may lose 3 to 4 liters of water to the atmosphere, giant ragweed (*Ambrosia trifida* L.) may transpire 6 to 7 liters per day, while the loss per day from a full-grown tree may be hundreds of liters.

The Driving Force of Water Movement. The words "driving force" with their connotation of a pushing upward and onward are not particularly apt in referring to transpiration. The upward movement of water in a plant is not the result of a pressure developed in the roots, but rather a pulling force developed in the cells of the leaf mesophyll when water passes from the leaves to the outside air.

When values for the water potential gradient from the soil solution, through the transpiring plant, to the atmosphere are determined, it is found that there is a relatively gradual and constant decrease in water potential from perhaps < -1 bar for the soil solution and -5 bar for the root, to say -20 to -30 bar for the leaf. When the water potential difference between the leaf interior and the outside air is determined, however, a sudden and dramatic discontinuity in this gradient becomes apparent, because the water potential value of the outside atmosphere can be 1,000 bar or more lower than that of the leaf atmosphere (i.e., inside leaf: outside leaf $= > -1{,}000$ bar).

To understand the origin of such a large $\Delta\psi$ value, we have to examine the effects of water content and temperature on the ψ value of air. We are most used to thinking of the water content of air in terms of a value that we call the relative humidity (RH). There is direct proportionality between the concentration of water molecules in the atmosphere and vapor pressure; relative humidity expresses the ratio between the actual vapor pressure of water in air relative to the saturation vapor pressure (the vapor pressure when air at that temperature is saturated with water). Vapor pressure is usually expressed in millibars. For example, the saturation vapor pressure (100% humidity) of air at 30 °C is 42 mbars. If air at 30 °C has a vapor pressure of 21 mbar then it is at 50% RH, while 30 °C air at 25% RH would have a vapor pressure value of 10.5 mbar. Because the total amount of water that air can hold increases considerably with increasing temperature (Table 2.1), relative humidity as a measure of absolute water content has meaning only if the temperature is also stated. A leaf and the surrounding air are frequently at different temperatures (due to the absorption and retention by the

Table 2.1. Relationship between vapor pressure of air, relative humidity, and air temperature

Temperature of air (°C)	Vapour-pressure values of air (mbar) at different relative humidities					
	0	10%	25%	50%	75%	100% (saturation vapor pressure)
0 →	0	0.60	1.5	3.0	4.5	6.0
20	0	2.32	5.80	11.60	17.40	23.2
25	0	3.14	7.85	15.70	23.55	31.4
30	0	4.20	10.50	21.00	31.50	42.0
35	0	5.57	13.93	27.85	41.78	55.7
40	0	7.30	18.25	36.50	54.75	73.0

leaf of radiant energy), hence, relative humidity data are of much less value in describing the change in gradient from leaf to air than are vapor-pressure values. However, RH values can be readily converted to water-potential values according to the following equation:

$$\psi(\text{bar}) = -10.7 \times T \times \log(100/RH),$$

where T is temperature in degrees absolute.

Thus, the ψ of air at 20 °C with an RH of 50% would be:

$$\psi = -10.7 \times 293 \times \log(100/50) = -943.7 \text{ bar}.$$

It is generally assumed that air within the large intercellular spaces of the spongy mesophyll of the leaf will be at saturation vapor pressure. For all practical purposes this assumption is true, though it would be strictly accurate only for those situations where the mesophyll cells were fully turgid, i.e., when $\psi = 0$. Cells of transpiring leaves may, in fact, have water potential values of -50 bars or less, but this turns out to be of little significance because quite large reductions in cell-water potential cause only small vapor-pressure changes (for a theoretical discussion, see Nobel 1983). For example, air equilibriated with cells with $\psi = -5$ bar would still have an RH value of 99.6%. As leaf temperature increases, the internal atmosphere is automatically adjusted to maintain 100% saturation vapor pressure by more water evaporating from the cell walls to enter the air.

While the vapor pressure of the leaf interior is determined by temperature and mesophyll-cell water potential, the vapor pressure of the external air is established by temperature and the absolute water content. Even when it is raining, the relative humidity of the air rarely exceeds 99%. If the interior of a leaf was at 100% RH (vapor pressure = 23.2 mbar at 20 °C) and the exterior atmosphere was at 99% RH (vapor pressure = 23.0 mbar at 20 °C), there would be a small (0.2 mbar) vapor pressure difference between the inside and the outside, and water would move from the leaf to the outside air. While this vapor pressure difference may appear very small, the difference inside to outside of a leaf becomes more apparent if we convert the two relative humidity values into water potential values. Using the equation presented earlier, 100% RH at 20 °C within the leaf equates with a leaf ψ value of 0 [log (100/100) = 0], while the 99% RH at 20 °C of the outside air corresponds to a ψ value for the air of -13.4 bar. Clearly water will move from inside the leaf to the outside down a gradient of $\Delta\psi = -13.4$ bar.

The atmosphere usually has a relative humidity of well below saturation; consequently, the difference in water potential between the inside and the outside of a leaf can be enormous. For example, at 20 °C, air at 50% RH has a water potential of approximately -940 bar, while at 20% RH the value is in excess of $-2{,}000$ bar. Because of the relationship between temperature and saturation vapor pressure, as the plant leaf and air are warmed during the day the numerical value of $\Delta\psi$ usually becomes even greater, since the absolute water content of the air does not necessarily increase with increasing temperature, while the air within the leaf is maintained at or around the saturation rapon pressure.

A small amount of water is lost directly through the leaf cuticle (cuticular transpiration), but in the majority of plants most of the water transpired ($>90\%$) passes outward through the stomata. Regulation of the size of the stomatal aperture is, therefore, of paramount importance in the water economy of a plant. Stomatal aperture is affected by a variety of internal and external factors. The two principal factors leading to stomatal closure are an increase in the CO_2 concentration of the air within the leaf, and a decrease in the water potential of the leaf mesophyll cells. When stomata close the loss of water from leaves is decreased dramatically.

Water Movement Through the Leaf. At each node of the stem where a leaf is attached, part of the vascular tissue, a leaf trace, branches off and passes through the petiole into the leaf blade. Within the leaf blade the pattern of the vascular tissue is highly variable but, characteristically, leaf traces branch repeatedly, frequently anastomose, and finally end up as single, open-ended xylem elements permeating the leaf mesophyll in such numbers that no cell of the leaf is more than two or three cells distant from a vein or vein-ending. It is through this intricate and extensive network of xylem elements that water and mineral nutrients move into the leaf.

Cell walls are composed of cellulose and other macromolecules. Because of their inherent properties and structural organization, these highly hydrophilic molecules confer on cell walls a very high matric potential. The water molecules transpired from a leaf enter the leaf atmosphere largely through evaporation from the microcapillary spaces within the walls of the cells bordering the intercellular air spaces of the spongy mesophyll. Reduction in the water content of these walls to below saturation level lowers the chemical potential of the remaining water; consequently, water flows into these cell walls from adjacent cells in response to their reduced water potential ($\psi = \psi_s + \psi_p + \psi_m$; in the cell wall $\psi_p = 0$ and $\psi_s = < -1$ bar; therefore the value of ψ will essentially be established by the value of ψ_m). In this way, a water-potential gradient becomes established across the mesophyll, the cells bordering the intercellular spaces being of lower water potential than those farther away. Water will move down this gradient until eventually the water lost in transpiration is replaced by water drawn from the free-water surfaces at the vein terminals. Removal of water from the xylem vessels creates a tension (a reduced pressure) in the xylem water columns which can be registered all the way down to the roots.

In describing the movement of water across the leaf mesophyll it has been assumed that the pathway is primarily through the microcapillary spaces within the cells walls. While there is considerable support for such a view, not all scientists believe this is so. We will return to this problem later when the transport of water across the root cortex is discussed.

Water Movement Through the Vascular System. Survival of land plants is dependent on sufficient water being absorbed by the root system and moved upward through the stem and branches to replace that lost by transpiration from the leaves. Moreover, the system through which water is moved has to accommodate a relatively rapid rate of transport. Diffusion of water from cell to cell down a water potential gradient as described earlier is too slow a process to supply sufficient water to the top of most land plants and, indeed, the existence of our present, tall land flora was dependent on the evolution of a water-conducting vascular system.

The suggestion that xylem was the tissue through which water ascended the stem was made first more than 250 years ago. Since that time, this has repeatedly been demonstrated to be true, using a wide variety of techniques. In angiosperms water is moved primarily through vessels, elements which during maturation lose their cytoplasm, develop perforated end walls, and fuse in files to form long tube-like structures. These vessels may be many meters in length and range from about 20 to 700 μm in diameter, depending on species. In gymnosperms, the conducting element is the tracheid. Tracheids are shorter (<5 mm) and narrower (<30 μm) than vessels; they have tapering end walls which overlap with the contiguous cells, and bordered pits in both lateral and end walls through which water flows. Because the cytoplasm disappears from both types of cell during their maturation, they are nonliving, with heavily lignified and relatively rigid cell walls. There are frequent connections between adjacent files of cells through pits in the lateral walls. The long, water-conducting conduits formed by vessels and tracheids extend through the stem and into the roots where they branch repeatedly, and also into the leaves where again they branch repeatedly to form the network of veins described earlier. Because xylem vessels and tracheids have lost their cytoplasm, resistance to water flow through the xylem is low. While it is convenient to think of xylem as a system of conduits through which water moves, the analogy to a pipe system must not be too closely drawn. Because of the presence of perforated crosswalls, which are numerous in vessels, we are dealing with pipes interrupted at frequent intervals by perforated septa. As will be explained later, these cross-walls appear, under certain circumstances, to be essential for the continued functioning of the transport system.

To be acceptable, any explanation of water ascent in plants must allow for movement of water to the tops of the tallest trees; the distances involved can, therefore, be considerable. The redwoods (*Sequoia spp.*) of California, Douglas-fir [*Pseudotsuga menziesii* (Mirb.) Franco] of the Pacific Northwest, and the blue gums (*Eucalyptus* spp.) of Australia can all attain heights of greater than 100 m. When the depth of penetration of roots into the soil is added to this height, it is clear that in extreme cases water has to be raised to a height approaching 120 m.

While the past 100 years has seem many mechanisms proposed to account for water ascent, today there is fairly general consensus that the explanation lies in a theory proposed in the 1890's and given more formal substance in the early years of this century; this is *the cohesion of water theory*. Water molecules show strong mutual attraction to each other (cohesive forces). When water is confined to narrow-diameter tubes with wettable walls (such as xylem vessels and tracheids) it can be subjected to a pull from the top and the stress will be transmitted through the water column without the water losing contact with the tube wall (adhesion forces). In other words, within such tubes the columns of water behave as if all of the molecules were connected and a stress applied to any part of the column will be transmitted throughout the column. If tension (a negative pressure) is applied to one end of such a column the water will move toward the source of the tension in exactly the same way that a drink is sucked through a straw.

The vessels and tracheids constitute such a system. These vascular elements enclose continuous, thread-like, frequently intermeshing columns of water which extend all the way from the leaf veins down to the xylem of the smallest roots. Transpiration from the leaves creates a water potential gradient across the leaf mesophyll which results in water being withdrawn from the free-water surfaces at the open ends of the leaf veins. Loss of water from the veins creates tension in the xylem water columns, the magnitude of which will depend on the rate of transpiration water loss. Xylem fluid is a very dilute solution of salts, the osmotic potential of which does not exceed -1 to -2 bar; therefore, water potential in the xylem will essentially reflect the applied tension. For example, if transpiration produces a ψ in the leaf mesophyll of -10 bar, then the ψ value of the xylem ($\psi = \psi_s + \psi_p$) will be: $\psi = -1 + (-9) = -10$ bar. A reduction in water potential at the leaf transpiring surface will, therefore, be transmitted through the xylem to the roots, where it will cause water to flow in from the soil, the rate of water uptake varying with the level of tension developed. Under such conditions there will be a continuous mass flow of water from the soil via the roots, stem, and leaves to the outside atmosphere. Because there is some resistance to water movement through the narrow, irregularly thickened lumina of the xylem elements, and resistance to the free movement of water across the root under conditions of high transpiration, the rate of flow of water through the xylem may be insufficient to keep up with the pace of transpiration loss. When this happens, considerable tensions can build up in the xylem and mesophyll cells. A reduction in mesophyll water potential is one of the most powerful triggers initiating stomatal closure. Following stomatal closure, the tensions are gradually released as water continues to flow into the system from the soil while water loss from the leaves is decreased. Thus, we have in the water-stressed transport system a finely tuned feedback system regulating and coordinating rates of water loss and uptake.

During the first half of this century, the cohesion of water theory was just one of a number of explanations advanced to account for xylem-fluid rise in plants. Each of the alternative proposals had its following of loyal supporters and the cohesion of water theory was subjected to considerable criticism from unbelievers. Most of the objections raised revolved around five key issues: (a) the adequacy of cohesive forces among water molecules to maintain water columns of the height involved, (b) the generation of the requisite tensions by structures as insubstantial as leaves, (c) the inherent instability of water columns under tension, (d) gas bubble formation (cavitation) in water columns interrupting water continuity, and (e) the observation that water movement is not halted by cuts in a stem which would sever water columns. Precise measurements and careful observations over the past 30 years have dealt with these concerns to the satisfaction of most scientists.

Under ideal conditions, the cohesive forces of water are very great and a stretched column of water could, in theory, withstand a tension of −1,000 to −2,000 bar before it would break. In xylem such ideal conditions do not pertain but, for tubes of the diameter and composition of vessels and tracheids, tensions as high as −350 bar have been applied before column breakage occurred. The larger the diameter of the column the less stable it is, and when the tension becomes too great cavitation occurs with gas-bubble formation. It has been estimated that the tension required to raise water in trees under conditions of maximum transpiration lies between −0.15 and −0.2 bar per meter of height (Zimmerman and Brown, 1971). This value accounts for the tension needed to balance the hydrostatic gravity effect, as well as that needed to overcome resistance to flow within the narrow xylem conduits. Hence, for the tallest trees known (<120 m) a tension of -0.2×120 or −24 bar would be sufficient to raise water from the soil to the tree canopy. Values of −5 to −10 bar are frequently recorded for the leaf water potential of herbaceous species, and values of −20 to −30 bar for modest-sized trees under conditions of high transpiration. For some desert species, values for leaf water potential of −100 bar and lower have been measured (Whiteman and Koller 1964; Milburn 1979). As xylem tension is increased, the walls of vessels and tracheids are drawn inward (because of the adhesive forces binding water to cell walls), superficial tissues lose water, and stem diameter may actually decrease. At night, when stomata close and transpiration loss is low, the stem returns to its original diameter as the tensions are released by water continuing to enter through the roots. Thus, it is clear that the forces necessary for water transport are compatible with the architecture of the transport system, and leaves can continue to function at tensions in excess of those required to satisfy the transpiration needs of even the tallest tree species.

When tension is applied to a water column there is a tendency for cavitation with the release of gas bubbles to occur. In larger-diameter vessels a considerable amount of cavitation probably occurs when xylem tension reaches −20 bar (D.A. Baker 1984). Another common source of cavitation bubbles is the gas evolved when water freezes. Freezing of the xylem sap of trees growing in northern latitudes is a common occurrence. Gas bubbles produced through cavitation or injury to individual vessels and tracheids will tend to remain in place and not spread throughout the xylem because, at the tensions involved, they are unable to escape through the very small perforations in the crosswalls and lateral walls of the individual cells.

In older trees, a considerable portion of the xylem may not be functioning in water transport because air embolisms or other materials in the xylem have interrupted water-column continuity. In a study of Canadian trees, J. Clark and Gibbs

(1957) showed that in certain species as much as 40% of the trunk volume could be occupied by gases. Fortunately, there is usually a great excess of xylem relative to the amount of water being transported. In ring-porous tree species, all water transport is essentially confined to the outermost (youngest) annual ring of secondary xylem, while in coniferous and diffuse-porous species several of the outer annual rings may be involved equally in water transport.

Because all of the water in a plant is connected, the cohesive forces of water in xylem are operative laterally as well as vertically. Hence, if some of the xylem elements become blocked by air bubbles, or if the vertical continuity of water columns should be interrupted by a deep cut, water will move laterally around the obstacle and then continue upward through uninjured xylem elements. Diversion of water traffic in this way leads to an increase in the resistance to water movement, but the rate of flow through the remaining, unblocked xylem is increased and, unless damage is sudden and very extensive, there may be little effect on the total amount of water that is moved. In many herbaceous species interconnections between individual elements of the vascular tissue, particularly at the nodes, may be so extensive that the occlusion of a percentage of the vascular tissue will have virtually no effect on total water movement.

With a knowledge of the mechanism and forces involved in water movement through the aerial parts of a plant, we are now in a position to see how water is absorbed by a plant-root system and moved radially to the vascular cylinder at the center of the root.

2.4.4 Movement of Water from Soil to the Root Xylem

The Absorbing Regions of the Root. Water enters roots in response to a water-potential gradient in the xylem established by transpiration. Clearly, therefore, it will enter most rapidly through those regions of the root which offer the least resistance to water movement. Precisely where those regions occur, however, depends on a variety of factors and is also species-dependent (Newman 1973).

Experiments performed with young, primary roots have shown that the apical, meristematic region is relatively impermeable to water. In this region the cells are nonvacuolated and contain dense cytoplasm. Also, the vascular tissue is not differentiated this close to the root apex and, hence, if water were able to enter freely it could not be transported backwards along the root. Maximum water absorption occurs in that root zone proximal to the meristematic region but distal to the region of extensive cutinization and suberization. Change in the rate of transpiration can cause a shift in that part of the root showing the maximum rate of uptake (Brouwer 1953), maximum absorption shifting further toward the base of the root as transpiration is increased.

The region of maximum absorption includes the root-hair zone. Root hairs can be extremely numerous and they greatly increase root surface area and the area of the root making contact with the soil (Dittmer 1937). It is part of established botanical folklore that root hairs penetrate the water-filled capillaries between soil particles and are important in relation to water and nutrient uptake. This idea may indeed be true, but there is remarkably little experimental evidence to support the contention (Newman 1974). Root hairs can absorb water (Cailloux 1972), but they are clearly not essential for water uptake because hydroponically grown plants, as well as aquatic plants, frequently lack root hairs, and they are not present on the adventitious roots of bulbs, corms, or rhizomes. However, root hairs serve to anchor young primary roots as they penetrate the soil, and as they age they secrete droplets of liquid. Both of these effects could contribute to and modify the rhizosphere and thus, indirectly, the total relationship of the root to soil water and nutrients.

Most research on the absorption of water by roots has concentrated on young, unsuberized roots which are more permeable to water and solutes than older, heavily suberized roots. In trees and other woody perennial species, however, by far the greatest part of the root system consists of older, suberized roots. Much of the water absorbed by these roots probably enters through the numerous fissures that develop with aging in the protective phellem and phelloderm surrounding the secondary vascular tissue. Kramer and Bullock (1966), in a study of roots under stands of loblolly pine (*Pinus taeda* L.) and yellow poplar (*Liriodendron tulipifera* L.), found that in midsummer about 99% of all of the roots were suberized. As midsummer would be the time of maximum transpiration, this must mean that most of the water was entering through suberized roots. For many plants, absorption through suberized roots may be of particular significance in winter or during periods of drought when there is little new root development and the growth of existing roots is restricted.

Radial Movement of Water Across the Root. Under most circumstances, the ψ_s value of the soil solution rarely exceeds -1 bar. As transpiration may easily create a ψ value for root xylem of -20 to -30 bar, a very large water potential gradient can exist between the soil solution and the xylem elements. Water will move across the root in response to the gradient established.

Water entering a primary root has first to traverse the epidermis and hypodermis (where present) and a parenchymatous cortical layer 5 to 15 cells in thickness with numerous, air-filled intercellular spaces. It must then cross the endodermis with its Casparian strip and, frequently, extensive suberization. Once inside the endodermis water only has to move through a few-cell-thick, parenchymatous pericycle layer before reaching the vascular tissue (see Sect. 2.3 and Fig. 2.6).

In describing the pathway of water movement across the leaf-mesophyll tissue, it was assumed that water was moving as surface water films and in the microcapillary spaces of the cell walls. While this would be the opinion of very many investigators it is by no means a consensus viewpoint, and the same diversity of opinion exists in describing the radial path of water movement across the root cortex.

Theoretically, there are three anatomically distinct pathways through which the water could move. Firstly, we have the route external to the living cytoplasm, i.e., the cell-wall continuum outside the cell membrane (plasmalemma). This is the so-called *apoplast* phase of the plant (Lauchli 1976). The root apoplast would represent somewhere in the region of 10% of the root volume and would be roughly equivalent to the *apparent free space* of Briggs and Robertson (1957). Secondly, there is the possibility that water crosses the cell wall and plasmalemma to enter the cytoplasm. It would then move through the interconnected, three-dimensional cytoplasmic continuum, moving through the plasmodesmata which connect the cytoplasm of adjacent cells. This route, representing the living portion of the cell, is termed the *symplast* pathway. Finally, there exists the possibility that water crosses both the plasmalemma and the tonoplast so that the vacuole becomes an integral part of the transport pathway. The relative rate of flow through each of the possible pathways will vary inversely with the level of resistance encountered.

Experimental procedures, as well as predictions based upon mathematical and computer modeling, have been used in attempting to assess the most likely route. In mathematical treatments of water movement, Molz and Ikenberry (1974), Molz (1976), and Molz and Ferrier (1982), have used a variety of circuit-analog models and have been unable to favor either the apoplast or the symplast-vacuole route. Symplast resistance might be expected to be high for two principal reasons. Firstly, symplastic movement requires transport of water through the plasmodesmata, which represent only about 1% of the surface area of adjacent cells and, secondly, the living cytoplasm is relatively viscous. However, Spanswick (1976) has suggested that movement through cytoplasm could be aided by the constant streaming of cytoplasm (cyclosis) within the cells. Newman (1974, 1976) favored the symplast pathway because he concluded from the data available that the resistance imposed in moving through the apoplast and through vacuoles were both too great to support the flow rates recorded. The experimental procedures used have provided us with no more definitive answers than have the theoretical studies and, at present, it is impossible to say that water is moved exclusively through any route. Most scientists today, however, would appear to favor the apoplastic pathway as quantitatively the most important route, though it is very likely that some of the water moves through each of the pathways. Because the water in a plant moves from regions of high activity to regions of lower activity, clearly there will be a considerable amount of water movement between each of these compartments as the water potential values in different parts of the system achieve equilibrium (Weatherley 1970). As pointed out by Molz and Ferrier (1982):

> "In a quantitative sense, the concept of the vacuolar pathway as distinct from the cell wall pathway is somewhat artificial (although perhaps quite useful). This is because water must enter the cell walls in order to get from vacuole to vacuole. If one allowed the resistance of the cell wall pathway to approach infinity, water transport would be halted also in the vacuolar pathway. Thus the two pathways are not really distinct."

While opinion may be divided regarding the pathway of water movement through the cortex, there is general agreement that water movement across the endodermis must be via the symplast route. Movement of water through the apoplast at the level of the endodermis is precluded by the lignin- and suberin-impregnated Casparian strip of the endodermal cells which forms an integral part of the radial and transverse walls (Fig. 2.7). Furthermore, in older endodermal cells, suberin lamellae cover the surface of the inner walls, and the tangential inner walls of the endodermal cells also may show extensive thickening. There are pores in these thickened walls through which plasmodesmata pass to connect the endodermal cytoplasm with the cytoplasm of the pericycle. In many roots unthickened endodermal cells, passage cells, occur opposite the protoxylem poles of the vascular cylinder. However, even in these there is an intact Casparian strip preventing apoplastic water transfer between the cortex and stele.

Like movement across the leaf, movement across the root is a major component of the total resistance to water movement through the plant. This is easily demonstrated by cutting off the root of a plant beneath water when the rate of transpiration will frequently increase. Movement of water through the symplast pathway at the endodermis contributes significantly to root resistance, and killing roots leads to a reduction in the resistance to water movement (Stoker and Weatherley 1971).

Once water has crossed the endodermal barrier and entered the stele, the resistances encountered will be comparable to those of the cortex; water will continue to move across the stele until it finally enters the xylem vessels and tracheids to be moved to the shoot system.

Root Pressure. If a plant is growing under conditions of very low transpiration with the root system in a moist, warm, well-aerated medium, droplets of liquid will frequently appear at the tips and occasionally along the margins of the leaves of many species. These droplets, which are termed guttation drops, are forced out of the leaf through hydathodes, which are specialized pores in the leaf epidermis overlying a region of loosely packed parenchyma cells (the epithem). The force responsible for guttation arises in the roots and is termed root pressure. Another expression of the same phenomenon can be seen when the stems of certain species, particularly some vines, are excised and water is forced out of the stumps remaining. The pressure that can develop in roots is usually in the range of 1 to 3 bar, but values as high as 6 bar have been measured (White 1938).

A number of explanations have been offered to account for the development of root pressure. The most likely explanation is that it is due to an increase in the ψ_s value of the root xylem sap, brought about by the active translocation of inorganic ions and organic materials into the xylem vessels and tracheids from the surrounding xylem parenchyma. Water will then move across the root in response to the water potential gradient developed. In other words, the root behaves essentially like the osmometer illustrated in Fig. 2.12, the water entering the xylem vessels creating a hydrostatic pressure within the stele.

Numerous attempts have been made to ascribe an important function to root pressure. A pressure of 1 bar could raise water almost 10 m and so, naturally, many scientists have proposed root pressure as a mechanism to account for the ascent of water through the plant. Several lines of evidence argue strongly against this hypothesis, however. Firstly, root pressure has not been shown in all species. Secondly, it frequently exhibits a diurnal periodicity, being maximal during the day and minimal at night; moreover it is usually seasonal, being at its greatest during the spring and lowest or not measurable in the summer when transpiration loss is maximal. Finally, it can be clearly demonstrated that during periods of active transpiration xylem water columns are under tension and never exhibit a positive pressure. Roots under cold or anaerobic conditions, or roots killed by metabolic poisons, do not show root pressure, presumably because under such conditions solute transport to the xylem ceases. Transpiration, however, will continue essentially unaffected for a considerable time after the roots of a plant have been killed and may even increase if the root system is excised.

One suggested function for root pressure which may have more validity is the idea that it could reconnect, and thereby reestablish, the continuity of xylem water columns in which cavitation has occurred due to rapid transpiration, freezing, or some other insult. Overall, however, it is probable that root pressure is a purely fortuitous consequence of the accumulation of salts in the xylem with no essential role in the total water economy of the plant.

2.5 The Root-Soil Boundary

At the beginning of this chapter reference was made to a commentary by Robert Hooke in which he equated the functions of the shoot and root systems of plants as organs obtaining nourishment from their respective environments and, in turn, modifying those environments by *"seasoning"* them with by-products of their growth activities. This remarkably astute suggestion, for it was little more than that at the time, has, in the broadest sense, proved to be correct. Today, the relationship existing between the shoot system and the atmosphere is very well understood but, unfortunately, there are still a great many gaps in our knowledge and understanding of the relationship between roots and the soil under natural conditions.

Most of our information on root-soil relationships has come from experiments in which usually small plants have been grown in containers of fixed dimensions containing a relatively homogeneous planting medium under controlled environmental conditions. However, it is difficult to extrapolate data obtained from such carefully regulated systems to the situation pertaining in the field. Plants growing in the field show enormous variation in root morphology both between and within species; also, field-grown plants are in soils which may show wide variation in composition, structure, and water status throughout the rooting depth. When one considers that this inherent variation in both roots and soil may be subject to further considerable modification throughout the growing season depending on environmental conditions, and that it is essentially impossible to devise satisfactory means of investigating many root-soil interrelationships without changing the system, then perhaps it is not surprising that we know so little.

Roots obtain their water and mineral nutrients from a very narrow zone of soil immediately adjacent to the roots. As the water and nutrients are withdrawn from the rhizosphere they are replaced from the mass of soil lying some distance away which is not itself directly exploited by the growing root. While a great deal is known about factors involved in the regulation of water and nutrient movement through soil, and the mechanism of uptake of water and ions and their movement within roots, precisely how the rhizosphere might affect water and ion movement between soil and root is very imperfectly understood.

The boundary zone between the mass of soil and the young, growing root consists of the outer, frequently cutinized, surface of the epidermal cells, and root hairs where present, together with the mucigel (Sect. 2.3), root exudates, microbial populations characteristic of the rhizoplane and rhizosphere, and those soil particles adhering to the viscous, fibrillar components of the mucigel. As one moves backward from the root tip toward the older regions of the roots, the mucigel component disappears, but there are likely to be more and different organic materials contributed to the zone through the death and decay of older epidermal and cortical cells. Because of the presence of such a diverse variety of energy-rich organic compounds in the root exudates (Table 3.2), and the high metabolic activity of many of the microbial species present, this boundary zone must be a veritable cauldron of chemical syntheses and interconversions which, it might reasonably be anticipated, must influence the transfer of materials between root and soil.

Because many microbial species show specific substrate requirements, and the nature of the organic constituents of the rhizosphere will vary at different positions along the root (Egeraat 1979), it is quite conceivable that the diversity and nature of the microbial components associated with the rhizosphere will also vary along the length of the roots. This is an area of inquiry which has not so far been fully exploited and which deserves further attention, as it could be of great importance in understanding some of the differences observed in the physiology of roots at different distances from the apex.

Much of our recent knowledge of the distribution and growth of roots under more natural conditions has been obtained from experiments conducted in rhizotrons. A rhizotron is a root-observation laboratory which consists basically of a subterranean chamber with slightly sloping glass walls some 3 to 4 m in height. The area behind the glass walls is divided into a series of compartments by means of concrete partitions. Soil profiles can be constructed in these compartments and planted with the species under study. As the roots grow vertically downward, many of them become pressed against the sloping glass walls and their growth can be observed. With the addition of sophisticated monitoring equipment, variables such as plant and soil water potentials, oxygen tension in the soil atmosphere, and the amount of solar radiation received by the plant can be measured continuously. The rate and direction of growth of individual roots can be followed at the glass surface, and a permanent visual record can be obtained through the use of time-lapse cinematography.

Superficial examination of the root systems of potted, greenhouse-grown plants in soil maintained at or near field capacity might suggest that growing roots maintain close contact with the soil particles, but a closer examination shows that this is not necessarily so. Frequently, roots can be seen to follow quite large gaps or channels between the soil particles. Using the more natural conditions provided by a rhizotron, it can be seen that plant roots often tend to follow pathways in the soil which present the least resistance to growth and so they frequently grow selectively along channels which are much wider than the diameter of the roots. Soils, particularly those with a high clay content, shrink and expand depending on the water content. Hence, the diameter of the channels through which roots grow may vary with the soil water status. Nevertheless, it is reasonable to assume that a considerable portion of the root system of a plant growing in the field is not at any one time making direct contact with the mass of soil. A further factor which has been found to influence root contact with the soil particles is the change in root diameter which can occur in water-stressed plants. Working in the rhizotron at Auburn, Alabama, Huck et al. (1970) showed that when cotton roots are growing in a drying soil diurnal changes in the diameter of the roots occur and, during the afternoon hours, the roots may shrink to about 60% of their original diameter.

It might be anticipated that the gaps between roots and soil, particularly under conditions of water stress when root diameter is decreased, would contribute significantly to the resistance to movement of water from soil to roots. This viewpoint has been disputed, however, by some scientists who believe that soil-root gaps may have little effect because water may be transferred in the vapor phase.

At this point we should again consider the possible role of root hairs in the interchange of materials between root and soil. The root hairs have been assigned by many writers an important role in both water and ion absorption. According to Newman (1974), however, there is little direct evidence to suggest that root hairs enter and extract water from pores in the soil structure. Indeed, rhizotron observations have shown that root hairs are frequently most obvious on those roots growing in soil channels where they do not make direct contact with the mass of soil. Two features of root hairs that are frequently ignored are that they anchor the growing root to soil particles and, as they age, they produce a viscous cuticle-bounded mucilage (Foster 1981) which is colonized by large populations of microorganisms. R. S. Russell (1977) suggests that while we cannot at present discount the possibility that root hairs are important in ion translocation and uptake, they could be functioning primarily as a source of organic substrates for rhizosphere organisms. He points out that those observations indicating a more rapid rate of ion uptake from soil in the root-hair zone could be explained equally well if diffusion of ions was more rapid in soil with extensive rhizosphere populations. Indeed, D. A. Barber et al. (1976) showed that the rhizoplane microflora increased the uptake and translocation of phosphate in barley.

One feature of root physiology which could have a marked effect on the rhizosphere flora is differential ion uptake. For example, the rapid absorption of phosphate by roots may lead to a local depletion of phosphate in the rhizosphere if the rate of diffusion of phosphate in the soil is low. On the other hand, certain ions, e.g., Al and Ca, may only be absorbed by roots very slowly, leading to their accumulation in the rhizosphere (Bowen 1981). Similarly, the form of nitrogen used as fertilizer may lead to a local change in pH of the rhizosphere, which may in turn bring about a change in the rhizoplane microflora (Smiley 1975).

The foregoing brief discussion of the root-soil boundary should be sufficient to indicate that we are dealing with a highly complex and poorly understood system. Many of the observations on which our theories of water and ion translocation and uptake are based have been made on pot-grown plants, and there is little evidence to suggest that for plants growing in the field the situations would be the same. Indeed, because of the enormous variations which can occur in both plant and soil, it is unlikely that any general conclusions reached will apply in all situations. One thing, however, which is very clear is that until recently little thought has been given to the possibility that the behavior of roots might be modified by their associated microbial flora. As our knowledge of the complex interactions occurring in the rhizosphere increases, it is quite likely that we will have to revise and modify some of the currently held precepts relating to the mechanism of root functioning.

2.6 Root Exudation

The existence around roots of a narrow zone of soil densely populated by microorganisms is directly due to the roots releasing a wide variety of organic materials which serve as a food source for the microorganisms. Because of their variety and complexity, it is convenient to adopt some classification of the types of materials known to be released by roots. We use here the classification of Hale et al. (1981), which was adapted from the one devised by Rovira et al. (1979).

Root exudates: Chemicals and elaborated metabolites released to the surface of the root or released into the root environment.

Root exudation: A process involving several pathways and biochemical mechanisms.

Leakages: Compounds of low molecular weight which diffuse into the apoplast and, via the apoplast, move to the root surface, or leak directly from epidermal or cortical cells.

Secretions: Compounds which cross membrane barriers as a result of expenditure of metabolic energy.

Mucilages: Four sources contributing to organic materials in the rhizosphere:

A. Those originating in root-cap cells and secreted by Golgi bodies.
B. Polysaccharide hydrolysates of primary cell walls between the epidermis and sloughed root-cap cells.
C. Those secreted by epidermal cells with only primary walls; includes root hairs.
D. Those produced by bacterial degradation of walls of old, dead cells.

Mucigel: The gelatinous material at the surface of roots: includes natural and modified mucilages, bacterial cells, metabolic products, colloidal mineral and organic matter.

Lysates: From autolysis of older sloughed cells which become heavily colonized or from released microbial metabolites.

A number of the low-molecular weight compounds released by roots are volatile at the temperatures and pressures occurring in nature. Because of this they are able to diffuse through the soil and affect the activities of microorganisms lying at some distance outside the rhizosphere (Linderman and Gilbert 1975). In this respect, these volatile compounds (alcohols, fatty acids, alkyl sulfides, etc.) differ from the other exudates, which tend to be concentrated in the rhizosphere and are frequently metabolized by the rhizosphere organisms before they can diffuse outward (Minchin and McNaughton 1984). For this reason, it would seem to us that in Hale's classification there should be a subdivision under leakages specifically to include volatile compounds which may well be more significant in relation to microbial activity in the mass of soil than they are in relation to microbial activity in the rhizosphere.

The range of organic materials known to be released by roots is extensive and will not be considered further here. These compounds are, however, considered in depth in Chap. 3.

Most of our information about root exudates has come from studies of plants growing under sterile conditions in nutrient solutions. The results of such investigations may, however, bear little relationship to the system pertaining for plants growing in soil under field conditions, because it is known that the presence of microorganisms in the rhizosphere increases root exudation (e.g., see Barber and Martin 1976; Prikřyl and Vančura 1980). The reasons for this are not entirely clear. It could be due to physical damage to the root tissue by the bacteria, or it may be due to the release from the microorganisms of metabolites which affect root physiology leading to increased exudation. Exudates are produced from plant metabolites, largely carbohydrates, which are primarily synthesized in the shoot system during the process of photosynthesis and then translocated to the root system. Clearly, therefore, any process affecting plant growth or physiology could have an affect on the quality and quantity of materials exuded by the roots. Many agricultural chemicals which have been applied as foliar sprays to plants have subsequently appeared in the exudate from roots (see Chap. 3). Some of the microorganisms occurring in the rhizosphere are known to release toxins as well as plant-growth regulators such as IAA, gibberellins, and cytokinins. It might be anticipated that these compounds would influence the release of exudates concomitantly with their effects on plant metabolism and growth (Hale et al. 1981). Research has generally tended to show that change in any biological or physical factor which affects plant growth also affects the quantity, and sometimes type, of exudate released by roots.

The mechanism by which the rhizosphere becomes enriched with these organic molecules is fairly well understood for some of the materials classed as exudates. For example, the origin of the mucigel in the root-cap tissue and its movement to the outside of the roots in vesicles derived from dictyosomes was described earlier (Sect. 2.3). It is also easy to see how necrotic and dead cells, abraded or sloughed off as the root grows through the soil, become lysed and release their nutritionally rich cytoplasmic contents. Similarly, the region of the root where lateral roots emerge is particularly rich in organic materials because the secondary roots, in forcing their way through the cortical tissue of primary roots, cause a considerable amount of tissue damage and the contents of these damaged cells then leak to the outside (Sect. 2.3, and Fig. 2.9). However, very little is known about how many of the materials found in the rhizosphere leave the symplast of the root to enter the apoplast and subsequently leak to the outside. The amount of such materials can be considerable. J. K. Martin (1977), working with a nonsterile root system, estimated that as much as 20% of the material assimilated by the plant could be lost in the form of root exudates.

Information which may prove to be very pertinent to an understanding of the release of exudates from living root cells has been obtained in studies conducted by Simon (1974, 1981) and others on the leakage of solutes that occurs when dry seeds are placed in water. In Simon's experiments, leakage was measured in terms of the conductivity of the solution into which the seeds were placed. He found that intact pea (*Pisum sativum* L.) seeds lost up to 30% of their conducting materials within 12 to 24 h, and when the testas were removed from the seeds some two to three times the amount of electrolytes leaked out. When the imbibing seeds or embryos had reached a water content of about 30% leakage was essentially halted. Simon believes that most of the solutes leaking from dry or slightly hydrated seeds are lost because, in the absence of sufficient water, cell membranes are disorganized. When sufficient water has been imbibed, the membranes rapidly assume a normal structure and further leakage is prevented. Simon derives theoretical support for his contention that at low water levels membranes become disorganized from the X-ray-diffraction studies of phospholipid-water mixtures made by Luzzati and Husson (1962). In their work, Luzzati and Husson found a variety of arrangements were possible in phospholipid-water mixtures and, at water concentrations below 20%, an arrangement described as a hexagonal mesophase was found. This is a predominantly hydrophobic phase interspersed by long water-filled channels. Such an arrangement in the cell membrane would permit the free exchange of soluble materials between the cell interior and the external environment. While Simon does not believe that the organization of membranes in a dry seed is likely to correspond exactly to the hexagonal mesophase arrangement, he believes that in dry seeds plasmalemma architecture is disorganized in some manner. Evidence similar to that of Simon's indicating a relationship between leakage of plant solutes and degree of tissue hydration has not been obtained for roots but, like the best of scientific hypotheses, the view of leakage he presents immediately suggests and invites experiments which could be extremely informative. In the meantime, there is some indirect evidence suggesting that root membrane permeability may be a factor in leakage. For example, leakage from roots is greater when the bathing medium contains only low concentrations of calcium (Shay and Hale 1973; Hale et al. 1981), which is known to be important in the maintenance of cell-membrane integrity. Leakage is also increased in the presence of respiratory poisons, such as azide and cyanide, and in the absence of oxygen, conditions under which maintenance of membrane organization would be expected to diminish due to a failure to produce ATP. Finally, one of the environmental factors known to affect leakage is the drying of roots.

As the foregoing indicates, the study of root exudation in nonsterile systems is highly complex because it is exceedingly difficult to differentiate between the intrinsic properties of the roots and the superimposed effects that the associated microorganisms may have on root growth and physiology. Many careful, balance-sheet studies of the contribution made by each component to the total system, similar to the recent study of Minchin and McNaughton (1984), will need to be conducted before any clear conclusions are likely to be reached.

Chapter 3 Root Exudates

Introduction

Hiltner's rhizosphere could not exist without the underlying phenomenon now most commonly referred to as root exudation, which is associated with all higher plants. Other terms that have been applied to the release or movement of substances from living roots into the surrounding soil are excretion, secretion, diffusion, liberation, and leakage. This varied terminology may reflect some confusion created by the fact that the mechanisms responsible for release of organic substances from roots have not been entirely clarified (see Chap. 2.2.6).

The liberation of compounds from roots in most experimental systems probably occurs by diffusion along electrochemical potential gradients ending in leakage. The exudation of ^{14}C-labeled photosynthates probably occurs by this process. While diffusion and leakage of cell contents, controlled by cell-membrane permeability, may be the fundamental process of exudate release, the more selective process of secretion also must be involved. Secretion, which requires the expending of metabolic energy, is a common process of living plant cells. The dictyosomes of root-cap cells secrete polysaccharides which, in experimental plant-culture systems, may appear as viscous droplets on the root tips. Hale et al. (1978, 1981) have discussed this topic in more detail and suggested a number of plant and environmental factors which may affect cell-membrane permeability and the kinds or quantities of exudates released.

Materials sloughed off during plant growth, such as root caps, epidermal cells and root hairs, do not comprise part of exudation, though they are extremely important in relation to microbial activity in the rhizosphere and must be considered along with the exuded materials. It is well known that injuries to roots, such as broken root hairs, incurred during growth through soil, can result in a higher amount of exudate; thus the difficulty of distinguishing between true cell exudate and combined exudate plus other cell contents can be appreciated. Root exudation also must not be confused with the metabolites synthesized through microbial activity, which may contain many of the same compounds found in exudates. Researchers seeking answers to fundamental exudation questions attempt to avoid the injury and microbial factors by employing special methods of axenic or gnotobiotic culture of experimental plants.

We begin our study with Hiltner because he gave us the term rhizosphere. But Börner's review (1960) on liberated organic substances in relation to the "soil sickness" problem toxic theory) reminds us that the ideas and philosophies of others long before Hiltner contributed toward the eventual recognition of root "excretions"; among these are such notables as Plenk in 1795, followed by De Candolle, Daubeny, and Liebig. Perhaps Micheli (1723) even earlier was referring to a root exudate effect when he observed that seeds of *Orobanche* germinated only in the presence of host roots. More directly related to the rhizosphere effect was the observation of Halsted (1888) that bacteria proliferated around the root tips of clover, probably, he surmised, due to the exudation of stimulatory substances from root cells. Russian scientists have been active for many years in the study of root "excretions", particularly in relation to their effects on *Azotobacter* and other bacteria which through their activities enhance plant growth (Krasilnikov 1958). Investigations which demonstrated that nodulated roots of legumes liberate nitrogenous substances have contributed to our basic knowledge of the exudation phenomenon, but this process, primarily a function of nodule development, is now considered to be outside the realm of true root exudation. Interest in plant-growth inhibitors released by roots goes back to the suggestion of De Candolle (1832) that substances from plant roots may retard the growth of other higher plants (allelopathy); this topic was reviewed by Börner (1960) and Rice (1979). Though Lyon and Wilson in 1920 had found that plants grown in sterile-water culture liberated nitrogenous substances and other organic matter, the study of root exudation in relation to what we now term the rhizosphere effect on microbial activity and plant health did not gain momentum and attract widespread attention until about 1955. From that time to the present, perhaps no individuals have contributed more to our knowledge of the subject than A. D. Rovira and his associates in Australia, and V. Vančura of the Czechoslovak Academy of Science.

It is evident from world-wide contributions that interest in root exudation in relation to rhizosphere phenomena is more than academic. Exudation relates to ecological investigations dealing with plant and microbial succession; root-cell permeability; and metabolism, growth, reproduction, and survival of components of the microflora and fauna. From the practical agriculture viewpoint, exudates are involved in microbial inhibition or stimulation of the growth of plants. A better understanding of the nature of exudates may provide the much-sought-after key to biological control of certain root-disease organisms (Baker and Cook 1982). A number of international symposia on the subjects of plant-root environment and microbial ecology relating to plant growth and root disease have included discussions on the role of root exudates.

3.2 Methods of Collection and Analysis

Manorik and Belima (1969) summarized and compared various methods that had been used up to 1968 for obtaining and expressing amounts of root exudates. No revolutionary new methods have been devised subsequent to that time, but existing techniques have been modified, biochemical procedures of analysis have been refined, and more sophisticated instrumentation has been developed. Increased use of radioactive materials has aided in detecting sites on roots where exudate components are released. Generally, the procedures used for collecting and analyzing exudates can be grouped as follows:

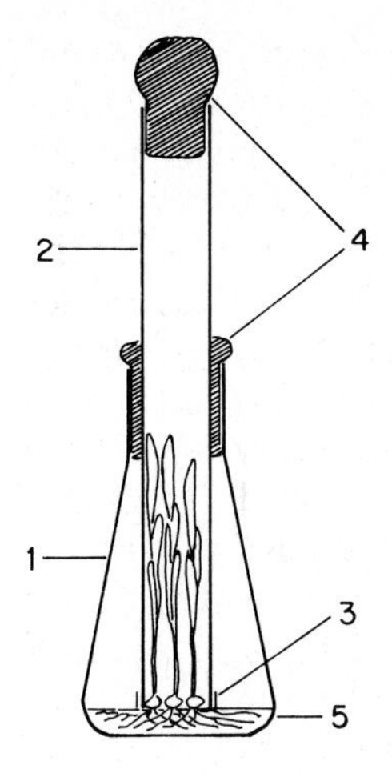

Fig. 3.1. Experimental system for the cultivation of plants under axenic conditions. *1* Wide-neck Erlenmeyer flask (500 ml); *2* glass tube (length 400 mm, diameter 35 mm); *3* gauze; *4* cotton stopper; *5* nutrient medium. (Prikřyl and Vančura 1980)

1. Plants are grown in sterile water or nutrient solution culture and the solution is analyzed for substances considered to be released from roots into the solution.
2. Plants are grown in solid substrates (soil, sand, or simulated soil) and washings or leachates from roots and from the substrate are analyzed by standard chromatographic and biochemical procedures.
3. Plants are grown in soil or other solid substrate and aerial parts are exposed to radioactive isotopes; any radioactive material subsequently released by the roots is then detected in the substrate leachates.

These basic procedures are represented in Table 3.1, which shows chronologically the transition from very simple to more sophisticated methodology. Some of the procedures have remained rather constant with only slight alterations, governed by the purpose of the study and the whim of the experimenter. When seeds or very young seedlings are being studied, exudates can be collected in small amounts of a nutrient medium (Fig. 3.1; Prikřyl and Vančura 1980) or even in sterile water without added nutrients. Seed sterilization is required to assure exclusion of living microbes and their metabolites; this is usually through treatment with alcohol +7% (w/v) sodium hypochlorite solution or sodium hypochlorite + a wetting agent such as Tween-20 (polysorbate), or by treatment in 0.1–0.2% (w/v) mercuric chloride or silver nitrate solution. In all cases, treatment should be followed by thorough washing of the seed in sterile water. Seeds then may be germinated on a nutrient agar medium, any seedlings still contaminated being discarded; sterile seedlings are transferred to sterile distilled water or plant-nutrient solution. Sometimes, it may be desirable to separate seed from root exudates. This can be done by supporting the seed on a perforated platform with only the roots extending down into the collecting solution. Where plants are grown in a nutrient solution, the solution containing the exudates is usually desalted before analysis by percolating through cation exchange resins.

If one is attempting to simulate more natural field conditions for study of root exudates, plants must be grown in a soil or a simulated soil-like substrate. Presumably because of damage occasioned by the mechanical forces when roots

Table 3.1. Selected methods used for collection and analysis of root exudates

Reference	Subject	Collection procedure	Analysis or assay
West (1939)	Execretion of bacterial growth factors by roots of flax in relation to character of rhizosphere	Seed sterilized, germinated, seed coats removed, and seedlings planted in tubes of sterile nutrient solution. After periods of growth, solutions combined and concentrated to 10 ml	Plant-growth solution assayed in growth response of *Staphylococcus aureus* in a basal medium minus thiamin. *Rhizobium trifolii* used for biotin assay
Parkinson (1955)	Liberation of amino acids by oat seedlings	Oats surface sterilized and grown in sterile-sand culture in Lees-Quastel perfusion apparatus (Lees and Quastel, 1946). Perfusates collected, concentrated *in vacuo* and residue taken up in alcohol	Colorimetric tests for amino acids present, and chromatographic analysis for specific amino acids
Katznelson et al. (1955)	Liberation of amino acids and reducing compounds by roots of pea, soybean, wheat, barley, and tomato in relation to desiccation	Seeds surface sterilized by immersion in 95% ethyl alcohol then in 1:1000 $HgCl_2$ 10 min; washed, germinated on agar and transplanted to pots of sand or sand-soil; one lot kept moist, another allowed to dry to signs of wilting. Pots leached with water and leachates stored in deep-freeze. Other plants grown aseptically by method of Blanchard and Diller (1950), leachates pooled and freeze-dried	Samples concentrated by freeze drying, residue taken up in isopropyl alcohol, and subjected to ascending paper chromatography. Paper dried, sprayed with ninhydrin (0.2% in ethyl alcohol) and developed at 45 °C. Another set developed with isatin color reagent for amino acids. Third set sprayed with *p*. anisidine hydrochloride and developed at 105 °C. Some leachates dispensed in small tubes, glucose added (=0.1% solution), sterilized, and inoculated with amino acid-requiring bacteria
Rovira (1956)	Excretion product of oat and pea roots grown under seminatural conditions	Seeds sterilized by immersion in alcohol, then in 0.2% $HgCl_2$ + Tween 80 wetting agent; serially washed, germinated on potato dextrose agar (PDA) and uncontaminated seedlings transferred to sterile tubes with acid-washed quartz sand + modified Crone's nutrient solution. Roots harvested, rinsed in distilled water; sand washed, and root + sand washings centrifuged, concentrated under vacuum and desalted	Amino acids and sugars determined by descending chromatography, followed by appropriate differential developing sprays of chromatograms for specific components. Root-cell debris also was determined from centrifuged material, dried and weighed

Rovira (1959)	Influence of plant species, age, light, temperature, and calcium nutrition on exudation by tomato roots	Seed surface sterilized in 7% sodium hypochlorite, washed, and germinated on glucose-yeast extract agar; contaminant-free seedlings transferred to nutrient solution in tubes	Amino acids determined by the ninhydrin test; relative concentration assessed approximately by spot color intensity rating
Bowen (1961)	Toxicity of legume seed (clover and lucerne) exudate to rhizobia and other bacteria	Seeds shaken in alcohol + hydrogen peroxide, and washed; seeds soaked in sterile water 18 h at 20 °C and diffusate decanted	Tested for effect on legume bacteria and other organisms by agar-plate assay
Pearson and Parkinson (1961)	Sites of exudation from broad bean *(Vicia faba)* seed and young roots	Seeds surface sterilized in Nance's solution 10 min, washed in sterile water, and planted in póts. Germinated seed samples taken at intervals, washed in water and placed on filter paper in dishes. Second sheet of paper (Whatman No. 50) placed on top of beans and moist Kleenex tissue placed over and around this. After 24 h at 25 °C, position of seed and root marked with needle punctures. Beans transferred to other dishes of filter paper for observation of damage	Paper from beneath beans (original dishes) dried in warm air and sprayed with 0.25% ninhydrin in alcohol, then dried again. Coloration patterns indicated sites of exudation representing primarily amino acids
Rovira and Harris (1961)	Exudation of B-group vitamins in relation to rhizosphere effect	Seeds of tomato and selected legumes surface sterilized with calcium hypochlorite and germinated on water agar. Plants grown by method of Rovira (1959) in either distilled water or plant nutrient solution. Solutions supporting roots were changed at intervals and the collected solutions concentrated	B-group vitamins of root-exudate concentrate determined by bioassay using organisms with essential growth-factor requirements: species of *Neurospora* and *Lactobacillus*
Harmon and Jager (1962)	Carbon and nitrogen in the rhizosphere of young wheat, spinach, and vetch plants	Roots from disinfected, germinated seed grown in perforated nickel-plated growth chamber inside larger container (glass pot) with artificial soil (sand, potassium-feldspar, kaolinite) wetted with Crone's solution. Soil from small chamber washed and washings dried	Nitrogen determined by micro-Kjeldahl method; carbon by dry combustion

Table 3.1 (continued)

Reference	Subject	Collection procedure	Analysis or assay
Sulochana (1962a)	Amino acid patterns of root exudates from cotton	Acid-delinted seed surface sterilized with $HgCl_2$ and washed; germinated in sterilized washed sand, inoculated with *Fusarium oxysporum* f. sp. *var. infectum*, and grown in sterile soil-water culture in flasks. After 15 days, soil-root washings filtered, pooled, concentrated *in vacuo*, made up to 250 ml and refrignerated with surface layer of toluene	Amino acid bioassays with *Lactobacillus arabinosus* and *Leuconostoc mesenteroides;* growth measured colorimetrically
Vančura (1964)	Root exudates of barley and wheat in early growth phase	Seeds disinfected with $HgCl_2$ solution, washed, and transferred to washed silica sand. After 7–9 days dish lids removed and plants allowed to partially dry, then rewetted. After 12 h, sand + roots washed in Büchner funnel with sterile water. Solution with exudate centrifuged, supernatant decanted and water removed by freeze-drying. Powder ground in mortar and stored over potassium hydroxide	Sugars, amino acids, organic acids, and phenolic substances determined by chromatographic methods
Schroth et al. (1966)	Temperature effect on quantitative differences in exudates from germinating seeds of bean, pea, and cotton	Seed coats removed, seed disinfected in sodium hypochlorite + Tween-20 wetting agent, and rinsed; planted in sterile moist silica sand in Petri dishes and incubated at different temperatures; exudates recovered by washing sand with deionized water	Ninhydrin-positive compounds and soluble carbohydrates determined with a Spectronic 20 photometer at 570 mμ (glycine standard) and 620 mμ (glucose standard, anthrone method), respectively
Boulter et al. (1966)	Amino acids from pea seedlings	Seeds treated with calcium hypochlorite solution on shaker, germinated in semisolid glucose agar; uncontaminated seedlings washed and grown in tubes of quartz sand wetted with N-free nutrient solution, or seedlings made to rest on invaginations of tube walls with roots submerged in culture solution. Seedlings harvested, roots and culture sand washed, and washings concentrated on rotary evaporator	Samples of concentrated washings analyzed with a Technicon Amino Acid Autoanalyser

Vančura (1967)	Effects of temperature and "cold shock" on exudates from seeds and seedlings of maize and cucumber	Seeds washed in sterile distilled water and surface sterilized with 0.1% $HgCl_2$ solution 20 min. Again washed repeatedly and seeds placed on Whatman No. 1 filter paper in sterile dishes and germinated at low to high temperatures. After development of young roots, seeds were discarded and filter papers eluted in distilled water on shaking machine. Eluates filtered through G3 glass filter	Filtered eluates fractionated on ion exchangers and analyzed by paper partition chromatography
Ayers and Thornton (1968)	Exudation of amino acids by intact and damaged roots of wheat and pea	Seeds surface sterilized by washing in Triton X-100 (Alkylphenoxypolyethoxyethanol) then in $AgNO_3$, washed with sterile NaCl, and rinsed in water: germinated on water agar and vigorous seedlings transferred to plant-culture vessels with washed quartz sand + nutrient solution. Other plants grown with roots in nutrient solution. Culture vessels designed with a perlite layer to separate germinating seed from subsequent root system and for monitoring gas exchange. Water rinsings of sand-grown plants and the solution cultures were filtered, concentrated in rotary evaporators and analyzed or stored frozen	Amino acid N determined by colorimetric method of Rosen (1957); solutions desalted by electrolytic apparatus and subjected to descending chromatography followed by detection of amino compounds with ninhydrin spray in n-butanol. Identify of compounds confirmed by chromatography
Booth (1969)	Amino acid exudation from roots of cotton tolerant or susceptible to *Verticillium albo-atrum*	Acid-delinted, uniform seeds, surface-disinfected in 0.1% $HgCl_2$ and rinsed with water. Planted in acid-washed sand in large test tubes; uniform seedlings selected for study. These grown long-term in specially designed apparatus with sterile sand + nutrient solution. Exudates collected at intervals by flooding unit and collecting through tube. Stored at −15 °C	Samples evaporated *in vacuo* at 50 °C and brought to 250 ml with water. Amino acids separated by paper chromatography
Bowen (1969)	Nutrient status effects on amides and amino acids in pine exudates	Pine seed surface sterilized with hydrogen peroxide + alcohol, washed in sterile water, and germinated on agar; germinated seed planted on stainless steel mesh just above nutrient solution in tubes. After growth of roots, solution pooled and reduced by rotary-film evaporation, desalted through cation-exchange column, and concentrated to 1 ml/240 plants	Amides and amino acids separated by two-dimensional paper chromatography with known compounds and spectrophotometric analysis

Table 3.1 (continued)

Reference	Subject	Collection procedure	Analysis or assay
W. H. Smith (1969b)	Release of organic materials from roots of tree seedlings	Pine seed surface sterilized in $HgCl_2$ and sodium hypochlorite, and rinsed with water; germinated on malt agar and contaminant-free seedlings grown in test tubes with glass beads and nutrient solution filtered, deionized, and concentrated to 1/200th original volume in flash-evaporator	Amino acids, organic acids, and carbohydrates separated by conventional chromatographic procedures
Chang-Ho (1970)	Effect of pea root exudates on germination of *Pythium aphanidermatum* zoospore cysts	Seeds surface sterilized in H_2SO_4 2½ min then repeatedly washed in sterile distilled water; planted on PDA and sterile seedlings transferred into glass cylinders with acid-washed sand. After 3 weeks, water with exudates drained off through bottom tube and system flushed with water. Solution concentrated *in vacuo* at 40 °C to 1 ml/10 plants, or freeze-dried, redissolved in water	Crude exudate separated into cationic, anionic, and neutral fractions through Dowex ion exchange resin columns and stored at −20 °C. Analysis of fractions by thin layer chromatography on cellulose plates for identification of sugars, amino acids and organic acids. Solution then tested on zoospore cysts in deep-well slides
McDougall (1970)	Movement of ^{14}C photosynthates into roots and exudation of ^{14}C from roots of wheat	Seed surface sterilized with 7% hypochlorite, washed, and germinated on agar; transferred to test tubes of sterile Hoagland and Arnon nutrient solution and grown in alternating light and dark. $^{14}CO_2$ injected into tube air space for photosynthetic assimilation and translocation of ^{14}C to roots	Translocation of ^{14}C to roots and detection of radioactive substances exuded into the medium around roots performed by autoradiography with No-screen X-ray film, scanning dried roots with radiochromatogram strip scanner, and ethanol extraction of soluble fraction from root tissue
Juo and Stotzky (1970)	Separation of proteins from root exudates of corn, pea, and sunflower	Plant species grown with roots maintained aseptically according to Stotzky et al. (1962) in a controlled environment, and soluble organic compounds liberated were collected at intervals. Collections plus washings were dialyzed against tris-glycine buffer (pH 8.3) and concentrated by evaporation from the dialysis bags in a stream of air to a volume of 1 ml. Concentrated samples redialyzed and centrifuged, and supernatants retained for analysis	Supernatants were subjected to electrophoretic separation of proteins, and total protein analysis performed by the method of Lowry et al. (1951)

Reference	Objective	Collection method	Analysis
Kovaks (1971)	Identification of aliphatic and aromatic acids in root and seed exudates of pea, cotton, and barley	Pea and barley seeds surface sterilized in $AgNO_3$; cotton in Triton X-100 then sodium hypochlorite solution and water rinse. Seeds germinated in filter paper cones supported by glass tube inside a larger tube. Cones moistened with nutrient solution. After growth, filter papers were combined and growth solutions were combined for analysis	Seed and root exudates analyzed by thin-layer gas-liquid chromatography for aliphatic and aromatic acids
J. K. Martin (1971a)	^{14}C-labeled material leached from the rhizosphere of wheat, clover, and ryegrass exposed to ^{14}C	Plants grown in pots with bottom drainage tubes and podzolic sand in a chamber designed for $^{14}CO_2$ atmosphere. Potted 4-week-old plants in polyethylene bags treated by injection of $^{14}CO_2$. Bags removed after 6 h, pots leached with distilled water and leachates collected. Growth procedures designed to include interactions of natural microflora	Activity of ^{14}C leachates and fractions from resin and Sephadex columns measured with liquid scintillation spectrometer providing information on water soluble organic compounds
Rittenhouse and Hale (1971)	O_2 and CO_2 effects on release of sugars from peanut roots	Embryos removed from peanut seed and surface sterilized 5 min in 20% sodium hypochlorite; germinated on salts-sucrose medium, selected for uniformity and transferred to transplanting tubes inside sterile isolation chambers. When roots reached length of tubes, the tubes with plants were inserted into the neck of 500-ml flasks containing nutrient solution. O_2, CO_2, or mixtures with H_2 were bubbled through the nutrient solution intermittently for 6 weeks	Nutrient solution was collected from three plants at intervals and pooled, filtered, reduced in volume, and prepared for standard chromatographic analysis
Vančura and Hanzlíková (1972)	Differences in chemical composition of seed and seedling exudates from barley, wheat, cucumber and bean	Seeds serially washed, then surface sterilized with $HgCl_2$, washed again, and seed germinated on moist filter paper in dishes. Noncontaminated seeds removed, paper cut into pieces and these eluted with distilled water; the solution centrifuged and supernatant lyophilized. For root exudates, sterilized seed were planted in dishes of washed quartz sand, plants later allowed to wilt, and watered again. Sand with roots washed in Büchner funnel and eluate used to wash other sand + roots; final eluate centrifuged and freeze-dried	Total N, free amino acids, protein and peptide N, and sugar analyzed by standard procedures described by others as cited

Table 3.1 (continued)

Reference	Subject	Collection procedure	Analysis or assay
Hamlen et al. (1972)	Influence of age and stage of development of alfalfa on carbohydrates in root exudates in a gnotobiotic environment	Seeds surface disinfested with 0.1% $HgCl_2$ and germinated on PDA, then planted in washed sand and subirrigated with a modified Hoagland's solution No. 1. Plants grown under gnotobiotic conditions (Lukezic et al. 1969). Plants clipped weekly. Exudates obtained from plants of different ages by flushing sand-root system *in situ* with sterilized, glass-distilled water. Exudate solution filtered (Millipore filter mean pore size 0.45 μm), concentrated under vacuum to 10 ml and frozen	Concentrated root exudates were desalted, lyophilized, and analyzed by gas-liquid chromatography-mass spectrometry for carbohydrate components
Papavizas and Kovacs (1972)	Fatty acids from the rhizosphere of snapbean and stimulation of spore germination of *Thielaviopsis basicola*	Seeds surface sterilized by method of Ayers and Thornton (1968) and germinated on water agar. Deep storage dishes with half-strength Hoagland's solution covered with perforated aluminum foil, were autoclaved and seedlings planted on the foil with primary roots through the holes into nutrient solution. This arrangement was placed inside deeper jars with proper air exchange. After growth, nutrient solutions from 10 plants were combined and evaporated under vacuum to 50 ml. Other seedlings were grown in sterile sand wetted with nutrient solution and later leached and condensed	Fatty acids were extracted and purified by standard procedures and components determined by gas chromatography
Rovira and Ridge (1973)	Effects of nutrients, microorganisms, and buffering compounds on exudation of wheat roots	Seed surface sterilized, germinated, and planted on stainless steel guaze above 30-ml sterile plant nutrient solution in 3 × 20 cm glass tubes as detailed by Ridge and Rovira (1971). Seedlings grown 6 days in controlled environment chamber. From some tubes trace elements or P were omitted; some tubes received soil suspension, others remained sterile. Appropriate buffer reagents added on 6th day of plant growth. All tops received $^{14}CO_2$ by injection through polyethylene covering the tubes	On days 7 and 8 solutions were sampled for radioactivity

Shay and Hale (1973)	Effects of calcium levels on exudation of sugars from peanut roots	Seed testa removed and seed surface sterilized 5 min in 20% sodium hypochlorite solution; germinated on nutrient agar and 5-day seedlings transferred to screw-capped glass jars and transported to transplanting tubes in perlite saturated with Hoagland's solution. After 4 weeks, sterile (axenic) seedlings selected for uniformity and placed in separate 1-liter flasks of nutrient solution; sterility checked. Various nutrient solutions appropriate to purpose of experiment used. Solutions from 3 plants of same treatment pooled, freeze-concentrated, filtered, and concentrated *in vacuo*, freeze-dried and stored at 5 °C	Sugars measured by gas-liquid chromatography of the trimethyl-silyl derivatives
Whalley and Taylor (1973)	Pea-root exudates, effect on germination of conidia and chlamydospores of *Fusarium oxysporum* f. sp. *pisi*	Seed surface sterilized with calcium hypochlorite (3% available chlorine) 45 min. Germinated on nutrient agar and uncontaminated seedlings grown in growth apparatus with roots immersed in 250 ml sterile, distilled water in growth cabinet with day-night control. Solutions tested for sterility. Uncontaminated solutions concentrated by rotary evaporation to 10 ml/ seedling, sterilized by filtration and stored at −13 °C	Crude exudates tested directly for effect on *Fusarium* spore germination without further analysis
Reid (1974)	Effects of water stress on movement of ^{14}C-labeled compounds in tree seedlings and resultant exudation of ^{14}C from roots	Pine seedlings maintained either in aerated distilled water or modified Hoagland's solution; both groups previously grown in vermiculite-nutrient solution medium. Root system removed, rinsed, and cleaned of root debris. Each plant placed in aerated solution 24 h before treatment in controlled environment chamber. Radioactive CO_2 introduced to seedling shoot by acidifying $NaH^{14}CO_3$ in closed assimilation chamber. Water stress treatments consisted of decreasing water potential of root-bathing media with polyethylene glycol	^{14}C determined by liquid scintillation spectrometer. Root solutions separated into fractions by use of ion exchange resins; all fractions evaporated *in vacuo* at 40 °C and samples counted for ^{14}C by liquid scintillation for detection of sugars, amino acids, and organic acids

Table 3.1 (continued)

Reference	Subject	Collection procedure	Analysis or assay
Barber and Gunn (1974)	Effect of mechanical forces on exudation of organic substances from barley and maize plants under sterile conditions	Plants grown under rigidly sterile conditions in enclosed Pyrex containers (Barber, 1967). Some root chambers contained nutrient solution only, some glass ballotini (1–3 mm diam). A sterile seedling planted in each vessel and grown under controlled light and temperature. Plants harvested in a sterile cabinet. Ballotini washed and washings added to original solutions; concentrated by rotary evaporation, filtered, and made up to 10 ml with deionized water; stored in deep-freeze. Roots separated from ballotini and mucigel removed by sonic cleaning in beakers of $CaCl_2$ solution. Solution put in bottle of chloroform and treated as first solution. Roots and shoots separated and treated in boiling 70% ethanol, rinsed, and solutions concentrated and frozen. Dry root weights determined	Assays for amino acids in various fractions by ninhydrin reaction and measure of resulting optical density. Ammonium determined by standard procedures; total carbohydrates by chemical procedures and expressed as milligrams of glucose. Root measurements and weights indicated effect of ballotini on growth as related to exudation
Booth (1974)	Effect of cotton root exudate on growth and pectolytic enzyme production by *Verticillium albo-atrum*	Cotton cultivars grown as previously described (Booth, 1969). Exudate solutions collected at 8 and 18 weeks, concentrated *in vacuo* at 50 °C and brought to 100 ml with deionized water; stored at −15 °C	Appropriate dilutions made for analysis. Growth factors determined by microbiological assay with specific organisms and compared with standards. Enzyme assays by standard enzymatic procedures
Egeraat (1975a)	Sites of exudation and composition of exudates from pea	Seeds surface sterilized by shaking in 3% H_2O_2 solution + drop of detergent for 20 min; germinated on water agar. Used modification of technique of Pearson and Parkinson (1961). Seedlings in dishes on wet paper tissue, root tips just in contact with a filter paper disc. Another moist filter paper disc placed over first discs and the root. After 24 h in darkness, upper paper disc removed and seedlings transferred to another similar device. Lower paper disc air-dried and sprayed with ninhydrin in alcohol and 4 ml collidine to detect ninhydrin-positive compounds. For collection of root exudates, jars (800 ml capacity) of plant nutrient solution were wrapped in aluminum foil and top covered with perforated glass plate; this placed in larger jar, covered with cellophane. The solution was	Nutrient solution in which roots grew was filtered to remove debris and root caps, desalted by cation exchange resin. Ninhydrin-positive compounds eluted from resin with NH_4OH and eluate evaporated under vacuum to dryness. Residue redissolved in buffer solution. Ninhydrin-positive analysis made with amino acid analyzer

		rated plate with roots through holes and root tips in contact with solution; incubated in controlled chamber	
Vančura and Stanek (1975)	Bean root exudates, relation to reserve compounds in cotyledons and true leaves	Bean seeds washed and surface sterilized in 0.1% solution of $HgCl_2$ and again washed repeatedly. Seeds germinated on wet filter paper and planted in washed sterile silica sand. After 11 days, some plants were deprived of cotyledons and leaves. Exudates collected by washing the sand at 4-day intervals and plants were replanted in new sand medium. Washings centrifuged to remove sloughed cells, root hairs, or sand; concentrated *in vacuo* and exudates lyophilized	Amounts of exudates at 4-day intervals determined gravimetrically. Amino acids and sugars determined by paper chromatographic and spectrophotometric methods
Barber and Martin (1976)	Comparison of amounts of organic matter released by roots of wheat and barley and effects of microorganisms	Wheat surface sterilized by agitation in calcium hypochlorite solution, washed, and planted in tubes of agar medium. Barley seed sterilized in $HgCl_2$ and hydroxylamine hydrochloride, germinated on filter paper, and transferred into tubes of agar. Three-day-old sterile seedlings planted in pots of soil in Pyrex glass jars with air inlet and outlet for maintaining atmosphere of CO_2 labeled with ^{14}C. Roots maintained under sterile or non-sterile conditions in soil so that ^{14}C labeled CO_2 released by plant could be distinguished from microbial activity; soil was sterilized by 5 M rad γ-radiation	Outgoing gas was bubbled through 30% ethanolamine in 2-methoxy-ethanol to collect respired CO_2; the ^{14}C content determined by liquid scintillation spectrometry. At harvest, ^{14}C and total C content of roots and shoots, and water-soluble or insoluble material in soil, were measured
Schulb and Schmitthenner (1978)	Exudation from soybean seed with intact or cracked coats	Seeds with or without cracked coats were either leached or soaked. Leaching: seeds buried in a bed of 1-mm-diameter glass beads and covered with 3-mm-diameter beads in Büchner funnel with neck removed. Leached with 3 ml water/h. Leachates collected in antibiotic solution and stored frozen. Seeds in vials of distilled water soaked for 3 h, then 0.5 ml transferred to antibiotic solution, filtered and frozen	Total carbohydrates by anthrone test, absorbance determined colorimetrically. Amino acids by modified ninhydrin test (Yemm and Cocking, 1955); absorbance determined with colorimeter
Prikřyl and Vančura (1980)	Factors affecting exudation: root growth, concentration gradient of exudates, presence of bacteria	Wheat seeds surface-sterilized by 0.1% $HgCl_2$ plus 0.1% detergent Tween 80 (Lachema). Seeds pregerminated on sterile filter paper and plants grown under axenic conditions in a nutrient medium which was replaced with fresh medium at 2-day intervals. Plants were maintained in a phytotron under controlled light-dark periods. Some cultures were inoculated with *Pseudomonas putida*	Amount of exudate was determined by bichromate titration as the total amount of carbon liberated by the plant roots into the nutrient medium containing no exogenous carbon source. Carbon exuded in presence of bacteria was calculated from minimum amount of substrate necessary for a given growth of bacterial biomass

Table 3.1 (continued)

Reference	Subject	Collection procedure	Analysis or assay
C. S. Tang and Young (1982)	Allelopathic compounds from roots of *Hemarthria altissima*	Stolons of tropical grass were washed, treated with 5% (v/v) Chlorox 15 min, and established in sand cultures in 1-gal inverted glass bottles with bottoms removed. Nutrient solution was circulated through the sand-root system and through a connected column containing Amberlite XAD-4-(resin) which absorbed hydrophobic metabolites. The column was eluted with methanol and the eluate separated into neutral, acidic, and basic fractions. Slight modifications have been described for collection and study of exudates of guava (Brown et al. 1983) and papaya (C. S. Tang and Takenaka, 1983)	Qualitative nature of compounds determined by chromatographic analyses and by bioassays of trapped exudates using lettuce seed
Chaboud (1983)	Maize root cap mucigel	Maize seeds sterilized in 1% NaClO 30 min, washed and allowed to imbibe in sterile distilled water for 24 h, then aseptically germinated in dark. Roots positioned through a grid over water in a petri dish so that roots are bathed in water (200–300 seedlings in 100 ml water). After incubation in moist chamber, droplets from root tips collected, centrifuged and supernatant dialyzed against distilled water + bacteriostat agent. Concentrated under vacuum, lyophilized and stored	Standard analyses performed for protein and sugar content

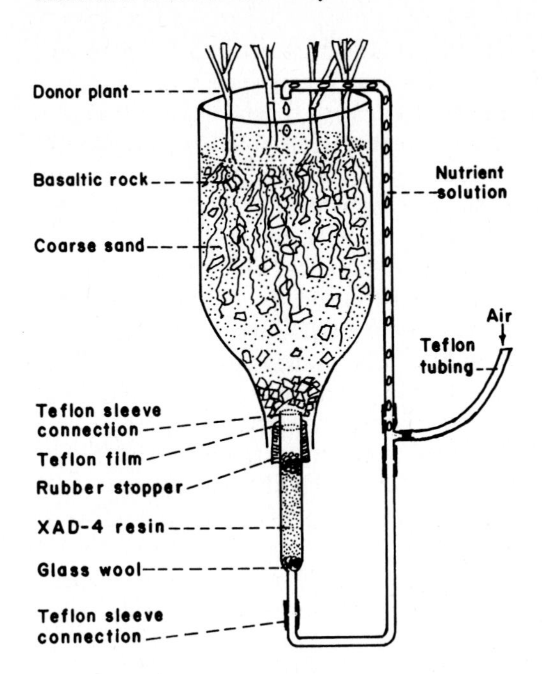

Fig. 3.2. Hydrophobic root exudate trapping system. The nutrient solution is continuously circulated through the root zone of living plants, eluting extracellular organic materials from the sand. Hydrophobic, or partially hydrophobic, exudates are selectively retained by the XAD-4 ion-exchange-resin column while the inorganic nutrients are unaffected. (Tang and Young 1982)

grow through substrates such as sand or ballotini beads (D. A. Barber and Gunn 1974), one finds increased exudation from growth in solid substrates as compared to that found in solution culture. Most frequently, washed quartz sand wetted with a nutrient solution is used under sterile conditions, or specific microorganisms may be added. A natural field soil, either sterilized or nonsterilized, may be desirable for certain studies, though the exact composition of such substrate may not be known. Plants have been grown in such substrates in simple glass tubes, followed by leaching or washing of substrate + roots to extract the exudates. C. S. Tang and Young (1982) developed a continuous trapping system (Fig. 3.2) for collecting quantities of exuded allelopathic chemicals from the undisturbed root system of a tropical grass, *Hemarthria altissima,* which had been observed to inhibit the growth of a legume, *Desmodium intortum* (see Table 3.1 for description). Where aseptic culture is not required, greenhouse pots with bottom drainage tubes have been used successfully. With the ready availability and ease of quantitative measurement of radioactive materials, radioactive labeling of plants has become increasingly popular in recent years as a tool for assessing the quantitative and qualitative nature of root exudates, for studying movement and transformation of compounds in plants, and for detecting exact sites of exudation in the root system. Most often used is CO_2 labeled with isotope ^{14}C; the gas is usually supplied to plants grown in glass tubes or jars, greenhouse pots, or specifically designed apparatus. Respired CO_2 then can be collected along with released root exudates which can be qualitatively analyzed for radioactive compounds and quantitatively analyzed for their ^{14}C content.

Examples of special apparatus employed along with the foregoing basic procedures are the Lees and Quastel (1946) perfusion apparatus used by Parkinson

(1955) to obtain oat-root perfusates, and the somewhat more complex plant-culture unit used by Stotzky et al. (1962) for collecting root exudates and CO_2. The most sophisticated installations for obtaining sterile root exudates are found in the many variations of gnotobiotic culture of plants (Hale et al. 1973). The term "gnotobiotic" is used to describe plants grown in the absence of microorganisms ("germ-free" or axenic culture), or to describe growth of plants in association with certain identified organisms (monoxenic or polyxenic). This type of experimentation does not simulate the complex environment of natural field soil – indeed presumably few, if any, controlled experiments can – but it is frequently essential to be able to distinguish between root exudates and those substances synthesized by any associated microorganisms. Elaborately engineered facilities and highly controlled environments are required for large-scale gnotobiotic research; however, many relatively simple modifications have been devised and found adequate for specific, usually short-term, experiments (Kreutzer and Baker 1975). Aseptic culture of plants for exudate collection in both washed quartz sand and a nutrient solution is well exemplified by the techniques of Ayers and Thornton (1968).

Many of the methods for exudate analysis require only standard chemical procedures, or modifications to fit the specific purpose of the experiment (Table 3.1). Procedures range from simple paper chromatography methods for separation and detection of sugars and amino acids, to the more expensive and sophisticated techniques of gas-liquid chromatography and liquid scintillation counting of the ^{14}C content released in exudates from labeled plants. Nitrogenous compounds, organic acids, phenolic substances, fatty acids, enzymes, and other compounds have been determined by a variety of such standard procedures. Chances of error occur even with these standard analytical techniques, and this probably is the reason why dissimilar results are often obtained by researchers studying the same plant species. An example of this was provided by Rovira (1956) in studies of the root exudates from oats and peas. Desalting of exudate material with acetone-HCl in preparation for chromatography caused partial breakdown of glutamine to glutamic acid and slight decomposition of asparagine to aspartic acid. Also, the relative quantity of some amino compounds (glutamic acid, glutamine, cystine, lysine, histidine, taurine and arginine) may be affected by their low solubility in acetone-HCl. Use of both this solvent and phenol as the extraction medium produced a wider range of amino acids from both pea and oat exudates.

When one is concerned with detecting minute quantities of exudate compounds, or determining patterns of exudation, bioassay methods can be used. *Neurospora* and *Lactobacillus*, which have essential growth factor requirements, have been used to detect certain B-group vitamins in exudates (Rovira and Harris 1961). Bioassays with *Lactobacillus arabinosus* and *Leuconostoc mesenteroides* have been useful in determining amino acid exudation patterns from roots of cotton infected by *Fusarium oxysporum* f.sp. *vasinfectum* (Sulochana 1962a). Simply testing crude exudates without analysis is appropriate in preliminary investigations for assessing total effects on germination of plant-pathogen resting spores and the subsequent growth or lysis of germ tubes.

Table 3.2. Substances detected in plant root exudates

Kind of compound	Exudate components	Plants most studied
Sugars	Glucose, fructose, sucrose, maltose, galactose, rhamnose, ribose, xylose, arabinose, raffinose, oligosccharide	*Triticum aestivum, Hordeum vulgare, Phaseolus vulgaris, Pinus* spp
Amino compounds	Asparagine, α-alanine, glutamine, aspartic acid, leucine/isoleucine, serine, -aminobutyric acid, glycine, cystine/cysteine, methionine, phenylalanine, tryosine, threonine, lysine, proline, tryptophane, β-alanine, arginine, homoserine, cystathionine	*Triticum aestivum, Zea mays, Avena sativa, Pisum sativum, Phalaris* spp., *Trifolium* spp., *Oryza sativa, Gossypium barbadense, Lycopersicon esculentum, Pinus* spp., *Robinia pseudo-acacia, Boutelova gracilis*
Organic acids	Tartaric, oxalic, citric, malic, acetic, propionic, butyric, succinic, fumaric, glycolic, valeric, malonic	*Triticum aestivum, Zea mays, Phaseolus vulgaris, Lycopersicon esculentum, Brassica* spp., *Pinus* spp., *Robinia pseudo-acacia*
Fatty acids and sterols	Palmitic, stearic, oleic, linoleic, linolenic acids; cholesterol, campesterol, stigmasterol, sitosterol	*Phaseolus vulgaris, Arachis hypogaea*
Growth factors	Biotin, thiamine, niacin, pantothenate, choline, inositol, pyridoxine, *p*-amino benzoic acid, n-methyl nicotinic acid	*Triticum aestivum, Phalaris* spp., *Phaseolus vulgaris, Pisum sativum, Trifolium* spp., *Medicago* spp., *Gossypium barbadense*
Nucleotides, flavonones and enzymes	Flavonone, adenine, guanine, uridine/cytidine, phosphatase, invertase, amylase, protease, polygalacturonase	*Triticum aestivum, Zea mays, Pisum sativum, Trifolium* spp.
Miscellaneous compounds	Auxins, scopoletin, fluorescent substances, hydrocyanic acid, glycosides, saponin (glucosides), organic phosphorus compounds, nematode cyst or egg-hatching factors, nematode attractants, fungal mycelium growth stimulants, mycelium-growth inhibitors, zoospore attractants, spore and sclerotium germination stimulants and inhibitors, bacterial stimulants and inhibitors, parasitic weed germination stimulators	*Avena sativa, Medicago* spp., *Trifolium* spp., *Pisum sativum, Lycopersicon esculentum, Lactuca* spp., *Fragaria vesca, Musa paradisiaca, Zea mays*

3.3 Qualitative and Quantitative Nature of Exudates

This topic was essentially summarized by Rovira (1965a) in tables prepared from his own work and from work of others up to that time. Table 3.2 shows the great diversity of chemical compounds that have been found in the root exudates of a variety of crop plants. Exudates have been most studied in wheat and other crops of high economic value. All substances noted in the table were supposedly detected in exudates of plants grown in microbe-free environments. Many of these same compounds have been extracted from natural rhizosphere soil where true exudates were not distinguished from microbial metabolites. Even with plants in gnotobiotic and aseptic culture, exuded material cannot always be distinguished from substances originating from sloughed root hairs and root-cap cells. Under natural growth conditions in soil, cell contents can be released from lesions due to disease, wounds caused by small insects and other soil animals, and by the abrasive action of soil particles.

Many studies have failed to include quantitative estimates of specific compounds. A greater effort seems to have been made by some of the early scientists, particularly Russian, German, and French, to obtain quantitative information on exudation than is evident in more recent studies. Keeping in mind that data from some of the work, both early and modern, may be based on experiments using questionable techniques, wide variation in the quantity of exuded compounds is indicated. A common problem in assessing exudate components quantitatively lies in the lack of standardized expression; various scientists have given this information in terms of mg or μg/plant, μg/unit dry weight of root, % of total material (i.e., reducing sugars, amino acids, etc.), or simply as number and intensity of spots on chromatograms. These then may vary further according to plant-culture techniques, whether in water, nutrient solution, quartz sand, soil (sterile or nonsterile), etc.

3.3.1 Carbohydrates and Amino Acids

One must be impressed with the large number of chemically different compounds exuded by a variety of plants and reflect on their potential influence on microbial activity around roots. Most of the chemicals identified are water-soluble, a fact that has been largely dictated by the techniques used for their extraction and analysis. Early recognition that root exudates contain a variety of sugars and amino acids, providing readily available carbon and nitrogen sources for microbial growth, lead to increased research attention being directed toward these compounds. Plant pathologists quickly associated such exudation with susceptibility or resistance of plants to root-infecting fungi. At least 10 sugars and 25 amino acids have been identified in the exudates of plants. Glucose and fructose are generally the most frequently reported. These and others listed in Table 3.2 are now known to be common in the exudation of many plants. Quantitatively, specific sugars in root exudates of two different plant species under the same cultural conditions may be similar or may vary considerably, as shown in comparisons between barley and wheat (Vančura 1964). Whereas oligosaccharides, arabinose,

Table 3.3. Sugars in exudates of barley and wheat (Vančura 1964)

Compound	% of reducing sugars	
	Barley	Wheat
Oligosaccharides	27.8	26.7
Maltose	5.4	3.1
Galactose	13.6	4.0
Glucose	9.5	16.8
Arabinose + fructose	19.0	17.7
Xylose	15.0	15.9
Ribose	1.3	0.9
Rhamnose	6.8	14.9
Deoxyribose	0.8	–
Deoxysugar	0.8	–

fructose, and xylose were only slightly different, galactose was three times higher in barley and glucose and rhamnose were approximately twice as high in wheat as in barley exudates (Table 3.3). That reducing compounds in general vary considerably between plant species is evidenced in further comparisons made by Vančura and Hanzlíková (1972). These compounds were markedly higher in root exudate of bean and wheat than in that of cucumber or barley. Where strictly aseptic plant-growth systems are not used to obtain exudates, secondary products may be mistaken for exudate components. Moghimi et al. (1978 b), using a variety of chromatographic techniques, determined that 2-ketogluconic acid was the only acid present in significant amounts in the rhizosphere of wheat seedlings. Since this acid is known to be a product of microbial action on glucose, it was suggested that it was produced from glucose in the rhizosphere.

Since Virtanen et al. (1936) reported that lysine and aspartic acid were "excreted" by root nodules of legumes, amino acids have become the most intensively studied group of compounds in the root exudates of plants. Exudation patterns, or spectra, of amino acids reported by different investigators for a single plant species often vary considerably due to the different growth conditions employed. This is to be expected if strict aseptic culture is not maintained, since bacteria may liberate amino compounds by synthesis or by decomposition of root cells and sloughed fragments. Exudates of different plants compared under identical conditions (Rovira 1956, 1959) may differ greatly in both kind and quantity of specific amino acids released from roots; seven times as much exudate from oats as from peas may be required to provide sufficient material for a single chromatogram. Principal compounds identified in pea-root exudate by paper-chromatographic methods were homoserine, threonine, α-alanine, glutamine, serine, and asparagine, whereas oat-root exudate yielded predominantly serine, lysine, and glycine. Using ion-exchange chromatography, Boulter et al. (1966) found that the spectrum of amino acids in exudate from 14-day-old pea seedlings was similar for plants grown in nutrient culture medium or quartz sand, but specific amino acids occurred in greater amounts in the sand medium (Table 3.4). Homoserine (1,412 $\mu g\ g^{-1}$ root dry wt.) was the most prominent amino acid in

Table 3.4. Amino acids (μg g^{-1} dry wt. root) in culture medium of 14-day-old pea seedlings. Average of 5 analyses (Boulter et al. 1966)

Acids	Culture solution	Quartz sand
Cysteic acid	Trace	Trace
Aspartic acid	280	740
Threonine	71	222
Serine[a]	68	232
Homoserine	190	1412
Glutamic acid	307	654
Proline	Trace	Trace
Glycine	41	160
Alanine	39	190
Valine	22	78
Cystine	14	N.L.
Isoleucine	13	43
Leucine	21	70
Tyrosine	19	54
Phenylalanine	17	67
γ-Aminobutyric acid	Trace	Trace
Ornithine	45	192
Lysine	29	123
Histidine	35	117
Arginine	61	219
Ammonia	1312	1540

[a] May contain asparagine and glutamine

quartz sand followed by aspartic acid (740 $\mu g\ g^{-1}$ and glutamic acid (654 $\mu g\ g^{-1}$). Such differences, along with differences in the many other exuded components released by plant roots, are responsible for the equally varied nature of microbial populations in the rhizosphere. Vančura and Hanzlíková (1972) found that nitrogen as part of free amino acids predominated over nitrogen in the form of protein or peptides in root exudates from seedlings of cucumber, bean, and wheat cultured in sterile quartz sand for 14 days.

3.3.2 Organic Acids and Lipids

Exudate compounds other than sugars and amino acids have received much less attention. Organic acids in particular deserve more attention because of their prominent role in cell metabolism and their effects on rhizosphere pH and microbial activity. Organic acids are good metal-chelating compounds and play an important role in the absorption and translocation of nutrient elements. They are found in relatively large amounts in many plants, but have been reported infrequently in root exudates. When Rivière (1960) cultured wheat plants in nutrient solution to the tillering stage (6 weeks) acetic acid was exuded in highest quantity (13 mg/plant). Propionic, butyric, valeric, and malic acids were found in smaller quantities and other organic acids only in trace amounts. Free fatty acids and sterols are important, often essential, in the activities of some microorganisms in the rhizosphere, yet they are rarely included in lists of exudate components. Palmitic, stearic, oleic, and linoleic acids have been found in bean (Papavizas and Kovacs 1972) and peanut (Thompson and Hale 1983) root exudates, and cholesterol, campesterol, stigmasterol, and sitosterol were found in peanut root exudate.

3.3.3 Growth Factors

The vitamins needed to satisfy the requirements of certain microorganisms in the rhizosphere may be provided from three sources: synthesis by other microorganisms, decomposition or autolysis of plant and animal tissues, and exudation from plant roots. Little was known about vitamins in root exudate before 1961, though West (1939) had reported much earlier that thiamine (vitamin B_1) and biotin were present in significant quantities in the exudate of flax grown aseptically in nutrient solution. Exudates were assayed for thiamine content by monitoring the growth of *Staphylococcus aureus* in a medium containing all essentials except thiamine. Biotin was assayed by a similar response of *Rhizobium trifolii*. Bioassay of thiamine with *S. aureus* may be of questionable accuracy since this organism also responds to the thiazole plus pyrimidine fraction of this vitamin. Nevertheless, West's observations brought to the attention of others the strong probability that essential vitamins may be released from higher plants in amounts sufficient to account in part for the recorded quantitative and qualitative differences in bacterial populations of the rhizosphere.

Rovira and Harris (1961), using specific vitamin-requiring bacteria for bioassay, determined the B-group vitamins in root exudates from seedlings of ten plant species grown in systems of controlled environment and nutrition. The vitamins and test organisms used for assay were: biotin, pantothenic acid, and niacin with *Lactobacillus plantarum* on different assay media; riboflavin (vitamin B_2) with *L. casei*; thiamine with *L. fermentum*; and pyridoxine (vitamin B_6) with *Neurospora sitophila*. Only biotin was found consistently in significant quantities in exudates from lucerne, white clover, field pea, tomato, and *Phalaris*. All others except pyridoxine were present in some exudates only in small to trace quantities; pyridoxine was not detected. Biotin in amounts of 0.1 to 16 $\mu g\ ml^{-1}$ of exudate and pantothenate, 0.01 to 0.32 $\mu g\ ml^{-1}$, were detected in the exudate from each species. Considerably higher amounts of biotin were released from field pea than from other plants. Bioassay with a mutant of *N. crassa* for *p*-aminobenzoic acid, thiamine, biotin, choline, and inositol revealed a total B-vitamin level of 46.6 μg/250 ml in root exudate of healthy cotton plants (Sulochana 1962b). Thus, plant roots, along with microbial synthesis, add to the supply of vitamins in soil which contribute to the requirements of heterotrophic microorganisms in the rhizosphere.

3.3.4 Enzymes

Some conflicting reports on the subject of enzymes in root exudate were made before 1920, but Knudson and Smith (1919) might have been the first to seek specific experimental evidence for enzyme exudation in strictly aseptic culture. When corn and field peas were grown from surface-disinfested seed in a nutrient solution containing soluble starch, the culture solution upon analysis showed only a slight increase in reducing sugars, not sufficient to indicate exudation of amylase. Similar results were obtained by Knudson (1920) with regard to invertase; he concluded that any increase in reducing sugars in a plant culture solution with field pea seedlings must be due to exudation of glucose and fructose following hydrolysis of sucrose absorbed by roots from the culture medium. These early efforts emphasize the need for caution when interpreting results even under aseptic conditions. Subsequent investigations (Krasilnikov 1958) have provided both convincing and questionable data to suggest root exudation of enzymes into the rhizosphere, as determined by the breakdown of starch, glycerophosphate, glucose phosphate, ribonucleic acid, and other substrates. Some of these studies have involved plant culture procedures which offered a high degree of opportunity for microbial contamination, which of course would accelerate the decomposition of substances supposedly hydrolyzed by exuded root enzymes. A common procedure has been to suspend plant roots in solutions containing starch, organic phosphates, etc., and to determine the extent of substrate decomposition by released amylase, phosphatase, etc. Such work by H. T. Rogers et al. (1942) suggested that phosphatase activity from corn and tomato roots may actually originate from sloughed root-cap cells, which in fact remain alive for some time. Miskovic et al. (1977) reported that a higher dehydrogenase activity occurred in the rhizosphere of corn than in root-free soil, and the application of increasing concentrations of nitrogen resulted in a further increase in enzyme activity. Dehydrogenase activity was positively correlated with crop yield.

3.3.5 Miscellaneous Compounds

Most of the exudate components discussed so far would be expected to supply nutrient sources making the few millimeters of soil around roots a virtual cultural medium for microbial growth. A variety of other, miscellaneous, defined and undefined compounds also are released by roots. Some of these, such as hydrocyanic acid and saponins (glucosides) remind us that exudates and the rhizosphere contain substances that are inhibitory as well as those that are stimulatory to microbial growth. A number of chemically undefined substances have been simply termed nematode or fungal stimulators, attractors, and inhibitors.

Methodology used for determining the nature of exudate compounds has been selective for water-soluble and nonvolatile materials while water-insoluble and volatile materials may in fact be extremly significant under natural field conditions. A brief look at volatiles in general may help to emphasize the potential importance and need of more information on this subject in relation to rhizosphere ecology in argricultural soils. Volatiles, being composed of compounds of relatively high vapor pressure, can move through soil in both the liquid and gas phases, thus affecting growth of microbes on the surface of organic particles, and aerial mycelia of fungi in the soil-pore spaces. Fries (1973) has reviewed the major sources of volatile organic compounds and discussed their effects on growth and development of fungi. Such compounds have been associated with all parts of plants: flowers, fruits, wood, decomposing residues, and living roots. Add to these the volatile compounds produced by a relatively large number of fungi and bacteria and the widespread occurrence of these compounds seems well estabished.

Russian workers, beginning about 1926, are given credit for first demonstrating that the soil atmosphere contains volatile organic constituents that should be considered equally with soil solution and the solid soil phase in their action upon fungi. Though we are concerned here primarily with root exudates, we are in fact persuaded to consider the nature of volatile components emanating from roots, decomposing organic matter, and microbial metabolites. Under natural conditions the contributions of each of these sources to the rhizosphere effect can hardly be separated one from the other.

Linderman and Gilbert (1975), drawing primarily from the U.S. Department of Agriculture investigations, reviewed the work on volatiles of plant origin and discussed their relationship to soil-borne plant pathogens. Sclerotia of *Sclerotium cepivorum* are stimulated to germinate in the presence of *Allium* spp. volatiles (Coley-Smith and King 1970), which contain organic sulfides, primarily methyl and propyl sulfides in onion and methyl and allyl sulfides in garlic. Sulfur-containing compounds (isothiocyanate and others) emanating from decomposing crucifer amendments (*Brassica oleracea* incorporated into soil) suppress all developmental stages of *Aphanomyces euteiches* (J.A. Lewis and Papavizas 1970, 1971). Seeds release gaseous and volatile compounds during the water-imbibing and germination stage. Gas chromatographic and chemical analyses have revealed that germinating seeds and seedlings of pine and a number of vegetable crops liberate ethanol, methanol, formaldehyde, acetaldehyde, propionaldehyde, formic acid, acetone, ethylene, and propylene (Vančura and Stotzky 1976). Somewhat less is known about the volatile compounds produced by developing roots other than volatile organic acids, such as propionic and valeric, and the common production of ethyl alcohol found under anaerobic conditions. Volatiles released from the ectomycorrhizal association between *Pinus sylvestris* and *Boletus variegatus* have fungistatic properties affecting growth of both the fungal symbiont and other fungi in the rhizosphere (Krupa and Fries 1971; Krupa and Nylund 1972). The opposite effect also may occur in which volatiles emanating from roots stimulate the fungal symbiont. Germ tubes from spores of the VA mycorrhizal fungus, *Gigaspora gigantea,* were able to grow through an air space in response to a volatile stimulus and contact roots of bean or corn plants (Koske 1982). The evolution of acetaldehyde from root exudates of wheat (Nance and Cunningham 1951; Plhak and Urbankova 1969) potentially could affect the physiology of both higher plants and soil microorganisms. We will consider volatiles again in Chapter 7 in relation to rhizosphere ecology and root-infecting pathogens.

3.4 Sites of Exudate Release

Though R. Brown and Edwards (1944) were not seeking information on sites of exudation, their observation that germination of seed of the parasitic plant *Striga lutea* occurred mainly in the root tip region of sorghum probably provided an early lead to determining the principal site of exudate release. The fact that nematodes (A. F. Bird 1959) and Phycomycete zoospores (Zentmyer 1961) tend to congregate in an area behind the root cap also indicates a region of high exudate concentration. That the meristematic zone immediately behind the root cap is a major source of exudates was clearly shown by Pearson and Parkinson (1961) in simple tests with broadbean (*Vicia faba*). Seed were surface-sterilized, germinated in soil, then washed and placed between absorbent filter papers for 24 h. When the germinated seed and young roots were removed and the filter paper sprayed with ninhydrin solution, the color pattern produced revealed the primary areas of exudation of ninhydrin-positive substances. Using similar methods, Schroth and Snyder (1961) showed that exudation of sugars and amino acids was most abundant from germinating bean seeds and root tips. While other investigations continued to show the apical region as the major site of root exudation, significant amounts of exudate also have been noted to occur from older plant roots. Frenzel (1960) demonstrated this fact with sunflower (*Helianthus annuus*). In his experiments he used, in addition to paper chromatography, bioassay with mutants of *Neurospora* having specific nutritional requirements for certain organic compounds. This revealed that threonine and asparagine came from root tips, whereas leucine, valine, phenylalanine, and glutamic acid came from the root-hair zone. Root hairs are involved in exudation, but lateral roots without root hairs also exude substances. Labeling of plants with radioactive materials such as ^{14}C-carbon dioxide, ^{32}P-phosphate, and ^{36}Cl, with subsequent radioautographs, has facilitated the precise location of exudation sites along roots (Rovira and Davey 1974). An advantage of labeling with $^{14}CO_2$, for example, is that total radioactive material representing all exuded carbon compounds is measured, rather than just that detectable by specific sprays. However, perhaps a disadvantage is that sloughed root cells are labeled along with the true exudate components in the form of water-soluble, water-insoluble, and volatile compounds. McDougall and Rovira (1970), employing a combination of autoradiography and a filter-paper technique with ^{14}C-labeled wheat plants, showed that non diffusible material was released from the apices of primary and lateral roots, whereas diffusible materials were released along the entire root length. A higher amount of non diffusible material was detected. Where exudation occurs along main roots it should be understood that organic compounds may be released during the formation of lateral roots, a process which damages the main root at the point of emergence. Egeraat (1975b) found that such wounds in pea-seedling roots primarily released homoserine.

Finally, it should be mentioned that exudates not only vary between specific sites on roots but also differ in quantity and chemical composition according to whether they are released from germinating seeds or from developing seedlings. For example, the spectrum of reducing sugars in bean seed as compared to seedling-root exudates differed greatly (Vančura and Hanzlíková 1972), and the quantity generally was greater from seedlings. Seed exudation is important in relation to the incidence of pre-emergence damping-off by certain pathogenic fungi which are favored by nutritive substances in exudate. Exudation is increased from poor quality seed with blemishes and cracked seed coats. Schroth and Cook

(1964) observed much higher incidence of damping-off of bean with cracked seed coats than from seeds with intact coats. Most exudation from undamaged seeds occurred from the hilum.

3.5 Factors Affecting Exudation

The principal factors that affect the kinds and quantity of substances released by roots into the rhizosphere soil include the species and developmental stage of plants, various soil physical-stress factors, plant nutrition, mechanical or disease injury, microbial activities, and foliar-applied chemicals (Hale and Moore 1979). Like most other phases of root-exudate investigation, the bulk of information on this subject has been derived from seedling plants grown in axenic culture for short duration. The relatively small effort to determine exudation in the field has involved the application of $^{14}CO_2$ to plants and subsequent testing for labeled metabolites released into the soil. Measuring microbial populations in the rhizosphere in response to various factors indirectly marks the occurrence of exudation, but does not eliminate the contribution of sloughed organic materials and microbial metabolites.

3.5.1 Plant Species and Developmental Stage

We now know that many different compounds are released by all plant species that have been tested and at least certain exudate components are released at all developmental stages from the seed, seedling, and maturing root system.

Numerous controlled experiments with plants grown in nutrient solutions or in sterile sand have provided convincing evidence that differences in the quantity and composition of exudates are likely to be greater among plants that are not closely related phylogenetically. Some of the earliest comprehensive work was done by Rovira (1956) who found both quantitative and qualitative differences between exudates of peas and oats grown aseptically in quartz sand. Over a 21-day period, peas released 22 amino compounds and oats only 14. Also, the proportions of the various amino acids differed between peas and oats.

Subsequently, other studies have shown differences among different plants for amino acids, sugars, the B-group vitamins, and other compounds in root exudate. Amino-acid exudation by *Phalaris* and tomato exceeded that of clover growing in nutrient solution (Rovira 1959). Young wheat and barley plants cultured in silica sand have released different kinds and amounts of specific compounds (Vančura 1964). Certain amino acids were found in wheat exudate but not in that of barley, and vice versa. For the sugars, galactose comprised 13.6% of those in barley exudate as compared to 4.0% in wheat exudate; glucose and rhamnose were approximately 8% higher in wheat. Field-pea plants grown aseptically in distilled water produced more of the B-group vitamins than did lucerne, white clover, canarygrass, or tomato (Rovira and Harris 1961). Further contributions to our knowledge about the differential exudation among plants have been made

Table 3.5. Quantities of seed and root exudate. (Vančura and Hanslíková 1972)

Plant	Seed exudates		Root exudates	
	mg/1000 seed	Number of seeds needed for 1 g of seed exudate	mg/1000 plants	Number of plants needed for 1 g of root exudate
Barley	242.6	4122	434.3	2417
Wheat	198.0	5050	502.6	2253
Cucumber	108.0	9259	620.0	1614
Bean	517.7	1943	557.8	1792

by Vančura and Hovadik (1965), Kovacs (1971), Vančura and Hanzlíková (1972), and others. Frequent conflicts in results obtained by different scientists studying the same plant species suggest that the extent of variation may be influenced by such factors as stage of plant growth (germinating seed or established seedlings) when exudates are collected, the specific exudate components being assessed, physical or chemical differences in the plant-culture environment, and slight differences in methods of analysis.

While the nature of exudates varies between plant species, changes also occur with plant age or stage of development. The colonization of seeds and seedling roots by microorganisms depends to a large extent on the nutrients in exudates provided by these sources. Thus, kinds and numbers of organisms change with plant growth as the chemical constituents of exudate change. A direct correlation can be found between seed size (reflecting storage material) and amount of seed exudate released (Table 3.5), or the number of seeds needed to produce 1 g of exudate; for example, nearly five times as many cucumber seed as bean seed are required to release the same quantity of exudate (Vančura and Hanzlíková (1972). Specific exudate components (sugars, amino acids, etc.) also vary from seed to seedling and the spectrum within these groups may differ considerably. Seed exudates of barley and wheat yielded a lower spectrum of sugars than did the exudates of seedlings. Seeds are most likely to evolve oxidizable volatile compounds, such as aliphatic aldehydes, during the first 3 or 4 days after imbibition (Stotzky and Schenck 1976). Active seed metabolism seems to be required for production of volatiles.

Few studies have been made of the exudates of mature plants, largely due to technical difficulties in maintaining aseptic plants for extended periods. Vančura and Hovadik (1965) grew red pepper and tomato plants hydroponically in gravel until they reached the fruiting stage, then washed the root systems and cultured the plants in nutrient solution, and finally in sterile distilled water from which exudates were obtained. Eighteen amino acids were identified in exudate from the initial growth phase of red pepper and 15 from the fruiting stage. Specific identified amino acids were approximately equal for the two phases of tomato culture. For both plants, chromatographic analysis indicated generally a higher concentration of exudates during the initial growth phase. More carbohydrates were found during the initial growth phase of both plants than during fruiting, oligosaccharides and fructose being the most prominent. W.H. Smith (1970) compared root exudates from 3-week-old sugar maple seedlings and a 55-year-old tree. Seedlings were cultured axenically in nutrient solution with glass beads, and a technique of root-pruning and air-layering was used to obtain new roots with unlignified tips from a mature tree; these new roots, 8 cm in length, were then cultured in nutrient solution. Analysis of exudates showed that carbohydrates from seedlings were more diverse and in higher quantity than those from the mature tree. Amino acids/amides and organic acids released from the new root tips of a mature tree were more diverse and abundant than those from seedlings. That age and stage of development may significantly influence the nature of root exudate was further demonstrated by Hamlen et al. (1972), who observed a decrease in total carbohydrates from alfalfa plants with increasing plant age over a period of 16 weeks in gnotobiotic culture; flowering plants released more materials than clipped plants. Many other examples could be cited to confirm that quantitative and qualitative changes occur as plants develop from the germinated seed to maturity and fruiting. Some exudate constituents increase and others decrease with time, depending upon the species of plant, cultural method, and many environmental factors. We will see later (Chap. 4) that microbial activities in the rhizosphere are closely linked with these same factors that influence exudation.

3.5.2 Temperature and Light

Reviews of factors that affect root exudation (Hale et al. 1971; Hale and Moore 1979) show that, while the rate of exudation generally increases for most plants with an increase of temperature above that which is optimal for plant growth, some plant species release greater quantitites at low temperatures. Further, as we noted earlier, the composition of exudates with regard to specific amino acids, carbohydrates, organic acids, vitamins, etc. varies between species. Different species of plants respond differently to temperatures and therefore reveal different rates of photosynthesis and of translocation of photosynthates to the roots, different enzymatic activities that synthesize or degrade photosynthates, and variations in membrane permeability (Hale and Moore 1979). Therefore, it is not surprising that a pattern of exudation in response to temperature has not been established. Relatively high temperatures (31 °C) induced an increase in amino acids exuded by tomato and clover roots in axenic nutrient solution (Rovira 1959). Tomato, cultured in distilled water, released exudate of greater vitamin content (biotin and traces of thiamine, riboflavin and pantothenate) at lower temperatures of 16°–30 °C (Rovira and Harris 1961). Clover released more biotin and niacin at higher temperatures. Other evidence of temperature effect is provided by findings that amino-acid exudation from strawberry roots is favored at 5°–10 °C (Hussain and McKeen 1963), cotton and bean exudate increased at 37 °C, while that of peas decreased (Schroth et al. 1966), and carbohydrate from cotton seed was exuded faster and in larger amounts at 12° and 18 °C than at 24°–36 °C (Hayman 1970); in the latter case amino-acid exudation increased at 36 °C. Some plants, as in the case of maize and cucumber (Vančura 1967), can be induced to release high quantities of exudate by subjecting the plants to a period of low temperature, creating a "cold shock".

Light intensity, by affecting photosynthesis, would be expected to influence both the qualitative and quantitative nature of root exudates. Exudation of serine, glutamic acid, and α-alanine by clover roots in nutrient solution decreased when daylight intensity was reduced 40% by shading (Rovira 1959). From tomato roots, aspartic acid, glutamic acid, phenylalanine and leucine were lower in quantity under reduced light, whereas serine and asparagine increased. Subsequently, little effect of light was found on vitamin exudation, though clover exuded more biotin at high light intensity (Rovira and Harris 1961).

3.5.3 Soil Moisture and Atmosphere

Methods for obtaining maximal quantities of root exudate often require subjecting young plants to a period of partial desiccation followed by rewetting and continued growth. Stress created by low soil moisture in many cases increases exudation of various compounds and results in a higher microbial population density than under conditions of optimal moisture content. Under aspetic cultural conditions, when pea, soybean, wheat, barley, or tomato in sand or a sand-soil mixture were allowed to dry to the wilting point and then were remoistened, the roots liberated higher amounts of amino acids than when kept constantly moist (Katznelson et al. 1955). In these early experiments no attempt was made to distinguish between true exudation from intact roots and the liberation of amino acids from autolysis of sloughed epidermal and root-cap cells, and root hairs. It has been suggested that an increase in α-amino acids in plants as a result of proteolysis during wilting (Kemble and Macpherson 1954) may contribute to an amino-acid imbalance resulting in an increased tendency to move outward. Reid (1974) studied the effect of specific levels of induced water stress over a period of 6 days on exudation of ^{14}C from *Pinus ponderosa* seedlings maintained in aerated nutrient solution. Water-stress treatments involved decreasing the water potential of the root-bathing medium in increments from 0 to -11.9 bar by using polyethylene glycol as the osmoticant. $^{14}CO_2$ was supplied to the seedling shoots and foliage. Assimilation of $^{14}CO_2$ with subsequent translocation of the ^{14}C label to the roots was inhibited by a decrease in water potential. Root exudation of ^{14}C-labeled sugars, amino acids, and organic acids increased as water potential decreased from 0 to -1.9 bar, then exudation declined between -1.9 and -5.5 bar, but increased again from -5.5 to -11.9 bar. As substrate water potential decreased, sugars as a percentage of the total exudate increased, organic acids decreased, and percentage of amino acids decreased slightly (Fig. 3.3). Thus, specific compounds may increase or decrease in the exudate depending on level of water stress. The extent to which experimental results of this type may relate to root exudation under natural soil conditions cannot be determined from the relatively little information available. It is evident, however, that soil moisture stress sufficient to interfere with water absorption and photosynthesis can be expected to induce changes in exudation patterns. In addition to the direct effect on the exudation process, soil moisture also influences the extent to which exudates can diffuse in the soil and affect microbial bebavior, particularly nematode migration and the movement of motile spores of fungi, as in the lower Phycomycetes.

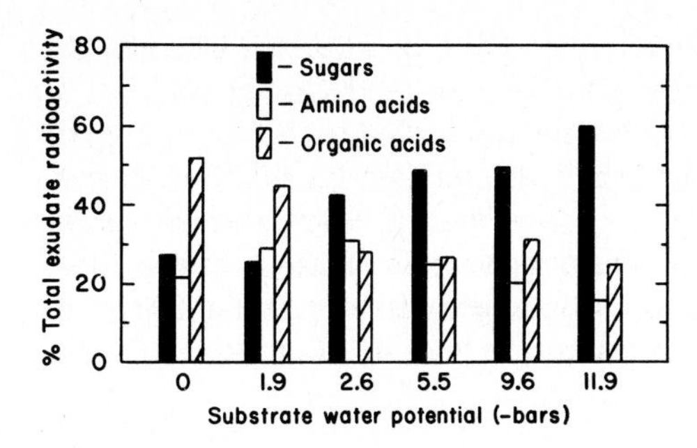

Fig. 3.3 Sugars, amino acids, and organic acids as percentages of total radioactivity of root exudates from 12-month-old ponderosa pine seedlings subjected to substrate water stress for 6 days. $^{14}CO_2$ was introduced to the foliage 4 days before induction of stress. (Reid 1974)

Most of the soil environmental factors that affect root exudation are closely interrelated. Gas exchange, which is closely linked with soil type and moisture, affects both root development and microbial activities. Most root systems grow profusely in well-aerated sandy soil but become thicker, shorter, and less branched in unfavorable heavy clay soils of high moisture and poor aeration. Soil aeration refers to oxygen, carbon dioxide, and the content of by-products of anaerobic decomposition, such as hydrogen sulfide, methane, and hydrogen that accumulate in soil (E. W. Russell 1973). Oxygen demand and sensitivity to CO_2 may increase with a rise in soil temperature.

Grineva (1962) subjected roots of young corn and sunflower plants to periods of anaerobiosis by immersion in water freed of oxygen. Exudate for analysis was collected from beakers of the water in which plants had been held for 6 h. Release of organic compounds (sugars, amino acids, and organic acids) was promoted under these anaerobic conditions. Glucose and fructose were "execreted" in greater amounts along with alanine, proline, glutamic acid, serine, threonine, aspartic acid and histidine as compared with plants in oxygenated water. Chief organic acids that increased under anaerobic conditions were oxalic, citric, malic, succinic, and fumaric. The effect was more pronounced for sunflower than for corn. Though Grineva concluded that the blocking of aerobic respiration changed metabolism, resulting in active excretion of compounds that were not metabolized, the experimental methods used probably also contributed to an increased yield of exudates.

Ayers and Thornton (1968) tested the effects of several gas mixtures on root exudation of pea growing in sterile sand towers. The sand towers consisted of a column of washed quartz sand moistened with sterile nutrient solution and fitted with lower and upper inlet-outlet ports through which filtered, moist air or gas mixtures of known composition were continuously passed at the rate of 2 liters h^{-1}. Rinsings from these sand-grown plants were reduced in volume and analyzed for amino acids using colorimetric and chromatographic methods. Qualitatively and quantitatively more ninhydrin-reactive material was released under the "soil air" series (CO_2-enriched air) than under other gas mixtures. When Rittenhouse and Hale (1971) grew peanut plants (*Arachis hypogaea* L.) in nutrient solution subjected to varying O_2 and CO_2 tensions under axenic conditions, root exudation was affected. Greater amounts of galactose and dihydroxyacetone were excreted under the most aerobic conditions after 2 weeks of plant growth, but with a longer growth period this trend appeared to be reversed. Usually, environmental conditions that impose stress upon plant growth can be expected to induce greater total exudation. Various concentrations of CO_2 above 1% may be injurious to roots, but rather high levels usually are required; or inhibition may occur when the O_2 concentration is too low for proper root development. Root growth usually is adversely affected when the oxygen content falls below 9–12%. The most evident fact presently is that too little information on soil atmosphere versus root exudation is available for interpretation in terms of natural soil conditions.

3.5.4 Plant Nutrition

The availability of mineral nutrients and nutrient uptake by root systems are logically related to the quantitative and qualitative release of nutrients into the soil surrounding roots. Knowing that mineral nutrition influences the levels of unbound amino acids in plant shoots, Rovira (1959) attempted to test calcium level effects on root exudation of plants grown in nutrient solutions, but found no consistent effect on amino-acid release from tomato, clover, and canarygrass with the calcium levels applied (0.05–0.0005 M). Bowen (1969), working with aseptically grown *Pinus radiata* seedlings, compared a complete nutrient solution and phosphate-deficient or nitrogen-deficient nutrient solutions. Loss of amides and amino acids was 2.5 times greater from roots grown in phosphate-deficient nutrient solution than in complete nutrient solution and 10 times greater than from roots in nitrogen-deficient solution (Table 3.6). Apparently, phosphate deficiency leads to excess free amide/amino nitrogen in the roots with increased exudation as opposed to low exudation by nitrogen-deficient plants. Such conditions will in-

Table 3.6. Amide and amino acid exudates from *Pinus radiata* seedlings. Moles $\times 10^{-9}$ per plant. Roots were of similar length in all treatments. (Bowen 1969)

Amide or amino acid	Nutrient solution		
	Complete	Phosphate-deficient	Nitrogen-deficient
Asparagine	10.9	32.5	3.0
Glutamine	23.6[a]	52.0	2.8
γ-Amino-butyric acid	5.2	13.8	1.0
α-Alanine	1.6	2.8	1.2
Asparatic acid	4.4	9.6	2.0
Glutamic acid	6.0	19.7	2.0
Glycine	7.3	14.0	3.4
Leucine	3.0	5.0	1.8
Serine	4.8	8.0	2.0
Threonine	1.4	2.0	–
Valine	1.8	4.0	0.1
Total amide-/amino nitrogen[b]	104.6	248.5	25.1

[a] Some arginine was also present but only in small amounts
[b] Including the two NH_2 groups of asparagine and glutamine

fluence microbial acitivty in general and also root colonization and infection by mycorrhizal and pathogenic fungi.

The peanut or groundnut plant requires a relatively large amount of available calcium for peanut seed development, and for this purpose $CaSO_4 \cdot 2H_2O$ (gypsum) is widely used in the cultural practices for peanuts in the southern United States; soil rot or peanut pod breakdown also apparently is less severe in soils with adequate calcium (Garren 1964; Hallock and Garren 1968). Calcium is known to affect membrane permeability, and changes in permeability may be related to the mechanism of root exudation (Vančura 1967). These facts led Shay and Hale (1973) to examine the effects of calcium levels on exudation of sugars and sugar derivatives from peanut roots grown under axenic conditions. Plants were grown in nutrient solutions with varied calcium levels, and sugars in the solutions were measured at weekly intervals by gas-liquid chromatography. Four times more sugar was exuded at 10 mg than at 50 mg of Ca^{2+} per liter. Ion influx measurements indicated that low levels of Ca^{2+} increased the root-cell-membrane permeability, resulting in a quantitative increase in exudation of sugars from peanut roots.

Nutrient effect is reflected in some methods used for assessing exudation of organic compounds from roots in solution culture. When phosphate was omitted from a sterile growth medium containing wheat seedlings, and the tops were treated with $^{14}CO_2$, the radioactivity of the root solution was reduced to 47% of the control (Rovira and Ridge 1973). Yet, we have already mentioned that Bowen (1969) found that pine seedlings grown in phosphate-deficient solution exuded much more amino-acid and amide material than seedlings in a complete nutrient solution. The question remains whether greater exudation in this case resulted from increased amino-nitrogen inside the roots or from an increase in "leakiness" due to phosphate deficiency. In the experiments of Rovira and Ridge, the addition of acetate buffer to the culture solution at pH 5.0 greatly increased exudation, whereas pH adjustment with H_2SO_4 or universal buffer, or addition of sodium acetate at pH 7.0, significantly reduced exudation.

One must keep in mind that, under a certain set of conditions, total exudation may increase, while at the same time release of a specific type of compound may decrease. In some cases nutrient factors may affect cell permeability, whereas in other cases cell metabolism may be affected, resulting in increased levels of compounds in the cytoplasm followed by increased loss from the roots. The importance of standardization in methodology among workers is quite evident.

3.5.5 Plant-Injury Factors

The amount of substances liberated from roots varies widely according to the conditions in which the root system is developing (Richter et al. 1968; Rovira and Davey 1974). Under controlled experimental conditions the plant-culture method itself may promote sufficient damage to the root surface to increase the quantity of exudation. Two of the most popular cultural media for collecting exudates in a controlled environment have been nutrient solution and sterilized sand. The contrastingly opposing results frequently obtained with these two media again emphasize the importance of considering methods used when comparing data. Boulter et al. (1966) found that greater amounts of various amino acids were exuded by pea seedlings in quartz sand than in a culture solution. Undisturbed solution-grown wheat roots held on chromatographic paper (Ayers and Thornton 1968) failed to yield detectable ninhydrin-reactive spots. But when these roots were gently drawn across filter paper and retested, a ninhydrin reaction was observed, and the amino acid pattern on developed chromatograms of these slightly damaged roots was similar to that of sand-grown roots. D. A. Barber and Gunn (1974) presented further evidence of this kind, showing that roots of barley and maize plants grown in 1-mm diameter glass ballotini beads bathed in nutrient solution released greater amounts of both amino acids and carbohydrates than roots of plants grown in nutrient solution in the absence of ballotini. Since the ballotini are free to move as roots grow in the interstices between beads, it is evident that only a very slight pressure or resistance is required to influence exudation. This probably represents a minimal effect, far less than would be encountered by roots growing through a natural field soil.

Other instances verifying the effects of inconspicuous root damage on release of material into the growth media have been cited (Rovira 1969). Injury can be chemical as well as physical, a fact well illuminated by Toussoun and Patrick (1963), who found that water-soluble substances from residues of several decomposing crop plants in field soil were highly phytotoxic in lettuce bioassay tests. When droplets of the extracts were placed on excised bean stems, chromatographic analysis showed that exudation of ninhydrin-positive substances was greatly increased, probably due to alteration of plant-cell permeability. This type of injury and induced exudation can contribute to the predisposition of plants to infection by pathogenic organisms.

Aside from the use of radioactive labeling methods, the fog-box technique (Went 1957) used by Clayton and Lamberton (1964) may provide the best means of collecting exudates from the injury-free roots of maturing plants. Plants are grown in an opaque box with roots extending free into a lower box where they are automatically sprayed with nutrient solution. Root drip is then collected and analyzed. Certain compounds which are commonly found in exudates from roots grown in sand are not detected in root drippings. The fog-box technique also has the advantage of reducing the possibility of reabsorption and utilization of compounds released from roots; this may occur in other types of plant-growth systems, resulting in a quantitatively false measure of exudation.

Since extremely minor abrasions of the root surface can induce greater exudate release, it might be expected that more serious injuries would magnify this effect. Increased exudation promoted by natural phenomena, such as cracking of seed coats and damage by lateral root emergence, has already been alluded to in relation to sites of exudation along roots. The probability that nematode action on roots may increase exudation has often been suggested as being related to fungus-incited disease, but experimentation does not always verify such an assumption. Rovira (1969), using tomato labeled with $^{14}CO_2$ and growing in agar or sand, detected no increase in radioactivity of exudate following inoculation with larvae of *Meloidogyne javanica*. On the other hand, Wang and Bergeson (1974), using a monoxenic culture technique on filter paper and in silica sand, found that sugars in root exudate from *M. incognita*-infected tomato plants increased significantly, while amino acids were somewhat reduced. They proposed that changes in total sugars and amino acids in the xylem sap and root exudate of nematode-infected plants contribute to *Fusarium* wilt. Obviously, this and other kinds of root wounding that disrupt the cellular structure must be recognized as contributing more to direct release of vascular tissue contents than to the diffusion outward of exudates through alteration of cell permeability. Projecting our thoughts to the natural environment of field soil, we can visualize the virtual inseparability of wounded vascular tissue exudation and true cellular leakage as they influence microbial activity at the root surface.

General knowledge of plant physiology tells us that any foliar destruction of plants, such as may be caused by clipping (defoliation), severe foliage disease, and herbicidal injury, interfering with the photosynthetic process, can be expected to influence root exudation. The logic of this assumption is suggested by the observations of Vančura and Stanek (1975) in which changes in exudation of intact bean plants (grown in nutrient-wet sand) were compared with those of plants deprived of cotyledons or of true leaves. The results indicated that the plant roots were nourished primarily by reserve compounds in the cotyledons for the first 24 days of growth and a decrease in root exudates between days 3 and 15 was associated with formation of true leaves, which are also initially fed by these reserve compounds. When true leaves were removed, the nutrient supply from the cotyledons to the roots increased, and the quantity of root exudates also increased. When the cotyledons (and thus the reserve compounds) were removed, root exudation decreased. The amount of reducing compounds in exudates decreased for plants deprived of true leaves or cotyledons, whereas some amino acids increased and others decreased. Under normal agricultural practices, such as frequent clipping and grazing of forage crops, it can be expected that root exudates and microbial activity in the rhizosphere will fluctuate accordingly. Indeed, J. K. Martin (1971 a) found that defoliation of ryegrass, cultured in sand and previously exposed to $^{14}CO_2$, resulted in an increase in amount of ^{14}C leachates from the sand.

Other forms of stress upon plants, not involving mechanical injury, can alter root exudation and the qualitative content of exudates. For example, a high sodium content (from NaCl) in a nutrient solution caused a marked decline in the free amino acids found in a barley root exudate (Polonenko et al. 1983). Salt-induced osmotic stress which occurs in high saline soils can be expected to affect membrane permeability, oxygen and nutrient uptake, and the movement of exudates from roots.

3.5.6 Microbial Effects

The first requirement for determining the true nature of root exudates is the exclusion of microorganisms from plant-culture systems to avoid confusion of their metabolic products with exudate components. Rovira and Davey (1974) named four ways in which microorganisms may affect the observed exudation: (a) by affecting permeability of root cells, (b) affecting metabolism of roots, (c) absorption of certain exuded, compounds, and (d) altering nutrient availability to the plant.

Since exudation of organic compounds from roots can be affected, presumably due to altered cell permeability, by the addition of certain chemicals such as buffers to the liquid cultures used in root exudation studies (Rovira and Ridge 1973), it is not surprising that some antibiotics also can affect membrane permeability and increase exudation (Norman 1955, 1961; P. Martin 1958). Increased exudation from bean roots induced by phytotoxic substances from decomposing plant residues (Toussoun and Patrick 1963) may be another example of an effect on cell permeability; we referred to this earlier also as possibly indicating a chemical wounding effect.

With its higher concentration of carbohydrates, amino acids, vitamins and other growth-promoting substances, the rhizosphere is a zone of intense microbiological activity where the metabolic interaction between the rhizosphere microflora and green plants is extremely complex (Nicholas 1965). The mucigel coating on the surface of roots provides a matrix in which microorganisms, soil nutrients, water, and gases are in intimate contact. In soil already low in nutrients, the microflora through competition can create a nutrient deficiency. On the other hand, plants may benefit from microbial action that releases nutrients, such as phosphates and trace metals, from insoluble sources making these available to the plants. In either case, growth of plants and the release of exudates from roots are affected. The mechanism of microbially induced exudation need not be related directly to plant metabolism. Significant increases in root exudation often observed in the presence of microorganisms as compared with exudation by axenically cultured plants in nutrient solution may result from an increased concentration gradient between root and nutrient solution created by microbial utilization of certain compounds in the exudate.

The great difficulty of assessing the true nature and quantity of exudates in the natural soil environment is attributable primarily to two actions of microorganisms: (a) absorption and utilization of exudate components and (b) synthesis of microbial metabolites in the rhizosphere. Since most exudate studies have been conducted in cultural systems that excluded microorganisms, the results do not represent what occurs in natural soil. Radioactive labeling techniques, however, have provided some information on the influence of rhizosphere microorganisms on exudation. D. A. Barber and Martin (1976) were able to assess separately the ^{14}C-labeled CO_2 produced in soil by plants and by microbial activity. Labeled wheat and barley plants were maintained in either sterilized or nonsterilized soil. Under sterile conditions 5–10% of the photosynthetically fixed carbon was released by roots, compared with the 12–18% released from roots in nonsterilized soil. More exudation was apparent from barley than from wheat roots and this was associated with higher numbers of bacteria in the rhizosphere of barley. Wheat plants cultivated in the presence of *Pseudomonas putida* released up to twice the amount of exudate released by axenically cultured plants (Prikřyl and Vančura 1980). Exudation may be affected by cytokinins produced by soil fungi and bacteria that colonize plant roots. Increased concentrations of free fatty acids and sterols were released by axenically grown peanut plants treated with kinetin (6-furfurylaminopurine) according to the studies of Thompson and Hale (1983).

J. K. Martin (1977) proposed that, in artificial systems of plant culture, a major loss of carbon from roots may result from autolysis of root hairs, sloughed root cap cells, epidermal cells, and cortex which would release a root lysate rather than true exudate. Root lysis can be increased by soil microorganisms, apparently without penetration of the plant cell walls.

Though little attention has been directed to the subject, enhanced exudation by roots growing in symbiotic association with rhizobia or mycorrhizal fungi would seem to be a logical assumption since plant growth is affected. Exudates from nodulated root systems are sometimes different from those in systems without *Rhizobium* nodules (Hale and Moore 1979). Inoculation of alfalfa with *Rhizobium* can result in an increase in nonreducing sugars, ortho-dihydroxyphenols, amino-N, polygalacturonase, and pectin methylesterase in the root exudate, whereas reducing sugars and total phenols decrease. *Rhizobium* itself produces indoleacetic acid in the rhizosphere.

The nutrient-absorbing capacity of mycorrhizal roots is enhanced with a consequent effect on plant growth that is quite contrasting with growth of plants with nonmycorrhizal roots. It should follow that root exudates would be quantitatively, and perhaps qualitatively, different also, but we know little about this. Carbon compounds are transferred from the host plant to the fungal symbiont and thus may temporarily curb the release of some compounds directly into the rhizosphere. It is difficult to distinguish between exudate components from mycorrhizal roots and the volatile or nonvolatile materials synthesized and released by the fungal symbiont. Therefore, they are often considered together as being derived from the mycorrhizosphere. Monoterpenes and sesquiterpenes produced by the ectomycorrhizal (*Boletus variegatus*) root system of Scots pine are usually not found in nonmycorrhizal root systems (Krupa and Nylund 1972). These compounds are known to inhibit the growth of plant pathogens *Phytophthora cinnamomi* and *Fomes annosus*, and it is upon this possible relationship to plant disease that most interest has been centered.

Even less is known about the effects of endomycorrhizal fungi on root exudation. Effects are frequently inferred from the observations that phosphate-solubilizing bacteria, free-living nitrogen-fixing bacteria, and others maintain higher populations on roots with vesicular-arbuscular mycorrhiza than on nonmycorrhizal roots. Increasing interest in the relationship between mycorrhizal roots and the establishment of nitrogen-fixing bacteria in particular should in time provide greater insight into the mycorrhizal contribution to the process of root exudation.

Not to be overlooked are the effects of plant-pathogenic microorganisms on root exudation. We will deal with the subject of rhizosphere and plant-disease relationships in Chap. 7. It will suffice here to mention that both foliage-infecting microorganisms or viruses and root-infecting pathogens can bring about changes in the metabolism of carbohydrates, amino acids, proteins, lipids, nucleic acids, and natural growth regulators (Hale and Moore 1979), resulting in more root exudation and higher populations of microorganisms around roots. Increases in exudation induced by root pathogens can be attributed to direct effects in which root cells are injured and cell contents are released into the rhizosphere; nematodes are particularly active in this type of effect. However, we have already mentioned that this is not a true exudation process. On the other hand, when microbial metabolites alter cell membrane permeability, the release of compounds can occur along electrochemical potential gradients in a process of simple diffusion.

Like other studies of root exduation, microbial effects have been determined largely in highly controlled monoxenic plant-culture systems. In most instances, plants inoculated with pathogenic organisms have released greater total amounts of the various compounds common in exudates than were found in exudates of healthy plants. In a few cases, healthy plants have yielded higher quantities of some specific compounds.

3.5.7 Foliar Sprays

Root exudates can be altered by the application of some compounds to leaves (Rovira 1969; Hale et al. 1971, 1978; Rovira and Davey 1974). Through the use of radioactive materials, autoradiography, and bioassay, a number of substances, such as growth-regulating compounds, antibiotics, and nutrient elements applied to the shoot system have been shown to move from foliage into roots and out into a culture medium by exudation. Since molecular structure and other factors can affect the translocation from leaves to roots, all foliar-applied compounds do not reach the roots. This was demonstrated by Linder et al. (1964) who showed that 2,3,6–trichlorobenzoic acid applied to leaves of bean was released unaltered into aerated water surrounding the roots, whereas 2,5-dichlorobenzoic acid was translocated from the leaves to stem but could not be detected in roots or root exudates. Another compound in a different family, α-methoxyphenylacetic acid, was exuded from roots in quantities up to 10% of the amount applied. In addition to 2,3,6-trichlorobenzoic acid, picloram (4-amino-3,5,6-trichloropicolinic acid), dicamba (3,6-dichloro-*o*-anisic acid), and 2,4,5-T [(2,4,5-trichlorophenoxy) acetic acid] are released from roots after foliage treatment (Hurtt and Foy 1965; Reid and Hurtt 1970); the amounts exuded may be high enough to affect contiguous plants. Such exudation may be particularly significant for certain persistent herbicides like picloram which are biologically active in extremely low concentrations. Three-day-old sicklepod (*Cassia obtusifolia*) plants, aseptically cultured in water and foliar treated with the herbicide linuron [3-(3,4-dichlorophenyl)-1-methoxy-1-methylurea], released root exudate representing a 19.8% increase in dry matter over that from untreated plants (S. L. Brown and Curl 1979). Such induced exudation may have relevance to the stimulation or suppression of spore production and germination by fungal pathogens in field soil.

Other effects of plant growth-regulating compounds on root exudation have been cited by Hale and Moore (1979). Among these are increased exudation by sorghum and increased cholesterol release from peanut roots following foliar applications of 2,4-D [(2,4-dichlorophenoxy) acetic acid], increased exudation of amino acids and sugars from root and hypocotyls of bean after foliar treatment with EPTC (*S*-ethyl dipropylthiocarbamate) and dinoseb (2-*sec*-butyl-4,6-dinitrophenol), and increased exudation by soybean after treatment with chloramben (3-amino-2,5-dichlorobenzoic acid). In a few cases, herbicides have been shown to suppress exudation of principal compounds with the consequence of reduced microbial populations in the rhizosphere.

Root exudates have been modified by foliar application of antibiotics in efforts to alter and control the rhizosphere microflora. ^{14}C-labeled streptomycin moved from treated *Coleus blumei* leaves down to the root tips within 24 h (Davey and Papavizas 1961). This resulted in suppression of Gram-negative bacteria in the rhizosphere. Exudation of amino acids and sugars from roots of wheat increased considerably following foliar applications of another antibiotic chloramphenicol, fewer bacteria and increased numbers of fungi occurring in the rhizosphere of treated plants.

A number of researchers have shown that foliar applications of plant nutrients may bring about changes in rhizosphere populations, and this is attributed primarily to modification of root exudates. For example, species of *Aspergillus* in the rhizosphere of wheat were affected when foliar sprays with urea caused increases in exudation of glucose, fructose, glutamine, and alanine and decreases in organic acids (Agnihotri 1964). Foliar-applied nitrogen in the form of $NaNO_3$ and phosphorus as Na_2HPO_4 changed the root exudation pattern for different crop plants (Balasubramanian and Rangaswami 1969). The concentration of amino acids generally increased and the sugar content in exudates decreased after treatment of foliage with nitrogen, whereas amino acids decreased and sugars increased following phosphorus treatment. The mechanisms involved in altering the metabolism of plants leading to such differentials in exudation are complex and have not been clarified.

The implication and potential significance of altering root exudations by foliar applications are related to the influence of exudates on microbial activity in the rhizosphere. The feasibility of imposing some control upon certain rhizosphere phenomena that affect root pathogens, plant-nutrient uptake, and interactions between contiguous plants seems real, but much more basic information is presently needed.

3.6 Sloughed Organic Matter

We have seen from the foregoing discussion in this chapter that it may be impossible under most plant-culture systems to clearly distinguish between materials exuded from roots and the degradation products of sloughed root cells that are released as a plant grows. In ^{14}C-labeling experiments with wheat, J.K. Martin (1977) was able to distinguish cell degeneration material released into soil and assess the effects of microorganisms on the process. He referred to these materials as root lysate rather than exudate. Most exudate studies have been conducted with plants in solution culture primarily to avoid root injury and to minimize the sloughing of cells that would confuse assessment of the true exudation picture. Yet, even in solution culture, it has been clearly established by microscope observations that whole root caps, tissue fragments, and individual cells are released during growth (G.J. Griffin et al. 1976). For peanut plants cultured axenically in Hoagland's solution, tissue representing approximately 0.15% of the plant's carbon, nitrogen, and hydrogen was sloughed per week. The C/N ratios of sloughed organic matter varied from 9:1 for plants in quarter strength Hoagland's solution to 18. 9:1 for plants in nitrogen-free solution; the H/C ratios were usually about 2:1. The sloughing of root cap and cortical cells is illustrated in Fig. 3.4.

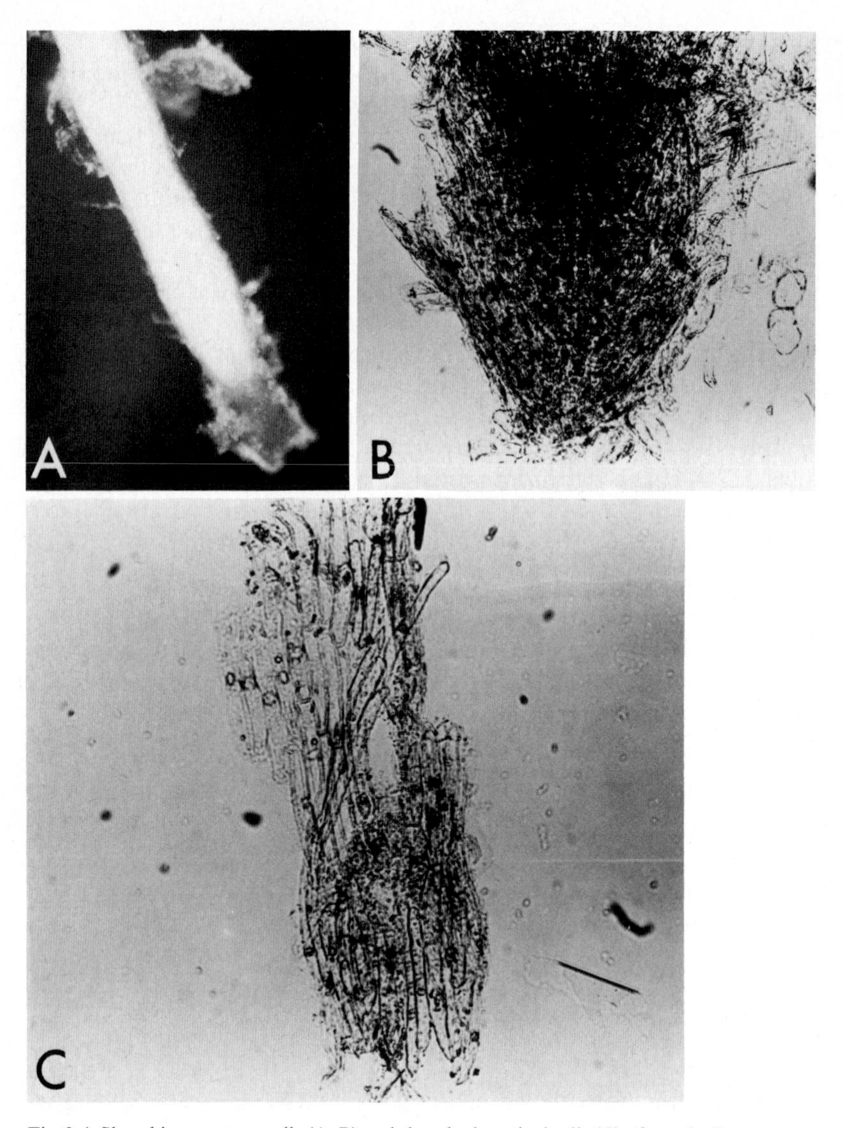

Fig. 3.4. Sloughing root cap cells (A, B) and sloughed cortical cells (C) of axenically grown peanut plant. (Griffin et al. 1976)

It would be expected that organic matter loss in the form of debris from roots growing in sand or natural soil would be many times greater than that in solution culture. Thus, while model systems of axenic culture are essential for the most accurate evaluation of the nature of exudates, a combination of exudates and sloughed matter actually provides the influencing principle affecting microbial populations and their many interactions. It bears repeating, therefore that exudate information obtained in controlled environments must be interpreted with caution when relating to natural conditions.

Chapter 4 Rhizosphere Populations

4.1 Introduction

The quantitative und qualitative nature of microbial populations in the rhizosphere and rhizoplane are related either directly or indirectly to root exudates and thus will vary somewhat according to the same environmental factors that influence exudation. These fluctuating populations constitute part of what we term the rhizosphere effect, this effect diminishing with increasing distance from the root surface. Populations are made up of components of both the microflora (bacteria, actinomycetes, fungi, and algae) and the micro- and mesofauna (protozoa, nematodes, mites, and insects). The microflora has received by far the greater research attention by soil microbiologists, plant pathologists, and ecologists; the fauna, except for nematodes, has been largely neglected.

A broad view of the microbial groups and other interacting agents relating to the rhizosphere effect are illustrated in Fig. 1.2. One can see that a large number of interactions are possible and thus appreciate the enormous complexity of this unique zone. From the time of Hiltner's (1904) investigations of root influence on bacterial activity, many controversies have been raised regarding the true nature of the microbial population; does the rhizosphere harbor a special flora or does the root system merely stimulate the general endemic population?

Research efforts toward solving the mysteries of the rhizosphere began long before Hiltner coined the term, but international focus upon the subject only came following a series of investigations beginning about 1929 by R. L. Starkey of the New Jersey Agricultural Experiment Station. More than 50 years of rhizosphere research worldwide has been documented in reviews (Starkey 1929; Katznelson et al. 1948; F. E. Clark 1949; Katznelson 1965; Rovira and Davey 1974; Bowen and Rovira 1976). These reviews show that a significant portion of the overall research effort has been applied to assessing microbial populations characteristic of various crop plants. Paralleling these investigations were the development and modification of microbiological techniques from the simple soil-dilution and plate-count methods, still in use, to direct observations with the stereoscan electron microsope. Results vary considerably with kinds of cultural media and the isolation or observation methods employed, thus creating a prime source of conflicting results and controversy among various researchers.

The great amount of time and effort spent on enumeration and characterization of microorganisms in the rhizosphere is justified by the essentiality of knowing something of the taxonomy and physiological behavior of the microbial components before their interspecific interactions and effects on plants can be studied.

4.2 Isolation and Enumeration of Microorganisms

4.2.1 The Microflora

Standardization of techniques for estimating populations of microorganisms in soil has not been developed to the advanced stage of other procedures such as chemical analysis. This is due to the extreme complexity and variability of soil environments. We sometimes refer to the "total" number of bacteria or fungi in a soil sample when in fact we mean the relative number as compared with other soil samples determined by the same method. The assessment of total numbers cannot be achieved with presently known cultural procedures because of the diverse growth requirements of the different physiological groups of organisms. Direct microscopy observation methods have revealed the presence of many microorganisms which will not grow on any of the nutrient media that have been used in soil-plating procedures. Therefore, direct counts usually show higher numbers and larger R/S values. For reasons not fully understood, however, direct microscopy counts sometimes reveal higher numbers of microorganisms in nonrhizosphere soil (lower R/S values). The populations recorded in rhizosphere soil of wheat seedlings were much higher than in root-free soil when the soil dilution-plate method was used, but the numbers were lower when fluorescence microsocpy was used (van Vuurde and Delange 1978).

Methods and media used for estimating numbers of bacteria, actinomycetes, and fungi in field samples have been described in some detail (Parkinson et al. 1971; L. F. Johnson and Curl 1972; Curl and Rodriguez-Kabana 1977). Tsao (1970) described selective media used specifically for isolating various plant pathogens and discussed the principles of media development. Since all organisms in a sample cannot be isolated by a single technique or by one culture medium, the ideal procedure would be to employ several different techniques and media. This, however, is rarely feasible when dealing with large numbers of soil samples from replicated plots. Therefore, one must select the method best suited to the purpose and one or more media that are most selective for the desired microbial group or individual, and clearly state the limitations under which the study is made.

The methods for quantitative and qualitative assessment of soil microorganisms are almost as diverse and numerous as the physiological groups of organisms to be studied. Most commonly used is the soil-dilution and plate-count method in which a soil sample is serially diluted in water and the appropriate final dilution is mixed in plates of molten agar medium before solidification, or is spread over a solid agar surface. Knowing the dry weight of soil used and the number of distinct microbial colonies appearing on the plates, one can calculate the relative numbers of propagules per gram of dry soil. Individual colonies can be transferred from such plates to pure culture on appropriate media for further study. The basic serial-dilution and plating procedure has been described in detail (L. F. Johnson and Curl 1972). A 25-g (dry-weight basis) sample of sieved soil is added to water in a 1-liter flask to provide a final volume of 250 ml. This is shaken and the serial dilutions prepared by pipetting the original 1:10 (soil:water) through a series of water blanks until appropriate dilutions for fungi (usually 1:10,000) and for bacteria (about 1:1,000,000) are reached. One milliliter of the end dilution is then mixed in a Petri dish with a suitable agar medium such as Ohio agar with sodium propionate and antibiotics for fungi, and a soil extract medium or Thornton's mannitol-asparagine agar for bacteria. The ingredients of many media are listed by Tuite (1969) and L.F. Johnson and Curl (1972). This basic soil-dilution procedure has been modified many times in the search for shorter, less cumbersome methods when large numbers of soil samples are to be processed. A short method that we have found suitable is one referred to as the dropper or modified dilution-plate method, which simply requires plating drops of the original soil-water suspension and mixing in a suitable agar medium. Other drops of the suspension are oven-dried and weighed so that numbers of colonies per plate can be converted to propagules per gram of soil. This method permits the processing of large numbers of soil samples with minimal glassware and also provides a more concentrated soil-water suspension in the agar medium, which increases chances of isolating nonsporulating fungi from mycelial fragments. Details of the procedure are outlined by Curl and Rodriguez-Kabana (1977). Other modifications of the dilution-plate procedure have been suggested, and various selective media developed which permit growth of individual genera or other groups.

Special treatment of soil suspensions and plating in selective media are necessary procedures when dealing with specific organisms such as the aerobic spore-forming, nitrifying, or nitrogen-fixing bacteria. For bacteria in general, the most-probable-number method for estimating population density is often employed (Alexander 1982).

Since soil-dilution and plating methods reveal only relative numbers of viable propagules (bacteria, fungal spores, mycelial fragments, microsclerotia) and also favor the isolation of abundantly sporulating genera of fungi such as *Penicillium, Aspergillus, Trichoderma,* and *Fusarium,* direct microscopy methods are desirable where precision is essential. A slide incubation chamber described by Polonenko and Mayfield (1979) provides a system in which plants can be grown from seed and long-term microscopy observations of root growth through soil can be made. The method was used to determine microbial colonization characteristics on the rhizoplane of *Pisum sativum* and to monitor population responses to root exudates. Such methods, along with improved staining techniques (Johnen 1978; van Vuurde and Elenbaas 1978), allow all cells to be counted and the morphology of organisms to be observed. Populations estimated from microscopy observations are usually many times higher than those recorded from dilution-plate procedures. Development of direct microscope techniques has progressed from use of the light microscope (Frederick 1965) to the highly specialized use of transmission and scanning electron microscopy (Dart 1971; Campbell and Rovira 1973; Old and Nicholson 1975; Foster and Rovira 1976) for studying spatial relationships between roots, soil, and microorganisms in a relatively undisturbed state. Numbers of microbial cells, whether determined by plate-count or by microscopy, do not provide information on weight of living material (biomass). This extra precision, which may be desirable in certain studies, can be attained through the use of appropriate techniques for conversion of numbers of bacterial cells, fungal spores, or lengths of fungal mycelium to biomass (Parkinson et al. 1971).

Soil algae have received much less attention than other members of the microflora, and recordings of their habitation have been more qualitative than quantitative. They have been studied mostly by three procedures. In the soil-block or enrichment-culture method, blocks or samples of soil are incubated in moist chambers and viable algae are picked off and identified. Serial dilution of soil followed by incubation of a sample of the suspension in Bristol's solution or other appropriate medium can provide cells for direct microscope examination and for application of the most-probable-number procedure. Or direct fluorescence microscopy may be used, a procedure in which a soil sample is irradiated with blue light through a yellow filter and the algae appear as red objects against a black background. These procedures have been described by F. E. Clark and Durrell (1965) and Shields (1982).

Since the rhizosphere is a zone of greater microbial activity than is root-free soil, the relation of microbial behavior to root development and plant health is logically studied most in this region. This zone of soil, lying under the influence of root exudation, is only a few millimeters wide; therefore, certain modifications of the generally used methods for soil-sampling and for microbial isolation and enumeration may be required. Relative populations of bacteria, actinomycetes, and fungi can be estimated by serial dilution and plating of rhizosphere soil in appropriate media. Numbers of microorganisms per gram of rhizosphere soil are computed from the average colony count, and transfers from individual colonies may be made to tubes of nutrient media for identification and further study. A numerical value, the R/S ratio (Katznelson 1946), representing the rhizosphere-soil number divided by the nonrhizosphere-soil number, is used as an index of rhizosphere effect on populations.

Rhizosphere soil is usually considered to be that soil which remains adhering to root surfaces after the roots have been shaken to dislodge the larger clumps of soil and organic matter. The amount of adhering soil remaining on roots will vary with soil type, soil moisture, vigor of shaking root systems, etc. A small amount of soil remaining on the roots will yield a higher R/S ratio than a larger amount of adhering soil. It follows then that expression of the rhizosphere effect as an R/S ratio is most accurate when comparing plants in one type of soil under similar cultural conditions.

A procedure described by Timonin (1940), with subsequent modifications of it, is most frequently used. Blocks of soil with root systems intact are cut out and roots gently removed with minimal damage. These are shaken briefly to remove loosely attached soil, then placed in weighed flasks containing 100 ml of sterile water. The flasks are uniformly shaken for 20 min and serial dilutions are prepared from the resultant soil suspensions. One-milliliter quantities of appropriate end dilutions for bacteria and for fungi are mixed in Petri dishes with molten agar media selective for these two groups and the cultures are incubated until colonies appear. The numbers of colonies are converted by calculations to organisms per gram of soil; this presumes that each colony represents a single propagule. The weight of rhizosphere soil is determined by removing the roots from the original flasks, evaporating the water, and obtaining the oven-dry weight of the residue. A number of variations of this procedure may be suitable for specific purposes (Parkinson et al. 1971; L. F. Johnson and Curl 1972). Typical dilution plates of

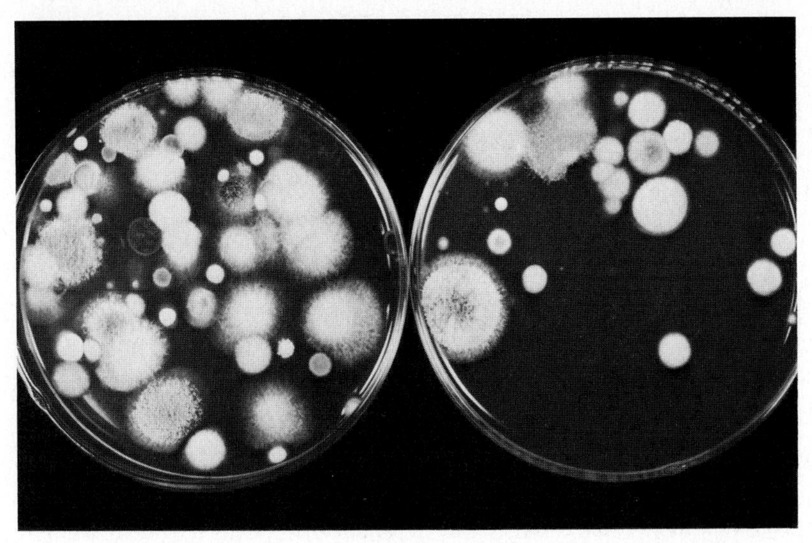

Fig. 4.1. Colonies on dilution plates representing fungi isolated from (*left*) rhizosphere soil of wheat plants and from (*right*) nonrhizosphere soil

fungi obtained by the short dropper method described earlier are shown in Fig. 4.1.

In some cases the simple Warcup soil plate (Warcup 1950) is useful, primarily to study the ecological distribution of various species of fungi in the rhizosphere soil of different plants where quantitative assessment is not needed. Rhizosphere soil obtained by shaking roots in a dry flask or plastic bag is transferred as small samples (5–15 mg) on a microspatula to sterile Petri dishes. A suitable agar medium is then poured and the soil particles are dispersed in it. Fungi appearing on the plates can be isolated, identified, and studied individually or in combination.

Most rhizosphere studies present population results that are based on both rhizosphere and rhizoplane organisms, since a distinction between the two is difficult or unlikely. Rovira and Davey (1974) have suggested that expression of numbers of organisms both on the basis of dry-soil weight and root weight would provide the most valid data where a distinction between the two zones is not made. Many studies, however, have been made of the rhizoplane (root surface) after all rhizosphere organisms presumably have been washed off. The technique of Harley and Waid (1955 a) is very effective for isolating fungi intimately associated with root surfaces. Root pieces are washed in 30 or more changes of sterile water and the efficiency of removal of microorganisms is determined by plating samples of the wash water in agar media. Figure 4.2 shows how plate-counts of fungal spores and mycelial fragments decline with successive washings. The washed roots are then cut into small segments and placed on agar media from which pure cultures of emerging fungi can be isolated for study. This is not a quantitative method, but it can be used to study the distribution of fungal species on different parts of the root system. Louw and Webley (1959 a) suggested that,

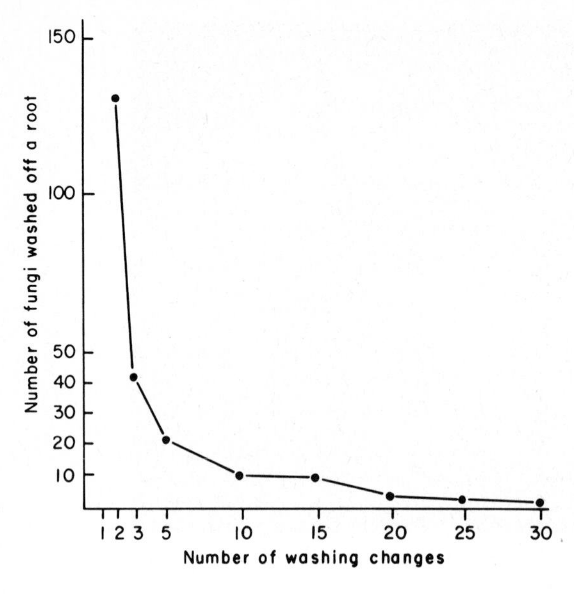

Fig. 4.2. Numbers of fungi washed off a root shaken in successive, similar volumes of sterile water. (Harley and Waid 1955a)

after rhizosphere soil has been removed, the roots may be immersed in flasks of sterile water plus glass beads. The pre-weighed flasks are then reweighed and shaken for 20 min, and dilutions plated on agar media as usual. Rhizoplane organisms are recorded as numbers per gram of wet weight of root tissue.

Root maceration techniques, such as the one used by Stover and Waite (1953) to isolate *Fusarium* spp., may yield a more representative variety of rhizoplane flora if we can assume that the species released from internal tissues are largely the same as those that inhabit the root surface and mucigel layer. Thoroughly washed roots are macerated in sterile water in a blender for 2 min, then diluted as appropriate and spread on the surface of agar media for production of colonies.

The qualitative nature of rhizoplane organisms and their distribution patterns can be determined most effectively either by light microscopy or stereoscan electron microscopy, with suitable staining of root preparations or by use of infrared color photography. Such techniques can reveal the intricate details of relationships between root surface and fungal mycelia or bacteria, or aid in identifying strains of *Rhizobium* responsible for nodule formation.

The feasibility of applying soil enzyme methodology to the assessment of microbial activity has been suggested (Rodríguez-Kábana and Truelove 1970, 1982) for situations in which it is not essential to identify microbial components or groups of organisms involved. Catalase activity, for example, has been correlated with bacterial and fungal counts, cation exchange capacity, dehydrogenase activity, and cotton yield. Catalase activity, measured polarographically, in soil samples from the approximate rhizospheres of several crops at the peak of the growth period varied considerably between plant species (Fig. 4.3). These differences were attributed to changes in biological activity, principally microbial, in the root zone. Similarly, shifts in soil microbial populations which occur in re-

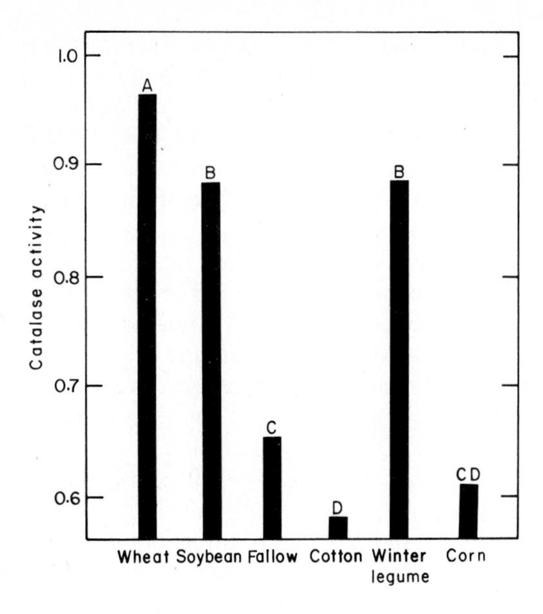

Fig. 4.3. Levels of catalase activity in soil collected from the root zones of crop plants at maturity. *Bars with the same letter* are not statistically different. (Rodríguez-Kábana and Truelove 1982)

sponse to crop rotations or fertilization can be detected by determining soil xylanase activity, reflecting changes in populations of xylanase-producing organisms (Rodríguez-Kábana 1982).

4.2.2 The Fauna

Nearly all of the animal phyla, except purely marine forms, are represented in the soil (Kevan 1965), but their numbers vary considerably. The most abundant forms (protozoa, nematodes, and microarthropods) can be expected to influence the activities of microflora components, or be influenced by the flora, since many of these animals are primarily bacterial and fungal feeders. Relatively little information on the activities of animals specifically in the rhizosphere has been acquired, except for the attraction of nematodes to roots and their relationships with plant pathogenic fungi and mycorrhizae. As with the microflora, populations of the soil fauna in the rhizosphere can be estimated by adaptations of methods used for soil populations in general.

Both cultivated and virgin soils harbor a heterogeneous abundance of protozoa (ciliates, flagellates, and rhizopods or amoeboids forms), which are of interest to ecologists mainly through their interrelationships with bacteria. An enrichment procedure is usually used when enumerating protozoa (F. E. Clark and Beard 1965) in which a washed-agar medium (nutrient deficient) is used. A bacterium, such as *Aerobacter aerogenes,* which is considered digestible by a large number of the protozoan fauna, is used as the food source in conjunction with a ring method (Singh 1946, 1955). Small glass rings are placed on the solidified agar surface in Petri dishes and a thick bacterial suspension is spread on the agar inside the rings. A soil suspension is diluted in series and 0.05-ml volumes of the desired end dilutions are applied inside the rings. After 2 weeks of incubation at 22 °C, the bacterial rings are examined with a microscope for presence of protozoa; the most-probable-number method (Alexander 1982) can be applied for estimating whole polulations, major subgroups, or individual species. T. R. Anderson and Patrick (1978) used spore suspensions of *Cochliobolus sativus* and *Thielaviopsis basicola* as food sources to increase numbers of spore-perforating amoebae. Other methods have been described (Stout et al. 1982).

Nematode counts in soil surrounding plant roots have been made but such counts usually are not compared with counts in root-free soil. The term rhizosphere is not always applicable, because nematodes can migrate into and out of the narrow soil zone that is influenced by roots. Most studies relating to the rhizosphere deal either with factors responsible for attraction of nematodes to roots, substances or stimuli of root origin that induce the hatching of larvae from cysts, or host–parasite relationships.

General extraction methods for free-living nematodes and for nematodes in plant tissues are well documented and described (Southey 1970; van Gundy 1982). Most methods employ some modification of the Baermann funnel in which the motility of nematodes aids in separating them from soil, debris, or macerated plant material. The material is held on muslin cloth or wet-strength paper tissue in water and active nematodes migrate to the bottom of the funnel stem where they can be collected. Depending on the purpose of specific studies, methods may be used involving simple flotation techniques for extracting cysts, or expensive, semi-automatic elutriation systems designed for the simultaneous processing of large numbers of soil samples.

Although many workers have observed the activities of nematodes and their orientation and attraction around roots, few have made population assessments specifically in the rhizosphere. Henderson and Katznelson (1961) used a modified Baermann funnel method with composite 33-g soil samples taken from 15–20 plants of several crops (wheat, barley, oat, soybean, and pea). Nematodes were extracted from both rhizosphere and nonrhizosphere soil and numbers were based on oven-dry weight of the soil. Comparative populations are sometimes reported in terms of the R/S ratio, as for the microflora.

Methods for estimating populations of soil animals other than protozoa and nematodes vary considerably according to the specific animal involved. Since microarthropods, particularly the Acarina (mites) and Collembola (springtails), make up the greatest numbers and are most likely to be affected by rhizosphere phenomena, we can logically concentrate on these. Like the microflora, they are not uniformly distributed through soil but tend to congregate at food-source sites. Many microarthropods are mycophagous, for example, and may be more abundant where fungi are plentiful, such as on organic matter or near root surfaces. Therefore, sampling procedures that will provide a representation of the animal's distribution are of prime importance. Though various flotation and screening techniques may be used for assessment of total dead and living populations, it is often desirable to extract living, undamaged animals. Woolley (1965) and Wallwork (1970) have described the procedures generally used for extracting mites and other microarthropods from soil and debris. These procedures are based on principles of the early Berlese-Tullgren extraction systems in which a properly col-

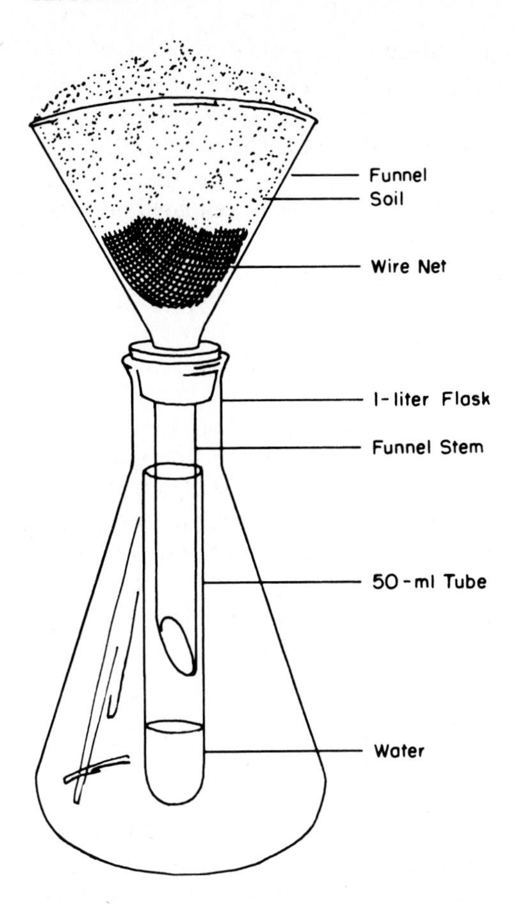

Fig. 4.4. Modified version of the Tullgren system for extracting live Collembola from soil. When placed under a 40-W incandescent light, the soil dries slowly, forcing downward migration of the insects, which drop onto the water surface in a collection tube. (Wiggins and Curl 1979)

lected soil sample is supported in a funnel and the sample is heated from above by a Mazda-type lamp or other suitable heat-light system. The animals, reacting negatively to heat, light, and desiccation, move slowly down the funnel over a 4- to 6-day period and fall into a collecting vial containing water or alcohol. Chemical repellants applied to the soil surface can be used where heat and light sources are not available, but the efficiency of extraction is usually lessened.

The usual method of collecting rhizosphere soil, by lifting whole root systems and shaking or washing off the soil, is not feasible for rapidly moving or jumping insects such as the Collembola. Wiggins and Curl (1979) used a conventional soil sampling tube of 2.2-cm inside diameter. The tube was inserted against a cotton plant parallel to the tap root to a depth of 20–25 cm and the sample quickly ejected into a suitable container. Obviously, this does not collect exclusively rhizosphere soil, as the definition of rhizosphere would require, but parts of the lateral root system are harvested with each probe sample. Thus, populations in samples containing some rhizosphere soil can be compared with populations in adjacent soil which is essentially root-free.

For extracting only living mites and Collembola, Wiggins and Curl employed a simple open modification (Fig. 4.4) of the Tullgren (1917) extraction apparatus.

Glass funnels with about 1,200 g of field soil were arranged under 40-W light bulbs maintaining a soil-surface temperature of 35 °C. The funnel stem extended into a 50-ml collecting tube standing inside a 1-liter Erlenmeyer flask, the stem end positioned a few millimeters above the surface of 10 ml of water in the tube. Insects which migrated in advance of the drying soil fell onto the water surface and were decanted to a microsieve. From here they could be refloated in a dish of water for counting or transferred for further study.

4.3 Populations of the Microflora

The rhizosphere effect of an actively growing crop plant is most pronounced with bacteria, followed by actinomycetes and fungi; the algae are least affected. Comparative estimates of populations for the rhizosphere soil (R) and nonrhizosphere soil (S), expressed as the R/S ratio, vary widely with different crop plants and with the sampling, isolation, and assessment procedures used. Since the earliest review by Starkey (1929), the magnitude of root influence on microbial populations has been documented many times worldwide, revealing increased intensity of investigation up to recent times.

4.3.1 Bacteria

Bacteria are the most abundant microorganisms in field soil, whether determined by plate count or by direct microscopy. Counts recorded have varied between 10^6 and 10^8 cells per cubic centimeter of soil (D. M. Griffin 1972). Most estimates, however, are based on numbers per gram dry weight of soil and these have been estimated to be in excess of 100 million g^{-1} according to plate counts and more than 3 billion g^{-1} by microscopy procedures. Based on direct microscope estimates of populations at 8 million g^{-} dry weight of soil, it has been estimated that bacteria make up between 0.03 and 0.28% of the total soil mass (F. E. Clark 1949). From population counts in Rothamsted field soils, E. W. Russel (1973) estimates that the bacteria in the top 15 cm of soil may weigh between 1,340 and 3,360 lb $acre^{-1}$ (1,500 to 3,750 kg ha^{-1}) live weight, or 250 to 650 lb dry weight $acre^{-1}$ (280 to 730 kg ha^{-1}).

The rhizosphere effect of actively growing crop plants usually corresponds to R/S ratio values of 2 to 20, but ratios of over 100 have been recorded. The wide variation in the value is because of the different rhizosphere effects of different plant species, and it also varies with the stage of plant growth. Thus, as we are reminded by Rovira and Davey (1974), R/S comparisons for a wide range of plants are valid only when such studies are made by the same methods under a constant set of conditions in one laboratory. Results of a study by Rouatt and Katznelson (1961), in which counts of bacteria in the rhizosphere of different crop plants were compared with root-free soil counts (Table 4.1), show the typical rhizosphere influence. The rhizosphere and control soil counts in this case were calculated from dilution-plate colonies and were based on the oven-dry weight of

Table 4.1. A comparison of the colony counts of bacteria in the rhizosphere and rhizoplane of different crop plants and in root-free soil. (Rouatt and Katznelson 1961)

	Colony count ($10^6\,g^{-1}$ of soil based on oven dry weight)		R/S	Colony count ($10^6\,g^{-1}$ of rhizoplane based on root dry weight)
	Rhizosphere	Root-free		
Red clover	3,255	134	24	3,844
Oats	1,090	184	6	3,588
Flax	1,015	184	5	2,450
Wheat	710	120	6	4,119
Maize	614	184	3	4,500
Barley	505	140	3	3,216

Table 4.2. Microorganisms (thousands g^{-1} of oven-dry soil) at various distances from the root surface of lupin seedlings grown in unamended Elsinboro sandy loam. (Papavizas and Davey 1961)

Distance from root surface (mm)	Bacteria	Streptomycetes	Fungi
0[a]	159,000	46,700	355
0– 3	49,700	15,500	176
3– 6	38,000	11,400	170
9–12	37,400	11,800	130
15–18	34,170	10,100	117
80[b]	27,300	9,100	91
R/S ratio[c]	5.8	5.1	3.9

[a] Rhizoplane
[b] Nonrhizosphere soil (control)
[c] $$\text{R/S ratio} = \frac{\text{No. of microorganisms 0 mm from root surface}}{\text{No. of microorganisms 80 mm from root surface}}$$

soil, hence they can be compared directly. The rhizoplane counts, however, were calculated on the basis of the dry weight of washed roots that were chopped in a blender before plating on media, and these values cannot be compared directly with the rhizosphere. Results obtained by Papavizas and Davey (1961), working with *Lupinus angustifolius,* clearly showed that the highest populations of bacteria, as well as those of streptomycetes and fungi, were in the rhizoplane and there was a decline in numbers with increasing distance from the primary root (Table 4.2). In this case a microsampler was used to take samples at 3-mm increments away from the root surface. The soil was serially diluted and plated on appropriate media and counts were based on the oven-dry weight of soil; rhizoplane counts were expressed as numbers of organisms per gram dry weight of roots plus adhering soil.

Some workers now believe that the rhizoplane population of microorganisms is more realistically estimated by direct microscopy examination of root samples, expressing the results in terms of numbers per root surface area or percent of root surface covered. Rovira et al. (1974) and Dangerfield et al. (1975) described and applied such methods to the assessment of the rhizoplane microflora of grassland species and of young and mature lodgepole pine. Eight grassland species had bacterial cover of the root surface within the range of 4–10%, while the calculated cover of pine roots ranged from 0.2–2.1%. For grasses, Rovira et al. (1974) estimated a tenfold difference between direct counts and plate counts, assuming that bacterial colonies on the roots are one cell deep and each cell covers 1 μm^2.

4.3.2 Actinomycetes

Most soil actinomycetes belong to three genera, *Streptomyces* and *Micromonospora* of the Streptomycetaceae and *Nocardia* of the Actinomycetaceae within the same taxonomic category as bacteria. Researchers often make reference to the streptomycetes of soil or rhizosphere, since most species isolated fall within the class Streptomycetaceae, largely in the genus *Streptomyces*. Hiltner and Störmer (1903) made the first quantitative enumeration of actinomycetes in soil from the flora developing on gelatin plates. Numbers varied between 13 and 30% of the total microbial flora, the variation being due to seasonal changes and decomposing organic matter effects. Much of our present knowledge of the distribution and nature of actinomycetes in soil is derived from the extensive studies of Waksman and associates (Waksman 1959).

Studies of actinomycetes in the rhizosphere have lagged far behind work with bacteria and fungi. Interest in these organisms has been primarily in their capabilities as bacterial and fungal antagonists and as producers of inhibitory metabolites. Their numbers are frequently recorded at the same time that bacteria are assessed and usually reveal R/S ratios that are considerably smaller than for bacteria, as shown for soil from wheat (Table 4.3) and yellow birch seedlings (Table 4.4). Plate-count estimates of microorganisms at the root surface of lupin seedlings (Papavizas and Davey 1961) showed populations of bacteria to be far greater than streptomycetes but these populations compared with numbers 80 mm distant from the roots yielded R/S values that were only slightly different.

4.3.3 Fungi

Populations of soil fungi in general fall in third place below those of bacteria and actinomycetes, based on dilution-plate counts. Fungi are more difficult to assess than are bacteria, because of their filamentous nature and the fact that the serial-dilution and plating procedure favors isolation of the abundantly sporulating species. Thus, comparisons of the rhizosphere effect upon bacteria and fungi should be made with caution. Depending on the kind of plant and whether soil samples are from the root surface or from the outer rhizosphere, R/S ratios may range from 3 to over 200, but most frequently they are around 10 to 20 for crop plants.

Serial root washing, as used by Harley and Waid (1955a), followed by direct microscope assessment of fungi on root surfaces, invariably reveals species of

Table 4.3. Comparison of the numbers of various groups of organisms g^{-1} dry soil in the rhizosphere of spring wheat and in control soil. (Rouatt et al. 1960)

Organisms	Rhizosphere soil	Control soil	R/S ratio
Bacteria	$1{,}200 \times 10^6$	53×10^6	23
Actinomycetes	46×10^6	7×10^6	7
Fungi	12×10^5	1×10^5	12
Algae	5×10^3	27×10^3	0.2

Table 4.4. The incidence of groups of soil microorganisms on the roots of yellow birch seedlings growing in two soil horizons. Numbers expressed as millions g^{-1} of oven-dry soil, determined by soil dilution and plate count. (Ivarson and Katznelson 1960)

Groups	A horizon			B horizon		
	R	S	R/S	R	S	R/S
Bacteria	353.6	23.4	15.1	179.1	3.1	57.8
Actinomycetes	13.0	2.5	5.2	15.7	3.3	4.8
Fungi	1.0	0.8	1.3	0.1	0.07	1.4

R = Rhizosphere soil
S = Nonrhizosphere soil (control)

$$R/S = \frac{\text{Numbers in rhizosphere}}{\text{Numbers in control soil}}$$

fungi not isolated in dilution plates. Therefore, since direct microscopy gives a more precise estimate of total fungi and bacteria, this procedure is currently being used more for both quantitative and qualitative studies of the rhizoplane, though most available data are still derived from the less tedious plate-count procedure.

Again, we can turn to Tables 4.2, 4.3, and 4.4 for representative comparisons of fungal populations with other groups of the rhizosphere flora. Whether dealing with crop plants or tree roots, fungal counts are nearly always lower than those for actinomycetes. The R/S ratio, however, is sometimes higher for fungi, as in the study with wheat shown in Table 4.3.

4.3.4 Algae

Soil algae are typically photoautotrophic and, therefore, would by expected to thrive best in the presence of a light-energy source and CO_2 for carbon. Some species of Chlorophyceae (green algae), Cyanophyceae (blue-green algae), and Bacillariaceae (diatoms) can also grow heterotrophically, utilizing the oxidation of organic compounds to replace a light-energy source (Alexander 1977). Although algal isolates have been obtained from soil depths of 50–100 cm, probably leached there by water, highest populations occur in the top 10 cm of soil. Soil-surface samples taken where a visible algal bloom has developed may yield several million cells g^{-1} of soil but subsurface numbers are more commonly less than 10,000.

Relatively little information is available to assess the effect of roots on algae. Being mostly autotrophic with some heterotrophic capabilities, they are weak competitors with other saprophytes and low populations would be expected in the rhizosphere. Katznelson (1946), however, did reveal a rhizosphere effect on green algae for mangels (a variety of beet) as the plants approached maturity; R/S values of 29 and 15 recorded in unfertilized and fertilized soil, respectively, were higher than for fungi. Rouatt et al. (1960) found lower numbers in the rhizosphere of wheat than in nonrhizosphere soil, thus yielding an R/S value of <1.0. Though nutrients in root exudates might be expected to affect the development of algae to some extent, it is more likely that they are primarily affected by physical changes in the environment.

4.4 Qualitative Changes in the Microflora

Effects of higher plants on the qualitative nature of soil microorganisms may be more important than the stimulation of greater populations in general. Since plant taxa differ in chemical composition and physiology, and in the quantity and quality of root exudates, they can be expected to have some selective effects on specific microorganisms. Even before Hiltner named the rhizosphere, controversy had begun concerning the kinds of microorganisms found on plant roots and whether the general soil microflora was simply stimulated en masse or whether certain types or groups were favored by roots. With the limited techniques available, some early workers believed that the root microflora was essentially like the general soil flora, while other workers recognized the probability of a specialized microbial population influenced by roots.

4.4.1 The Bacterial Flora

With the development of improved microbiological techniques, and extensive research on the subject, it is now well established that bacteria are more responsive to growth-stimulating factors from roots than are the actinomycetes, fungi, or algae. The qualitative nature of the bacterial flora has been investigated primarily by: (1) determining the numbers belonging to definite groups that grow on selective media used in dilution-plate procedures, and (2) isolating bacteria on a nonselective medium such as soil-extract agar and studying the nutritional requirements of these individually. A point to keep in mind is that no culture medium can provide all requirements for isolation or growth of all types of soil bacteria. Supportive information on morphological and taxonomic types is obtained by direct microscopy. Both plate-culture and microscopy techniques have revealed that roots of most plants favor Gram-negative, nonsporing, rod-shaped bacteria, whereas Gram-positive, nonsporing rods and cocci, pleomorphic rods, and aerobic spore-forming bacteria are relatively less abundant directly adjacent to plants (Lochhead 1959; Katznelson 1965; Rovira and Davey 1974). The rhizosphere harbors a higher percentage of motile bacteria and chromogenic types, as well as amylolytic, proteolytic, ammonifying, denitrifying, sugar-fermenting, and cellulose-decomposing bacteria, and those that can reduce methylene blue or resazurin. Most investigators agree that *Pseudomonas* spp. are among the most consistently abundant bacteria under the influence of roots. *Agrobacterium* and *Achromobacter* also are relatively common. While *Arthrobacter* has been found in large numbers in the rhizospheres of certain crops, species of this genus apparently are suppressed by the more competitive *Pseudomonas* (Katznelson 1965). Vančura (1980) found that the fluorescent pseudomonads were the most prevalent bacteria found in the rhizosphere and rhizoplane of a number of plants investigated. Their numbers varied with plant species and age. *Pseudomonas fluorescens* represented 60–93% of the population on roots. The pseudomonads apparently function best during periods of greatest root-exudate release, whereas organisms of the *Arthrobacter* type may be more efficient in nutrient utilization when the nutrient flux is less intense. Generalizations about the rhizosphere effect on specific bacteria are hazardous, since exceptions to any such statements can be found. The stage of plant growth and the soil conditions at time of sampling have varied greatly among research investigations and have often resulted in conflicting reports. Russian workers and others have reported benefits to crop growth and yield after seed inoculation with nonsymbiotic nitrogen-fixing bacteria (*Azotobacter, Bacillus, Clostridium*). Yet, when their numbers, which are indeed high in the rhizosphere of certain crops, are viewed in proportion to other bacteria, evidence is yet insufficient to proclaim a strong selective rhizosphere effect. When given an initial advantage over competing microorganisms by application to seed, *Azotobacter* can multiply rapidly and persist in the rhizosphere of developing roots (M. E. Brown et al. 1962). *Azospirillum,* another aerobic nitrogen-fixing bacterium, is commonly associated with the rhizosphere of certain tropical grasses in which it may also inhabit the cortical layers of plant tissue (Atlas and Bartha 1981).

The status of *Rhizobium* is difficult to assess from conflicting reports, although some early work (Rovira 1960) has shown a proliferation of numbers around roots of both legume and nonlegume plants. Some believe that rhizobia may be largely suppressed in the face of microbial antagonism and competition for nutrients at the root surface. However, it would seem essential for rhizobia to multiply extensively in the legume rhizosphere before effective symbiosis can be established. Scanning electron microscope studies of inoculated legume roots (e.g., Dart 1971) have clearly demonstrated the capacity of *Rhizobium* to colonize root surfaces (Fig. 4.5). V. G. Reyes and Schmidt (1981), with the aid of immunofluorescence, calculated the populations of native *Rhizobium japonicum* based on root surface area. The density per square centimeter of soybean root surface fluctuated between a few hundred to over a thousand and declined as the root volume expanded. In soil, the advantage for competitive colonization might be provided in part by substances such as biotin and thiamine exuded by roots (Graham 1963). Perhaps high populations are not necessary for effective symbiosis in view of the capacity of rhizobia for lectin-mediated attachment to root hairs, thereby permitting escape from some competitive influences.

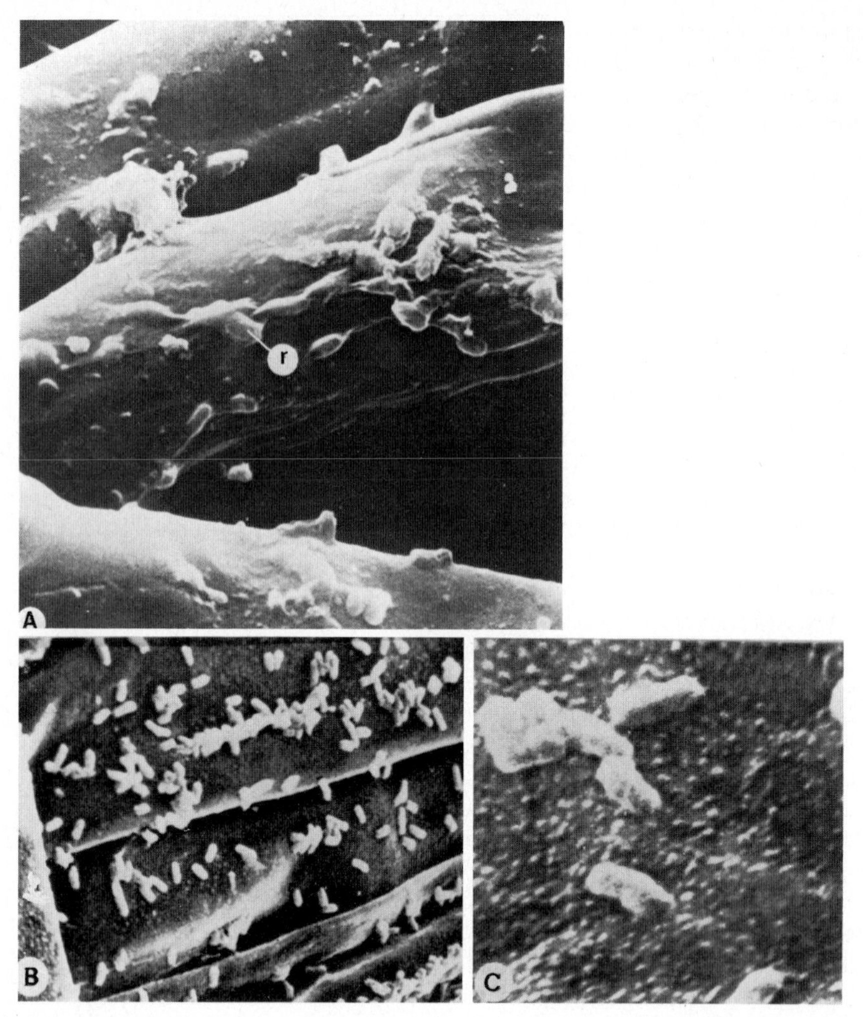

Fig. 4.5 A–C. *Rhizobium* cells on roots prepared for scanning electron microscopy by a glycol methacrylate infiltration and polymerization technique. (A) Root hairs of *Trifolium glomeratum* with attached *Rhizobium* cells (r); small granules also are prominent. × 3,100. (B, C) Root tip of *T. subterraneum* showing epidermis coated with bacteria and small granules. The longitudinal junction between epidermal cells is prominent. (B) × 550, (C) × 4,600. (Dart 1971)

Contradictory results, due largely to nonstandard procedures, also have confused the picture for the nitrifying bacteria, *Nitrosomonas* and *Nitrobacter*. Both were stimulated in the rhizosphere of corn and alfalfa in the early stage of growth (Molina and Rovira 1964), and in the inner rhizosphere of wheat at tillering and flowering (Rivière 1960).

Soil bacteria are regarded as relatively unstable physiologically, the incidence and activity of a particular group depending largely upon the availability of required nutrients. Therefore, it seems reasonable to expect that the physiological activities and nutrient requirements of bacteria in the rhizosphere, or at the root surface, may be somewhat different from those of bacteria in root-free soil. Within the rhizosphere the competitive interactions influencing the equilibrium of various groups depend on the availability of certain nutrients. Lochhead and Chase (1943) classified soil bacteria according to the complexity of their nutrient requirements. They isolated bacteria by plating on a nonselective soil-extract agar medium, and individual isolates were then tested for growth on other media varying from simple to very complex. In this way seven nutritional groups were established:

I. Bacteria with simple requirements provided by a glucose-nitrate-salts medium.
II. Bacteria requiring one or more amino acids.
III. Bacteria which require growth factors, especially cystine, thiamine, and biotin.
IV. Bacteria requiring amino acids plus growth factors.
V. Bacteria requiring substances in yeast extract.
VI. Bacteria requiring substances in soil extract.
VII. Bacteria requiring substances in both yeast extract and soil extract.

Though some microbiologists did not subscribe to this grouping scheme entirely, it served as a basis or source of reference for many other qualitative studies in succeeding years and contributed much to the clarification of the rhizosphere effect. Extensive work by Lochhead (1959), Katznelson (1965) and Canadian associates on the nutrition of soil and rhizosphere bacteria showed a higher percentage incidence in the root zone of bacteria requiring amino acids for maximal growth, and lower relative numbers of those requiring nutrients supplied in soil extract; the amino acid effect is even greater at the rhizoplane. There are also differences in individual nutritional requirements within a given group, as indicated by growth responses to specific nutrient components (amino acids and growth factors) in a basal medium (Lochhead and Chase 1943). The greater incidence of amino-acid-requiring bacteria (group II) in the rhizosphere of various crop plants is clearly evident in a 15-year summary prepared by Lochhead and Rouatt (1955).

The rhizosphere effect on nutritional groups of bacteria can be attributed both to substances released in root exudates and growth factors synthesized by other bacteria in the root zone. A higher percentage of bacteria capable of synthesizing growth factors required by other bacteria was found in the rhizoplane than in the rhizosphere of wheat (Lochhead, 1959). Over 50% of the rhizosphere and rhizoplane isolates tested were able to synthesize thiamine, pantothenic acid, and riboflavin, while significant but smaller percentages produced other essential factors such as biotin and vitamin B_{12}. Such evidence would seem to support the view held by some that growth factors available in the root zone may be largely products of microbial syntheses rather than of root exudation. The summary implication derived from all of this is that certain groups of bacteria are preferentially stimulated by essential nutrients released from living roots, or synthesized by other microorganisms, or occurring as decomposition products from sloughed epidermal cells. At the same time, another factor which may have a considerable influence upon the equilibrium of nutritional groups is the production of toxic metabolites by still other bacteria or fungi in the rhizosphere. Valuable information has been derived from the recognition of nutritional groups with pure cultures of rhizosphere or root-surface microorganisms, but presently available methodology precludes elucidation of the true balance between such groups in a natural soil environment.

Actinomycetes are frequently studied along with bacteria and sometimes isolated from soil on the same media. However, they have received less attention nutritionally and physiologically in relation to rhizosphere phenomena. They are best known for antibiotic production and inhibition of other soil microorganisms, especially root pathogens. The same species usually occur in both the rhizosphere and root-free soil, though proportions of antibacterial or antifungal types may be favored by certain crop plants. *Streptomyces* species usually predominate. Rouatt et al. (1951) found higher relative numbers of antibacterial actinomycetes in the rhizospheres of several crop plants than in root-free soil. Rhizosphere isolates from corn and soybean could be differentiated from nonrhizosphere actinomycetes on the basis of their ability to hydrolyze starch (Abraham and Herr 1964).

4.4.2 The Fungal Flora

Most of the information available on the qualitative nature of the rhizosphere mycoflora has been derived from indirect methods of counting and isolation on synthetic media. Therefore, the results reflect a degree of selectivity favoring fungi which grow rapidly on the media used and minimize the importance of those which grow slowly or require other essential nutrients or growth factors. In recent years the increased use of transmission and scanning electron microscopy has added significantly to our knowledge of the fungal flora in the outer rhizosphere, the rhizoplane, and in root tissues of healthy plants. At the same time, a better understanding has been acquired of the intimate relationships between fungal hyphae and root cells.

Unfortunately, the nutrition and physiology of soil fungi in relation to rhizosphere effects have not been studied to the same extent as for bacteria. However, the substrate-specific nature of soil fungi, proposed by Garrett (1970) as "competitive saprophytic ability" to colonize organic matter, involves some differences in nutrient utilization by different ecological groups, genera, or species. If we expand on the usual concept of "substrate" as dead plant debris or tissue and include the soluble substances from root exudates plus sloughed root cells, we may expect the "sugar fungi" (Burges 1939) to be favored in the initial colonization of the rhizosphere and rhizoplane. This would include primarily the Zygomycetes (*Absidia, Mucor, Rhizopus*) and many Hyphomycetes which rapidly utilize simple sugars and, to some extent, cellulose. Thrower (1954), by soil-dilution and plating procedures, obtained many genera of fungi from the rhizosphere of native heathland plants and found that a higher percentage of these were capable of maximal growth in a simple glucose-salts medium and in amino-acid-supplemented media than fungi from root-free soil; thus a similarity to the nutrition of rhizosphere bacteria was evident. *Penicillium* spp., *Cladosporium herbarum,* and species of Phycomycetes were well represented in the rhizosphere.

With modification of soil-dilution and plating procedures, greater use of selective media, and direct plating of rhizosphere soil (Warcup 1950) or root tissue, coupled with microscope observations, a more extensive and accurate listing of root-associated fungi is now possible. Many of them are common soil fungi which vary in preponderance according to the plant species, soil type, and prevailing environmental conditions. The logic of extending fungal assessments from the outer rhizosphere inward to root tissues is based on the fact that many of the same genera or species can be found throughout this range of root influence. A study of root–fungus relationships must deal with microbial interactions across this gradient of varied chemical and physical environments.

Employing a soil-dilution and plating procedure with V-8 juice-dextrose-yeast extract agar to study the rhizosphere and rhizoplane of *Lupinus angustifolius*, Papavizas and Davey (1961) isolated a wide variety of fungi, the dominant genera depending upon kinds of organic amendments applied to the soil. Overall, the abundantly sporulating Hyphomycetes made up a large portion of the population, some species occurring primarily in the rhizoplane and diminishing with distance from the root surface.

Since Harley and Waid (1955 a) demonstrated the value of serial root washing for studying fungal succession on roots, many studies have focused upon the flora of the rhizoplane and subepidermal tissues. Isolations from washed roots on Czapek-Dox + yeast-extract agar, along with microscopy (Parkinson et al. 1963), showed that a high percentage of the fungi associated with roots of dwarf bean, barley, and cabbage included species of *Fusarium, Cylindrocarpon, Mortierella, Gliocladium, Trichoderma, Penicillium,* and *Pythium,* as well as many sterile dark forms. A wide range of fungi may colonize roots initially, but with time a stable and qualitatively typical mycoflora tends to dominate. The dominance of *Fusarium* spp. among root colonizers has become well recognized largely through the development of selective media for isolation of the genus. Kreutzer (1972) using a modification of Warcup's direct soil-plating and selective media techniques showed that *Fusarium* spp. comprised 62–87% of isolates from the rhizosphere and rhizoplane of barley, 58–78% from all root zones of sugar beet (including inner root tissue), and 68–94% of isolates from all feeder-root zones of volunteer grasses. *F. roseum* was dominant in the root zones of barley and grasses, and *F. solani* was dominant in root zones of sugar beet, although each fungus was isolated more frequently from outer soil than from the rhizosphere or rhizoplane. Note the inverse relationship of these *Fusarium* spp. to nonfusarial fungi with regard to the direction of increase or decline in prevalence from outer soil to inner root of barley and sugar beet (Fig. 4.6). The prominence of *Fusarium* also extends to nonroot plant parts which lie on or under the soil surface, creating a habitat sometimes referred to as the laimosphere. Healthy clover stolons, for example, have been found to harbor a higher concentration of *Fusarium* species than of Mucorales, *Trichoderma, Aspergillus, Colletotrichum, Rhizoctonia,* or others (Hansen and Curl 1964).

Not to be overlooked as important components of rhizosphere and rhizoplane mycofloras are the fungal associates of ectomycorrhizae, consisting largely of Agaricales and Gasteromycetes on woody plants. According to the ecological classification of Garrett (1970), these are highly specialized root-infecting fungi

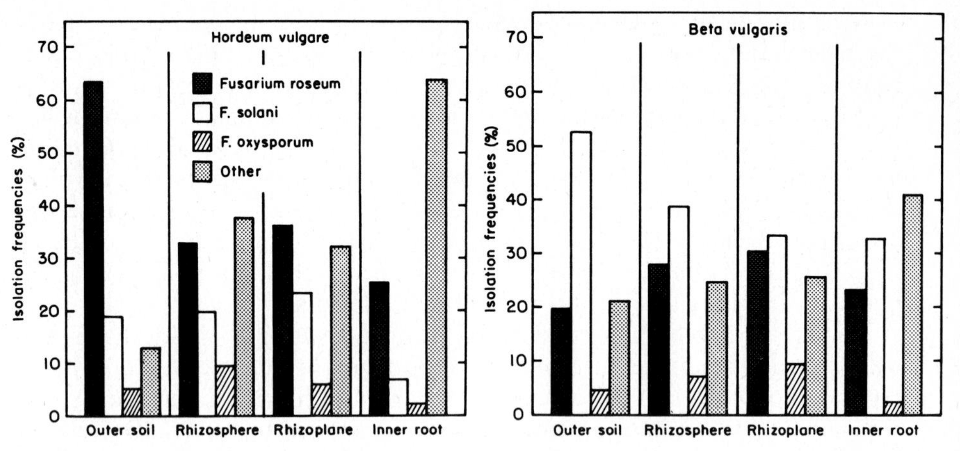

Fig. 4.6. Isolation frequencies of *Fusarium* spp. and other fungi from feeder-root zones of barley and sugar beet. (Kreutzer 1972)

which, prior to escaping into a state of mutualistic symbiosis with living host cells, are poor competitors with other microorganisms in the rhizosphere and therefore are said to be root-inhabiting. Some common ectomycorrhizal fungi are species of *Boletus, Elaphomyces, Lactarius, Suillus, Cortinarius, Laccaria, Pisolithus, Cantharellus, Corticium, Amanita, Cenococcum, Poria, Thelephora,* and *Rhizopogon.* Like the general rhizosphere flora, ectomycorrhizal fungi respond to the benefits of amino acids, sugars, and other factors from root exudates (Bowen and Theodorou 1973) before and during ectotrophic growth on roots, as the Hartig net becomes established around the outer cortical cells. Since mycorrhizal roots are known to either stimulate or inhibit other microorganisms in the rhizosphere-rhizoplane area, a "mycorrhizosphere" has become recognized (Katznelson et al. 1962; Rambelli 1973). A number of comparisons made of the microflora of mycorrhizal and nonmycorrhizal roots have shown both quantitative and qualitative differences in fungal and bacterial components (Slankis 1974). This effect may be especially significant with regard to the suppression of plant pathogenic fungi, such as species of *Phytophthora, Pythium,* and *Rhizoctonia,* at the surface of feeder roots; such inhibition has been attributed both to production of antibiotic substances by the mycorrhizal fungus and the formation of physical barriers by the Hartig net and mantle (Marx 1973, 1975).

Another mutualistically beneficial fungus–root association, the endomycorrhiza of the vesicular-arbuscular type, involves chlamydosporic, zygosporic, and azygosporic fungi in the family Endogonaceae (Gerdemann 1968), and occurs in a very large number of agricultural crops and forest trees. These fungi, represented by species of *Gigaspora, Acaulospora, Glomus,* and *Sclerocystis,* do not form a hyphal mantle over roots, as in the ectomycorrhizae, but spores and sporocarps can be found near mycorrhizal roots and an extensive network of aseptate hyphae may extend outward from roots for several centimeters. Therefore, they too are subjected to the competitive environment of the rhizosphere, but much is yet to be learned of the specific effects of exudates and rhizosphere microorganisms on the behavior of the fungal symbionts.

4.5 Factors Affecting Populations of the Microflora

The quantitative and qualitative nature of the rhizosphere-rhizoplane flora is subject to many influencing factors, as evidenced in a series of early papers by R. L. Starkey (1958) of the New Jersey Agricultural Experiment Station and in subsequent studies (Rovira 1965b; Katznelson 1965; Bowen and Rovira 1976; Dommergues and Krupa 1978). It should not be surprising that microbial populations are affected by many of the same factors that affect root exudation: age and developmental stage of plant, soil type and treatment, environmental factors (light, temperature, moisture, pH), foliar application of chemicals, and microbial interactions.

4.5.1 Plant Type, Developmental Stage, and Vigor

Different genera and species of plants are likely to differ to some degree in their metabolism and vigor of growth, depending upon the prevailing environmental conditions, including the kinds and quantities of nutrients released in root exudates. This might be expected to have some selective action upon species of the microflora, though little evidence is available to relate population changes to exudation patterns. Certain nutritional groups of microorganisms are affected by root exudates directly or by growth factors synthesized by other organisms, particularly bacteria at the root surface. Again, we should be reminded that reports on factors affecting populations vary somewhat according to the different sampling procedures, isolation media, and experimental plant-growth systems used, introducing considerable risk when making comparisons of the rhizosphere effect of different plants. From the time of Starkey's historical series of studies, it was evident that relative numbers of bacteria, actinomycetes, and fungi near roots varied with type of plant (apple, rye, corn, sugar beet, etc.). For most plants the term rhizosphere effect means increased microbial activity, and this becomes evident soon after seed germination. Common fungi such as *Penicillium, Aspergillus, Alternaria,* and the Mucorales, as well as bacteria, are associated with seed but these apparently contribute little to the established rhizosphere population. The rhizosphere effect generally increases with age of a plant, reaching a peak at the height of vegetative development and declining with root senescence. This is a generalization, however, and we shall see that specific plants may show a peak of microbial population in the early seedling state. Dominance of specific microorganisms has been associated with certain plants more than others but, overall, the rhizosphere effect of different plants is essentially one of magnitude rather than of specific kinds of organisms; this effect then diminishes with distance from the root surface.

Crop plants tend to show a greater rhizosphere effect than tree roots (Ivarson and Katznelson 1960; Dangerfield et al. 1978). Legumes usually induce a greater proliferation of bacteria than is observed with nonlegume plants. Rouatt and Katznelson (1961) found that on a soil-weight basis, the order of decreasing rhizosphere effect was clover > flax > oat > wheat > maize > barley, but rhizoplane counts on a root-weight basis showed red clover to be third in order below maize and wheat. Several workers (Katznelson 1965) have observed doubled R/S ratios of both bacteria and fungi for a number of plants after the third day of root development. The most pronounced increase often involves the amino-acid-requiring bacteria and other organisms with simple nutritional requirements and rapid growth rates. J.K. Martin (1971 b), using a different approach to bacterial assessment, collected water leachates from potted sand supporting wheat, clover, or ryegrass from the seedling stage to seed production and found by dilution plating that bacterial numbers reached a peak coinciding with flowering of each plant species. Bacterial numbers were 33–99 times greater than the numbers in leachates from control pots without plants.

A young seedling is first colonized by microorganisms which are by chance within the sphere of root influence, but as the root develops a species-stable population becomes established (Parkinson et al. 1963); on roots of barley, cabbage, and dwarf bean initial colonizing fungi were replaced by a typical flora of *Fusarium* spp., *Cylindrocarpon radicicola, Gliocladium* spp., and *Penicillium,* spp. Rivière (1960) observed the greatest stimulation of fungi in wheat rhizosphere at the tillering stage with a decline thereafter. For groundnut (peanut), the quantitative ratio between the mycoflora of rhizosphere and root-free soil is highest when plants are in the flowering stage. The population then declines and rises again at maturity (Rao 1962; Joffe 1969). Certain fungi may predominate in the early stage of root development, as Peterson (1959) showed with *Pythium* spp. isolated from washed roots of barley, flax, and wheat; these fungi then declined with plant age and dominance was taken over by *Phoma* and *Fusarium*. Actinomycetes follow the general pattern shown by bacteria and most fungi, increasing in density with plant age, though exceptions to this can be found.

Differences in microbial populations on roots of unrelated plant genera are to be expected, but it is especially interesting that differences between varieties or cultivars of a single species also occur. This has been observed mostly between plants susceptible or nonsusceptible to certain diseases, rhizosphere populations of bacteria or fungi in most cases being higher for disease-susceptible varieties (Katznelson 1965). Little is yet known of a possible genetic basis for differences in rhizosphere or rhizoplane populations between plant cultivars. Studies by Neal et al. (1973) have indicated that rhizosphere population characteristics of two wheat cultivars may be attributable to genetic differences that provide some degree of plant control over the microbial environment.

Microorganisms are not uniformly and randomly distributed over a root system, but occur in clumps or patches interspersed by areas where organisms are sparse or absent. Employing a modification of the Greig-Smith (1964) pattern-analysis technique for vegetation, Newman and Bowen (1974) confirmed that rhizoplane bacteria on several species of grassland plants (*Lolium perenne, Plantago lanceolata,* etc.) were not randomly distributed but aggregated. The growing root tips and root hairs of many plants are free of microorganisms; as the plant develops, bacteria tend to colonize sites of lateral root emergence from the main root. The production of homoserine at these sites has been suggested as a selective stimulant for growth of *Rhizobium leguminosarum* in the rhizosphere of pea seedlings (Egeraat 1975 b). Inhibitory effects also may occur, as in the case of *Rhizobium* suppression in the spermosphere of *Crotalaria juncea* by phenolic compounds released from the seed (Kandasamy and Prasad 1979). Successive lateral colonization of growing roots by microorganisms is more common than growth downward from the stem base parallel to the longitudinal axis of roots. By the use of a root-observation box model system, van Vuurde et al. (1979) were able to determine that the abundance of bacteria on the root surface of wheat has two periods of exponential increase, each followed by a decrease. The two peaks apparently were related to two periods of proportionately high increase in senescent cells between root tip and root base and the accompanying variations in substrate release.

Perhaps the most distinctly selective action of plants upon root surface fungi is seen in the associations between woody plants and members of the Hymenomycetes and Gasteromycetes which form ectomycorrhizae. This symbiotic association is most common in the families Pinaceae, Salicaceae, Betulaceae, and Fagaceae (Meyer 1973) and in certain genera of some other families. Little is known about the factors that govern the formation of ectomycorrhizae within distinct groups of plants, though specific chemical compounds synthesized in some plant taxa and not others may contribute to the regulatory mechanism. Herbaceous crop plants do not normally induce this form of symbiotic alliance, but serve as hosts instead for species of the Endogonaceae which form endomycorrhizae. Kruckelmann (1975) inoculated a soil : sand mixture with sievings from field soil containing *Endogone* (*Glomus*) chlamydospores and found that the response of a large number of plant species grown in the mixture varied widely in susceptibility to mycorrhizal infection. Highest percentages of mycorrhizal root segments were obtained from the crops *Vicia sativa, Trifolium repens,* and *Zea mays,* and from the weed *Amagallis arvensis.*

4.5.2 Soil Type

Soil type refers to a combination of properties relating to texture (size and distribution of mineral particles) and structure (arrangement or grouping of particles into aggregates). It is apparent then that soil conditions such as aeration, water-potential gradients, and heat transfer will vary with soil type and will be further altered as growing root systems displace soil particles and increase the bulk density in the rhizosphere. Early work (F. E. Clark 1949), primarily recorded in the Russian literature, suggests that microbial activity in the root zone contributes to soil aggregation and changes in soil structure. Research results that appear to show differential effects of soil types on microbial activity in the rhizosphere must be interpreted with caution, because growth rates of the same plant species will differ with soil type and any amendments applied. Growth rate, or vigor, influences root exudation and microbial activity.

Few studies have dealt with distinctly different soil types, but investigations often involve the microbiology of horizons in the soil profile and the influence of various organic or inorganic amendments on microbial populations. Large variations in absolute counts and in the R/S ratios of bacteria and fungi have been recorded for individual species of plants in different soils; for example, the R/S ratios for bacteria and fungi in soil with rye were 5 and 6 in a clay loam as com-

Table 4.5. Numbers and percentage of isolates belonging to different genera from the rhizosphere of radish plants grown in two soils. (Loutit et al. 1972)

Genera	Napier silt loam		Hastings sandy loam	
	Number of isolates	Percentage	Number of isolates	Percentage
Arthrobacter	86	22.9	22	6.5
Streptomyces	66	17.6	28	8.3
Achromobacter	39	10.4	66	19.6
Flavobacterium	30	8.0	28	8.3
Bacillus	29	7.7	45	13.35
Nocardia	27	7.2	2	0.6
Pseudomonas	21	5.6	44	13.1
Xanthomonas	16	4.3	10	3.0
Klebsiella	2	0.5	3	0.9
Alginomonas	1	0.3	0	0.0
Erwinia	1	0.3	5	1.5
Protaminobacter	1	0.3	2	0.6
Brevibacterium	0	0.0	1	0.3
Chromobacterium	0	0.0	0	0.0
Mycobacterium	0	0.0	1	0.3

pared with 16 and 15 in a fine sandy loam (Thom and Humfeld 1932). Bacterial counts on roots of yellow birch seedlings were 353×10^6 in the humus A horizon and 180×10^6 in the sandy B horizon (Ivarson and Katznelson 1960), but R/S ratios were 15 and 60 for the A and B horizons respectively. In addition to the soil effects on total populations and R/S ratios, differences also occur in the proportions of various genera in the rhizosphere of the same plant species in different soils (Loutit et al. 1972). Numbers of *Pseudomonas* and *Achromobacter* were much higher in the rhizosphere of radish grown in a sandy loam than when grown in a silt loam, whereas numbers of isolates of *Arthrobacter* and *Streptomyces* were higher in the silt loam (Table 4.5).

A comparison of soils by Peterson (1958), based on plate counts of fungi, revealed higher R/S values for wheat in an acid sandy loam than in neutral clay loam. A comparative study made in groundnut (peanut) fields in Israel (Joffe 1969) showed that 133 species of fungi occurred in rhizosphere soil, 96 in plant-free soil, and 86 in the "geocarposphere" (peanut-pod surface) soil. Species of *Aspergillus* were most numerous in heavy (light clay) soil, species of *Penicillium* in light (sandy) soil, and *Fusarium* in medium (loamy sand) soil. From the foregoing examples it can be stated that, in general, the rhizosphere effect on whole populations is most pronounced in sandy soils and least evident in heavy clay and humus soils, though numbers of certain species may deviate from this observation. The highest R/S ratios are found in desert and sand-dune soils, where nutrients and moisture are largely limited to the root-surface area.

4.5.3 Soil Treatment

More important than soil type per se are the kinds of treatment applied: inorganic fertilization and liming, organic amendments, and pesticides. The hydrogen-ion concentration in the rhizosphere changes along with these treatments as plant growth, root exudation, microbial activity, CO_2 evolution, and cation exchange capacity are altered by the soil treatments. The pH of soil is particularly a reflection of the history of nitrogen application and the extent of leaching; bases tend to be leached out and are replaced by hydrogen ions, resulting in an acid soil and often in nutrient deficiencies. The pH and nutrient status in a microenvironment at the root surface most probably will be different from that in soil generally. Treatment of soil with calcium oxide greatly increased the numbers of rhizobia in the rhizosphere of red clover (Rovira 1960); however, the R/S ratio actually was somewhat reduced because populations increased also in root-free soil.

There is convincing evidence that the application of mineral fertilizers increases microbial populations in the rhizosphere of many plants, although a number of instances can be cited in which fertilizers had little effect (Rovira 1965b; Katznelson 1965). The effect of fertilization upon microbial proliferation in the rhizosphere may be either direct or indirect. A change in the R/S value of a soil population reflects an indirect effect as a consequence of a fertilizer effect on plant growth and vigor accompanied by qualitative and quantitative changes in nutrients released in root exudates. An increase in the number of bacteria in the rhizosphere of oat plants has been observed following application of superphosphate to soil (Louw and Webley 1959b), yet microbial stimulation has been noted in phosphorus-deficient soil where cotton had been monocultured for many years (Blair 1978); in the latter case, the response to varied fertilization regimes was greater for bacteria than for actinomycetes or fungi. Bowen (1969) demonstrated a significantly greater increase in the exudation of amides and amino acids from pine roots grown in a P-deficient nutrient solution than from roots grown in a complete nutrient solution, thus suggesting a condition that might influence microorganism–plant interactions. Nitrogenous fertilizers often induce the greatest proliferation of both bacteria and fungi in the rhizoplane with less effect occurring in the rhizosphere, as observed for winter wheat 2 months after sowing (Samstevich and Borisova 1961).

Mineral fertilization not only affects total populations of fungi but specific fungi also. Though Kaufman and Williams (1964) did not work directly with the rhizosphere, an assay of soils treated in the laboratory revealed that NPK fertilization significantly affected the soil-fungus population, with N having the greatest effect on individual fungi. The response to N, however, was dependent on the P or K levels. Further studies in field soil showed that *Myrothecium* sp., *Verticillium* sp., and the *Penicillium purpurogenum* series were most affected by N fertilization. Several fungi also responded to P and K fertilization. Clear-cut effects of mineral fertilization are difficult to demonstrate in natural soil. Hydrogen-ion concentration determines to a large extent the availability of N, P, or K for fungal utilization. At the same time, both microorganisms and roots alter the pH of the medium in which they grow so that the rhizosphere may have a pH different from

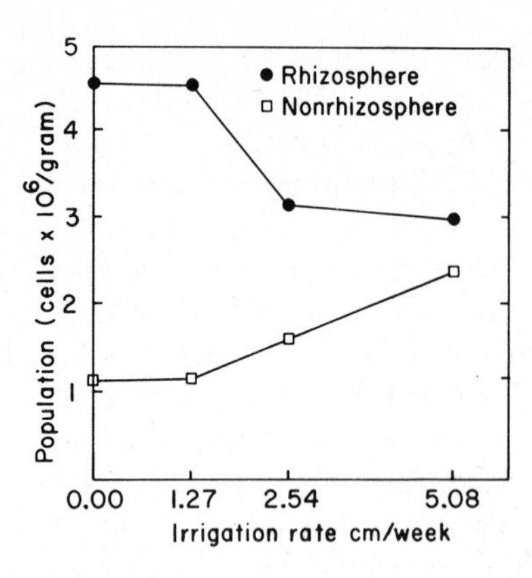

Fig. 4.7. The effect of irrigated cow-manure waste on proteolytic microorganism populations in millet rhizosphere and nonrhizosphere soil. (Dazzo et al. 1974)

root-free soil. Mineral fertilization of field soil then may be expected to affect the rhizosphere population differently from the general soil population.

Since the 1930's, numerous studies have been made of the effects of organic manures and crop-residue amendments on the rhizosphere microflora (F. E. Clark 1949). Opposing results of various workers in different regions make generalizations unwise perhaps, but most evidence suggests that organic manures do not greatly affect the rhizosphere flora. Where positive effects have been observed they can be attributed largely to direct effects on plant growth which lead indirectly to altered microbial activity; we have already seen that plant growth rate, vigor, and stage of development influence root exudation and rhizosphere microorganisms.

When a sandy soil is irrigated with cow-manure slurry, microbial activity responds to the increase in nutrients by a corresponding increase in metabolism of proteins, lipids, urea, and starch (Dazzo et al. 1974). The resulting change in bacterial populations is less pronounced in the rhizosphere soil (of millet) than in root-free soil, and consequently the R/S ratio declines with increased rates of irrigation. The effect of cow-manure waste on the proteolytic microorganism population is clearly shown in Fig. 4.7. Work with millet showed that, following manure-slurry treatment, bacteria requiring mostly amino acids increased and those requiring undefined nutrients in soil extract decreased in numbers. It has been further observed (Mishra and Das 1975) that bacteria with comparatively simple nutritional requirements are more abundant in the rhizosphere of rice plants grown in unamended soil, while bacteria with complex requirements were predominant in the rhizosphere of plants supplied with farmyard manure or mineral fertilizers.

Where organic amendments and nitrogen affect rhizosphere fungi, the effect usually depends on (a) the time between amendment incorporation and planting, (b) kind and maturity or condition of amendment, and (c) the individual species of fungi responding. Davey and Papavizas (1960) found that a 25-day fallow period for decomposition was required between amendment incorporation and planting of beans before a positive rhizosphere effect was obtained with green buckwheat, corn, oat, and sudangrass materials. Certain genera or species of fungi were stimulated by some plant materials and not others. A pronounced rhizosphere effect was observed for *Penicillium* with all amendments. Enrichment of the amendments with NH_4NO_3 further altered the rhizosphere picture, usually resulting in a higher R/S value than in unamended soil. This was due to the proportionately greater stimulation of fungi by the supplemental nitrogen when added to unamended soil. In terms of the R/S ratios of fungal populations, the rhizosphere effect of lupine seedlings was less pronounced in soil amended with various plant materials (barley straw, corn stover, or soybean meal) and nitrogen (NH_4NO_3) than in unamended soil (Papvizas and Davey 1961). The amendments apparently stimulated fungal proliferation outside the rhizoplane, resulting in reduced R/S values and less rhizoplane-rhizosphere effect. However, a definite rhizosphere effect was observed for individual fungal genera and species, depending on the kind of plant material added and the specific fungi involved. An increased frequency of occurrence of *Cephalosporium, Sporotrichum,* and *Verticillium* in particular was observed in soil amended with plant materials with or without nitrogen enrichment.

Many of the same soil factors that affect populations and activities of relatively unspecialized microorganisms also affect the specialized symbiotic types. Chlamydospores of some VA mycorrhizal fungi tend to be more prevalent in loamy soils than in sandy soils, and spore numbers are influenced by agricultural practices such as manuring, inorganic fertilization, and soil tillage (Mosse 1973; Kruckelmann 1975). There is good evidence that nitrogen may reduce spore formation, but interactions between NPK fertilization and soil type may result in opposing effects in different areas. Any treatments that affect the nutritional status and growth of the host plant might be expected to affect the fungal symbiont at the root surface and subsequent mycorrhizal infection.

Pesticides applied to soil may alter the rhizosphere microflora either directly or by affecting plant development. Most investigations, however, have dealt with effects of herbicides, insecticides, fungicides, and nematicides on general soil populations rather than in the rhizosphere specifically. Dissimilar results have been reported for the same chemicals in different areas worldwide; either stimulation, inhibition, or no effect on nontarget organisms has been recorded for various concentrations of pesticides. Herbicides have logically received most attention (F. E. Clark 1949; Fletcher 1960; Bollen 1961; Audus 1964; Curl and Rodríguez-Kábana 1977) because of their effects on the physiology of plants and the potential for predisposition of herbicide-resistant plants to secondary problems such as disease (Katan and Eshel 1973; Altman and Campbell 1977). Convincing evidence that pesticides affect the activities of nontarget microorganisms in soil is provided by the many documented cases of microbial degradation of herbicides. The correspondence between microbial populations and O_2 uptake, CO_2 evolution, or soil-enzyme activity is often used as a measure of a pesticide's effects on microorganisms. Kaufman and Kearney (1970) prepared an extensive list of bacteria and fungi that have been implicated in the degradation of a number of *S*-triazine herbicides. Similar lists could be prepared for other soil-applied pesticide families. From many conflicting reports it would be reasonable to state that the presently used soil pesticides may increase or reduce microbial populations temporarily, but in a relatively short time the population density again reaches an equilibrium approximating the original level. Shamiyeh and Johnson (1973) provided a good example of the variable effects of a commonly used insecticide, heptachlor, on all major microflora groups. Representatives of bacteria, actinomycetes, and fungi were severely inhibited on culture media containing heptachlor, but in treated field soil numbers of bacteria increased with increasing dosage of the chemical while fungal populations declined.

The significance of pesticide effects on nontarget microorganisms lies not so much in altered whole populations but in specific effects on individual organisms within a population, such as nitrifying bacteria, rhizobia, mycorrhizal fungi, and plant pathogens at the root surface. Some of the earliest studies (F. E. Clark 1949) with a plant-growth regulator (2,4-D) and an insecticide (DDT) showed highly inconsistent results (stimulation, inhibition, or no effect) and such confusion has persisted to current times with new pesticides.

While some important interactions between pesticides and the growth and reproduction of individual microorganisms have been demonstrated, these have rarely been related directly to the rhizosphere. There is little doubt that rhizosphere phenomena are involved in cases where plants exposed to herbicides suffer increased root-disease severity as in the case of *Rhizoctonia* in cotton (Neubauer and Avizohar-Hershenson 1973). Effects of herbicides on specific members of the root-surface flora may be due to stress upon plant growth and altered resistance, direct stimulation of the pathogen, or increased root exudation causing changes in the general microbial activity favoring a competitive advantage for the pathogen. Benson (1976) showed that trifluralin at levels approximating usual field rates either reduced or increased the overall microbial populations in cotton rhizosphere, depending upon the plant cultivar. An increase in the percentage germination of *Fusarium oxysporum* f. sp. *vasinfectum* chlamydospores occurred in trifluralin-treated nonsterilized soil, whereas an increase in new chlamydospore production occurred only in sterilized soil; increased lysis of germ tubes following chlamydospore germination in nonsterilized soil probably accounted for reduced spore production.

Studies of pesticide effects on populations in soil are usually accompanied by tests for effects on growth of specific organisms in pure culture. Depending on the organism, the compound, and the concentration used, such tests show varying degrees of stimulation or inhibition. Several triazine, urea, and carbamate herbicides stimulated the excretion of vitamins (riboflavin, niacin, or pantothenic acid) by the yeast genus *Rhodotorula* isolated from plant rhizospheres, and inhibited vitamin production by *Cryptococcus* (Leszczyński 1970). Such effects in the rhizosphere can be expected to influence other microbial activities as part of a chain reaction phenomenon.

A fact of little mystery is that fungicides may change the growth and morphology of microflora species which the products were not designed to control. Again, little evidence of their effects on the rhizosphere flora is available, other than in root-pathogen-related incidences, such as the increase in *Pythium* and *Fusarium* species frequently experienced where pentachloronitrobenzene (PCNB) is applied to control *Rhizoctonia* disease in cotton and other crops. One study directly concerning the rhizosphere is that of Bhaskaran (1973), who observed a marked reduction in populations of fungi and actinomycetes in rhizosphere soil of 12-day-old cotton plants following soil drenches with PCNB. A marked decrease in auxin (indole-3-acetic acid) synthesis in the soil accompanied microbial suppression. The selective action favoring *Pythium* and *Fusarium* probably stems from a reduction in competition from other microorganisms. In another study (de Bertoldi et al. 1978) the fungal population of the rhizosphere of *Allium cepa* was greatly inhibited by both captan and benomyl.

Various broad-spectrum soil fumigants can virtually eradicate ectomycorrhiza-forming fungi and lead to the production of abnormal pine seedlings with inadequate root systems and a deficiency of mycorrhizal short roots (Persidsky and Wilde, 1960; Iyer and Wilde 1965). Effective in this capacity were SMDC (Vapam) and a mixture of chlordane, thiosan, and allyl alcohol. Formation of vesicular-arbuscular mycorrhizae in crop plants also is affected. Eight fungitoxicants tested by Nesheim and Linn (1969), even at low concentrations, restricted the formation of the fungus-root relationship in corn. PCNB was most harmful to the fungus and captan least harmful; Vorlex was less injurious than Mylone (dazomet) and SMDC. As fungistatic compounds are degraded in the soil, mycorrhizal fungi resume growth rather rapidly, but highly fungicidal materials may cause permanent eradication. Stunting and chlorosis of citrus seedlings growing in California and Florida nurseries have been attributed to the absence of endomycorrhizae in soil after fumigation with a mixture of methyl bromide and chloropicrin (Kleinschmidt and Gerdemann 1972). Inoculation of such plants with *Endogone (Glomus) mosseae* in treated soil reestablished the mycorrhizal association and resulted in improved growth. Other effects of soil sterilants have been documented (Mosse 1973).

4.5.4 Foliar Treatments

Many chemicals applied to plant foliage can alter normal physiological processes of plants and modify root exudation and activities of the rhizosphere or rhizoplane flora (Hale and Moore 1979). Where leaf injury or defoliation occurs, reduced carbohydrate levels and simultaneous changes in the root-surface microflora of such stressed plants can be expected. Materials such as urea, antibiotics, and herbicides may, depending on the plant species and nature of the chemical compound, be translocated to the roots from which either the unchanged compound or a by-product is released into the rhizosphere where it directly affects microbial behavior. Applications of urea to leaves variously affects the rhizosphere flora of different plants. For example, populations of *Penicillium* spp. increased, whereas the bacterial and actinomycete floras decreased around rice roots (Ramachandra-Reddy 1959); the concentration of amino acids increased in treated wheat plants and numbers of *Aspergillus* spp. and bacteria increased in the wheat rhizosphere (Vrany et al. 1962); numbers of bacteria and fungi declined in the tea plant rhizosphere (Venkata-Ram 1960); nodulation in legumes was adversely affected (Cartwright and Snow 1962), suggesting inhibition of rhizobia; and the numbers of actinomycetes antagonistic to *Fusarium roseum* f. sp. *cerealis* greatly increased in corn rhizosphere (Horst and Herr 1962). Such results, however, are often modified by time, the effect usually declining with successive sampling periods. Davey and Papavizas (1961) showed that streptomycin applied to leaves of *Coleus blumei* was translocated laterally and downward to the roots where Gram-negative rhizosphere bacteria were suppressed by the antibiotic or some derivative of it.

A natural concern of biologists is the possible effect of various pesticides on root-surface flora and the physiological processes of plants. Herbicides and other growth regulators are of particular interest because of the proven translocation of some of these to roots from which they are released into the rhizosphere soil along with increased amounts of sugars, amino acids, and other nutrients that affect the microflora (Rovira 1969; Hale et al. 1971). Aside from the fact that radioactive materials have been traced from foliage to root exudates, the observed injury of untreated plants growing in close proximity to treated plants proves translocation either of unchanged compounds or derivatives. Of 31 compounds tested by Foy and Hurtt (1967) in foliar applications to bean plants, picloram, dicamba, and 2,3,6-trichlorobenzoic acid were readily transferred to the roots and significantly affected neighboring plants.

Fungicides might be expected to alter microbial behavior in the rhizosphere more than plant-growth regulators. A number of fungicidal and fungistatic materials applied to leaves of bean plants apparently were translocated in some form and affected numbers of bacteria in the rhizosphere (Halleck and Cochrane 1950). Bordeaux mixture, malachite green and Dithane Z-78 (zineb) reduced relative numbers of bacteria, whereas Spergon, Phygon-XL, and related compounds increased numbers.

The hypogeal fruit of the peanut plant provides a changing habitat for a community of microorganisms from early pod development (about 17 weeks after planting) to maturity and harvest. Prominent in the succession of fungi in the geocarposphere are species of *Aspergillus* which are known mycotoxin producers and which also cause pod rot and the deterioration of kernel quality. An extensive field study by Hancock (1981) showed that certain foliar-applied fungicides altered populations of the mycoflora of peanut rhizosphere and geocarposphere but had little effect on bacteria. The frequency-of-isolation diagrams in Fig. 4.8 were prepared from population data obtained when geocarposphere soil was processed through the soil-dilution and plate-count method using a high-salt medium selective for osmophilic fungi. The entire diameter of each diagram represents the total population of osmophilic fungi (mostly *Aspergillus* spp.) and the center, lightly shaded portion represents the percentage of this total that was comprised of *A. flavus*. The diameter, plus the perpendicular form of each diagram, indicates how populations changed between the first sampling at early pod development and the final sampling at harvest time (21 weeks after planting) in response to a foliar spray program with three fungicides, fentin hydroxide (FTH), benomyl (BEN), and thiabendazole (TBZ).

The diagram for untreated plants indicates that *Aspergillus* spp. comprised approximately 58% of the total osmophilic fungi isolated from geocarposphere soil at 19 weeks after planting, and approximately 25% were *A. flavus*. Foliar treatment with FTH (WP) 2X (wettable powder formulation used at twice the recommended field rate) resulted in a considerable increase in the percentage of *A. flavus* in the geocarposphere population. Also, increased populations of *A. terreus* (not shown) occurred when plants were treated with BEN (WP) 1X or TBZ

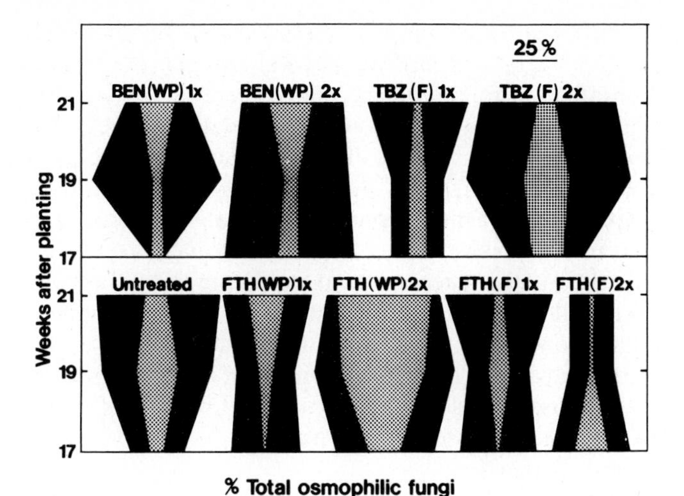

Fig. 4.8. Effects of foliar fungicide treatments on populations of *Aspergillus flavus* (width of lightly shaded portion of each diagram) expressed as percentages of the total osmophilic fungal populations of peanut geocarposphere in a drought-stressed field during a period of 17 weeks after planting to harvest (21 weeks). *BEN* benomyl; *TBZ* thiabendazole; *FTH* fentin hydroxide; *WP* wettable powder; *F* flowable formulation, *1*x and *2*x one and two times recommended field rates. (Hancock 1981)

(F) 2X (F = flowable formulation). While this interesting field study clearly demonstrates fungicide-induced fluctuations in populations of nontarget fungi, the mechanisms involved are not equally clear. Our common knowledge of microbial physiology and of microbial interactions (Chap. 5) permits safe conjecture that the observed effects on *Aspergillus* in the geocarposphere of the peanut could be either a direct effect of the fungicide treatments or an indirect effect as other microorganisms respond (are stimulated or inhibited) to the treatments and subsequently affect *Aspergillus* populations. The systemic fungicides, benomyl and thiabendazole, might be expected to translocate through the vascular system to developing peanut pods and there directly affect pod-surface microorganisms. Fentin hydroxide, however, is a contact fungicide not readily translocated from foliage to other parts of the plant. Its observed effect on geocarposphere fungi then must be explained through some other mechanism, such as foliage injury, that could induce physiological stress and an increase in root exudation. It is inevitable that foliar spraying of fungitoxicants in the field also will result in some run-off and deposition of the chemicals onto the soil surface followed by leaching and contact with the geocarposphere.

Leaf extracts from some plants have been tested for effects on the rhizosphere flora of other plants. Extracts from *Colotropsis procera* and *Datura metal*, used as foliar sprays on *Pennisetum typhoides*, reduced fungal populations in the rhizosphere, the effect decreasing with time after treatment (Mishra and Kanaujia 1972).

More information is needed about the effects of applied chemicals on specific, economically important organisms in relation to crop-plant development. Foliar treatment with the systemic compounds triforine and benomyl are known to inhibit the establishment of endomycorrhizal fungi, and several systemics have reduced chlamydospore numbers by 50% or more (Jalali and Domsch 1975). The potential significance of the effects of foliar applications of chemicals on soil microbiological phenomena at the root surface seems evident in relation to the physiology of plants and their susceptibility to root diseases. Much more study is needed in this complex area.

Finally, it should be mentioned that the mechanical removal of foliage from plants can result in changes in microbial numbers in the rhizoplane, as such imposed stress results in altered (usually increased) root exudation. For example, Sherwood and Klein (1981), using antibiotic-resistant species of *Pseudomonas* and *Arthrobacter* as markers, showed differential responses of these organisms to the defoliation of blue grama (*Bouteloua gracilis*) grown in natural soil. *Pseudomonas* showed a 2-log-unit increase over a 60-h period after defoliation, whereas *Arthrobacter* exhibited a 1-log-unit decrease in viable numbers.

4.5.5 Environmental Factors

An extensive discussion of this topic may be unwarranted in view of the known effects known of environmental factors on root exudates (Chap. 3) and subsequent effects on microbial activity. Evidence that there is a correspondence between amounts and kinds of amino acids or other exuded organic materials and populations of rhizosphere microorganisms has been adequately documented (Rovira and Davey 1974; Hale et al. 1978). Effects of major factors such as soil temperature and moisture are quite clearly reflected in population changes with changing seasons (Thrower 1954; Barclay and Crosse 1974). A certain amount of fluctuation, particularly for bacteria, occurs with time even when the soil environment is constant, numbers varying daily or even hourly. At the same time, bacteria are most sensitive to changes in soil environment.

Light would be expected to impose indirect effects upon the rhizosphere flora paralleling the effect on photosynthesis and plant growth, whereas soil environmental factors may provide both direct and indirect effects on microbial behavior. Rouatt and Katznelson (1960) found that total bacterial counts and numbers of glucose fermenters, methylene-blue reducers, amino-acid-requiring bacteria, and especially ammonifiers in wheat rhizoplane diminished as light intensity was reduced from approximately 11,000 to 3,200 klx. Different levels of daylight influence the root-surface mycoflora of forest trees. Genera of root surface fungi on beech seedlings varied considerably from full daylight through different levels of shading (Harley and Waid 1955b). Affected were mycorrhizal fungi as well as the plant pathogen *Rhizoctonia* and its common antagonist *Trichoderma*. Figure 4.9 illustrates the opposing effects of varying light intensities on numbers of colonies of the two fungi appearing on plated root fragments from beech seedlings. Peterson (1961) found essentially no differences in colonization of wheat or soybean roots exposed to two levels of illumination, though marked differences in plant development occurred. However, other work (Srivastava 1971) with wheat and barley showed considerable variation in the rhizosphere populations of fungi and bacteria in response to light and dark treatments. The favorable ef-

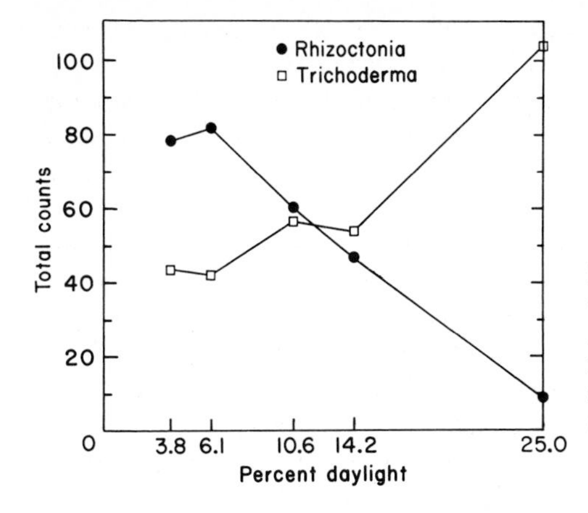

Fig. 4.9. The effect of varying light intensity upon the numbers (frequency of occurrence) of *Rhizoctonia* and *Trichoderma* on the root surface of beech seedlings, as determined by plating root fragments on a nutrient agar medium. (Harley and Waid 1955 b)

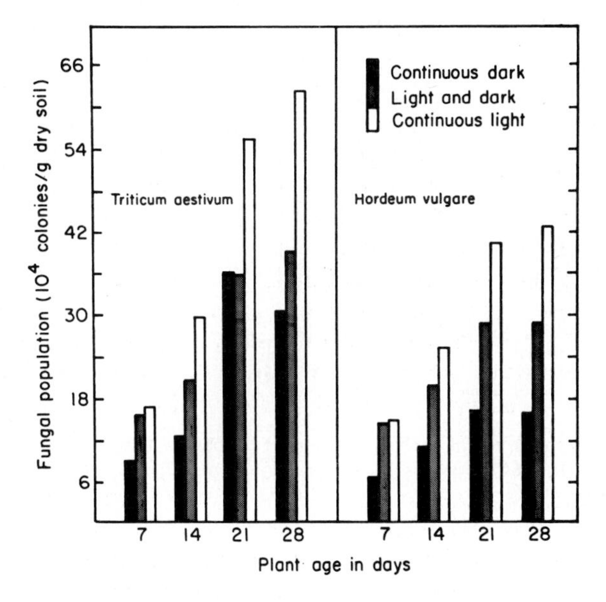

Fig. 4.10. Effect of light treatment on the fungal populations in the rhizospheres of wheat and barley at different plant ages. (Srivastava 1971)

fect of light on fungal populations as plant age increased is illustrated in Fig. 4.10. Liberation of amino acids from roots also was maximum for plants grown under continuous light.

Soil temperature is the most obvious, and perhaps the most biologically significant, physical variable in soil. Microorganisms, whether in root-free soil or the rhizosphere, are exposed to a wide range of temperatures that vary with plant cover, soil moisture, and soil depth (D. M. Griffin 1972). The effect of temperature on microbial activity in the root zone is most likely to be indirect by controlling plant growth and thereby influencing the nature and amount of root exudates. Since exudation also varies with plant species (Chap. 3), it might be expected that plants of different species impose different influences at the same temperature; this was demonstrated with wheat and soybean in greenhouse studies

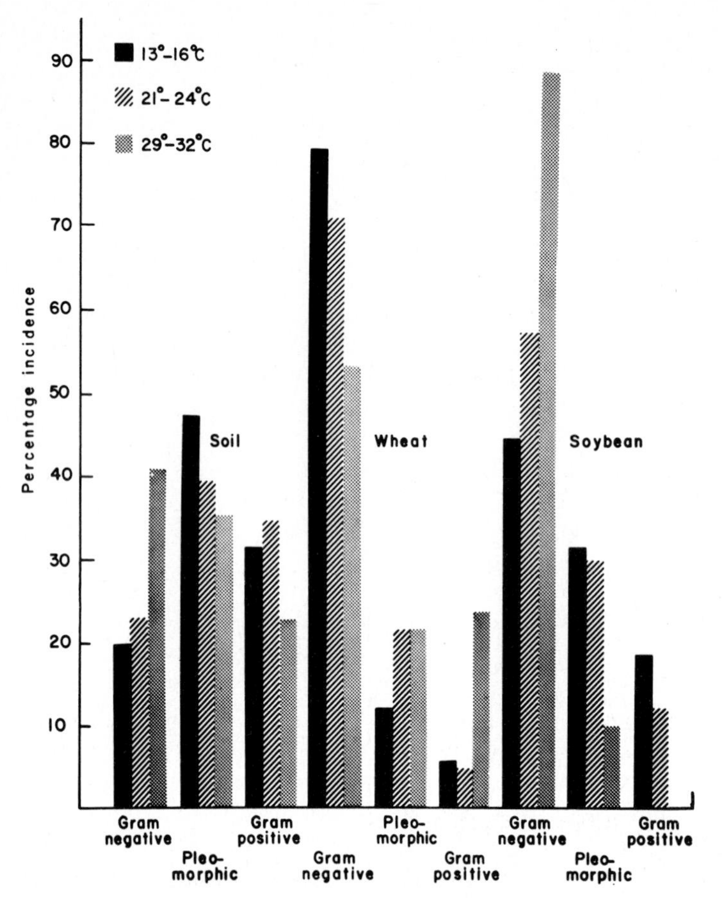

Fig. 4.11. Effects of temperature on the composition of the bacterial flora of rhizoplanes of wheat and soybean and of root-free soil. (Rouatt et al. 1963)

(Rouatt et al. 1963). Numbers of Gram-negative bacteria in the rhizoplane of wheat decreased with increased temperature (13°–32 °C), whereas numbers in root-free soil and in soybean rhizoplane increased with rising temperature (Fig. 4.11). Gram-positive and pleomorphic organisms, which were relatively low in total numbers in the rhizoplane, responded in a manner opposite to this. For fungi, a greater incidence of *Mucor, Rhizopus, Rhizoctonia,* and *Gliocladium* occurred on soybean roots at high temperatures, whereas *Fusarium* spp. and *Cylindrocarpon* spp. were favored by low temperature. On wheat, nonsporing dark species of fungi occurred at high temperatures and nonsporing hyaline types at low temperatures. In field studies, Ivarson and Mack (1972) showed that *Mucor* was much less abundant on soybean roots at 28 °C than at 20° or 12 °C. Low soil temperature favored *Pythium, Mortierella, Alternaria* and *Cladosporium* throughout the growing season. For onion plants, the rhizosphere effect on both bacteria and fungi was more pronounced at 25 °C than at 18° or 16 °C (Fenwick 1973), but the greatest rhizosphere effect did not occur at the optimum temperature (16 °C) for growth of onions. Ammonifying bacteria in particular were greatly increased at the higher temperature. Other examples could be cited, and comparisons would show frequent conflicting results, depending upon whether studies were performed under controlled greenhouse conditions or in field plots. Also, as usual, it must be remembered that media on which soil or root fragments are plated and the temperature used for incubation may be selective for specific microorganisms.

Soil moisture has a decided effect upon microbial behavior and proliferation in the root zone, the response of microorganisms being either directly related to a moisture gradient or indirectly related to plant growth. It would be helpful at this point for the reader to review the physical and chemical features of soil-water potential as described by D. M. Griffin (1972) or Cook and Papendick (1972). All field plants can be expected to undergo water stress at some time during growth, due to evaporation and transpiration that fluctuate with hot dry days and cool moist periods. The steepest water gradients occur at leaf and root surfaces. Thus the soil matric and osmotic potentials in the rhizosphere-rhizoplane region are likely to be quite different from root-free soil, and microbial behavior will respond according to the optimal and minimal requirements of specific organisms.

Soil after maximal wetting followed by drainage (field capacity) is at -0.3 to -0.5 bar. Water potential at the root-soil interface will fluctuate somewhere above these values, depending on stage of plant development, soil type, temperature, microbial activity, and several other factors. The water potential at permanent wilting is approximately -15 bar. Potentials of -10 to -15 bar prevent growth of most bacteria, whereas at least -30 to -100 bar is required to prevent the growth of many fungi and actinomycetes.

While decreasing soil moisture content from maximum water-holding capacity to the permanent wilting percentage is generally accompanied by decreasing bacterial numbers, higher populations are often encountered in the rhizosphere of drier soils, as Clark (1947) demonstrated nearly 40 years ago. We can recall at this point that water stress is one of the factors which induces increased exudation of amino acids and other nutrients from roots, providing a source of microbial stimulation. Diffusion of exudates outward and away from the root, however, is enhanced by high soil-water content.

Fungi also respond to root exudation and, therefore, indirectly to soil-moisture stress upon plants, though the response is usually not as rapid and readily evident as for bacteria. The precise measurement of soil-water potential is complicated, and even more uncertain is our ability to compare accurately the water potentials for optimal growth of different soil fungi. Pathogenic fungi vary in their capacity to incite root disease in either dry or wet soil because different species have different moisture requirements for spore production, spore germination, and vegetative growth at the root surface. The soil-dilution and plate-count procedure of assessing populations usually shows that the abundantly sporulating Hyphomycetes are present in quantity in dry soils. Many fungi reproduce rapidly under moisture stress, either in response to restricted vegetative growth or to increased availability of essential nutrients and growth factors diffusing from roots. It would be logical to expect that the lower taxa of Phycomycetes would not flourish in soils with water contents too low for the free movement and congregation of zoospores around roots.

Soil gases and volatiles are prominent among soil environmental factors that may be critical for microbial activity in the root zone. Oxygen diffusion, concentration, and uptake by roots, along with CO_2 production by roots and microorganisms, are obvious major factors and are interlinked with soil water, temperature, bulk density of the soil and other factors. Activities of the more competitive fungi and bacteria at the root surface create low-oxygen microenvironments that affect other microbes. As roots grow, soil particles are displaced according to root volume, creating a zone of maximal bulk density around the root referred to as the minimal-voids zone. The effect of this zone on microbial activity will vary with soil type, water potential, CO_2/O_2 ratio, and other factors. Oxygen supply becomes progressively lower as it diffuses across the rhizosphere to the rhizoplane, but diffusion of oxygen from plant shoots to roots probably maintains an adequate supply for essential microbial functions except in heavy, water-saturated soils. Where the oxygen supply is low, even temporarily, increased root-disease injury sometimes occurs, as demonstrated by Miller and Burke (1985) for *Fusarium solani* f. sp. *phaseoli* on bean plants. Carbon dioxide evolved by root cells and by microbial respiration increases in concentration toward the root surface and becomes a limiting factor for growth of certain microorganisms. However, it generally has little effect on most fungi until the partial pressure exceeds 0.1 atm (D. M. Griffin 1972). Growth, spore formation, or spore germination of some fungi are stimulated by CO_2, while other fungi are inhibited by high concentrations such as may occur around roots at considerable soil depth. Soil aeration is probably the most critical single environmental factor affecting *Phymatotrichum omnivorum* and cotton root rot in the southwestern United States (Lyda 1978). Carbon dioxide or the bicarbonate ion formed in alkaline soil favors growth and sclerotium production of the pathogen, while many associated microbial competitors, less tolerant of high CO_2 levels, are suppressed.

Organic volatile compounds that originate from plant roots, plant residues, and soil microorganisms move rapidly through the soil, and extremely small quantities can either stimulate or inhibit the germination or growth of specific microorganisms (Linderman and Gilbert 1975). Such chemicals are primarily aldehydes and alcohols. Since they have relatively high vapor pressures and diffuse rapidly through soil pores for greater distances than nonvolatiles, their activity is not uniquely confined to the rhizosphere. However, they undoubtedly influence the nature and behavior of the rhizosphere flora.

Volatiles have been studied with some intensity only recently and largely in relation to their effects on plant pathogens. Alkyl sulfides from root tips of *Allium* spp. stimulate sclerotia germination of *Sclerotium cepivorum* (Coley-Smith and King 1970). Volatiles, primarily monoterpenes and sesquiterpenes, produced by mycorrhizal roots of *Pinus sylvestris*, have been implicated in growth inhibition of mycorrhizal fungi (Melin and Krupa 1971) and the pathogenic fungi, *Phytophthora cinnamomi* and *Fomes annosus* (Krupa and Nylund 1972); ethanol vapors produced by *Boletus variegatus* in culture, however, stimulated these two pathogens. Compounds from decomposing crucifers are strongly inhibitory to all stages of *Aphanomyces euteiches* (Lewis and Papavizas 1971), whereas acetaldehyde and methanol from alfalfa distillate enhance general microbial respiration and either stimulate or inhibit *Verticillium dahliae*, depending upon the concentration and exposure time (Linderman and Gilbert 1975); germination of sclerotia of *Sclerotium rolfsii* can be triggered by alfalfa volatiles.

The influence of volatiles from plant residues generally distributed in soil is less likely to create a differential between rhizosphere populations than that of volatiles of root origin. A reasonable speculation, however, is that either may selectively stimulate microbial activity (bacteria in particular) at the root surface, resulting in microbial production of amino acids or growth factors required by other microbes.

Soil pH in the rhizosphere is another obvious influencing factor on microbial activity and populations. Though a common assumption is that the pH of soil near roots tends to be acid due to the release of CO_2 and organic acids by roots, the pH may in fact become either acid or alkaline, depending upon whether nitrogen is supplied as NH_4 or NO_3. Thus the accumulation of HCO_3 is not controlled solely by CO_2 evolution from roots but also in part by the relative uptake of cations and anions (Riley and Barber 1969; Nye 1981). It follows then that microbial activity at the root–soil interface also will fluctuate with these and other factors that affect plant growth.

The rhizosphere pH may differ from the pH of adjacent root-free soil by 1–2 units, which could have a significant effect on bacteria, actinomycetes, and some fungi. The fungi in general can tolerate a wider range of pH, and some species such as *Sclerotium rolfsii* can adjust the pH to their own optimal values.

4.5.6 Microbial Interrelationships

Associative activities of microorganisms in the rhizosphere will be dealt with in detail in Chapter 5. Remarks here are confined to interactions that may significantly influence populations of the microflora around roots. We have seen that populations of bacteria, actinomycetes and fungi flourish along growing roots, especially at principal sites of exudation and sloughing of root cells. As this initial population develops, it is rapidly modified by other microorganisms in the process of succession during rhizosphere colonization. Antagonism of one organism by another is a harmful association which may be imposed through three avenues: (1) competition, primarily for nutrients; (2) antibiosis, the inhibition of one organism by metabolites of another; and (3) parasitism or predation of one organism on another.

Fungistasis (mycostasis), the wide-spread factor in soils that prevents fungal spores from germinating, is due to a combination of antibiosis and nutrient deficiency, both of which may be created by competitive microbial activity. Lysis, the dissolution of cell walls and protoplast, is often closely associated with antibiosis. Parasitism involves bacterial viruses (phages), *Bdellovibrio* attacks on other bacteria, and the parasitic action of one fungus on another (mycoparasitism). Examples of predators are nematodes and protoza which consume bacteria, nematode-trapping fungi, and soil microarthropods which feed on bacteria and fungi. All of these processes can inhibit or displace some species of microbes, while allowing others to proliferate. As usual, the intensity of such activities is governed by the common soil environmental factors already discussed.

While antagonistic or inhibitory processes are at work in the rhizosphere, microbial activities that benefit individuals or groups of organisms are also going on. We have already pointed out that amino acids and a variety of growth factors synthesized by one group of bacteria may be stimulatory to other groups that require specific nutrients. Changes in the composition of the microflora around ectomycorrhizal rootlets, a zone referred to as the mycorrhizosphere, have been observed and attributed primarily to root exudates and to metabolites of mycorrhizal fungi (Rambelli 1973; Marx 1973). Distinct quantitative and qualitative differences have been found in the surface microflora of mycorrhizal and nonmycorrhizal roots of yellow birch (Katznelson et al. 1962). Higher bacterial counts and greater numbers of actinomycetes and of methylene-blue-reducing, sugar-fermenting, ammonifying, and fluorescent bacteria were found on mycorrhizal roots. By light and electron microscopy, large numbers of bacteria and of fungal species other than those forming the mycorrhiza were observed in the mycorrhizosphere of *Pinus radiata* (Foster and Marks 1967).

Endomycorrhizal fungi alone or in combination with nitrogen-fixing bacteria have been observed to alter microbial populations in the rhizosphere. For example, tomato roots inoculated with the VA mycorrhizal fungus, *Glomus fasciculatum*, and *Azotobacter chroococcum* yielded higher populations of bacteria and actinomycetes than noninoculated roots (Bagyaraj and Menge 1978). A synergistic effect was evident as plants inoculated with both the fungal symbiont and nitrogen-fixing bacterium had higher populations of bacteria and actinomycetes in the rhizosphere than plants inoculated with either organism alone. The mechanisms involved in this case are most probably related to the increased uptake of phosphorus and other mineral nutrients by mycorrhizal roots along with nitrogen fixation or the production of plant growth-promoting substances by *Azotobacter*. These influences on plant growth may be accompanied by changes in root exudation and microbial activity.

It is not unusual that plants stressed by disease may support microbial populations that are different from those in the rhizosphere of healthy plants, because the quantity or quality of root exudates released may be different. An interesting and more direct influence of a pathogen on bacterial populations has been observed in wheat rhizosphere (Bednarova et al. 1979). When soil was artificially infested with *Gaeumannomyces graminis*, the causal agent of take-all of wheat, both the number of bacteria in the rhizosphere and the bacteria/fungi ratio temporarily increased. *Pseudomonas fluorescens* and related species colonized the rhizosphere to a greater extent in the presence of the pathogen. It was suggested that the bacteria occurred on hyphal surfaces and accompanied the fungus in simultaneous colonization of the rhizosphere. As the bacterial population increased, a temporary stimulation of wheat growth was observed.

Microflora-grazing animals of the micro- and mesofauna may play a significant role in the dynamics of microbial populations in the rhizosphere (see Coleman et al. 1984). However, contrary to common expectation, the microflora density is not always reduced by faunal activity. Whereas some studies in nonrhizosphere soil have shown decreased numbers of bacteria in the presence of bacteriophagic nematodes, bacterial activity often is stimulated in the rhizosphere where both nematode and bacterial densities are initially high. Such stimulation probably is due primarily to a temporary reduction in the bacterial population which then releases certain, more aggressive species from the stress of competition, resulting ultimately in higher total numbers. Microarthropod grazing on fungi in the rhizosphere, particularly relating to plant pathogens (Curl et al. 1985) and VA mycorrhizal fungi (Warnock et al. 1982), has received limited attention, but laboratory demonstrations of the destruction of hyphae and chlamydospores by springtails (Insecta, Collembola) in the genera *Folsomia, Onychiurus,* and *Proisotoma* suggest a high potential for effects on root-colonizing fungi and consequent relevance to plant health and growth.

The reader should have reached the conclusion by now that the subject of microbial interactions in the rhizosphere is extremely complex and still largely undefined. Most of our knowledge is derived from in vitro experiments with pure cultures in model systems, and from work with whole populations in natural soils. However, our present level of understanding leaves little doubt that the quantitative and qualitative nature of the rhizosphere and rhizoplane flora must fluctuate considerably according to associative interactions.

4.6 Populations of the Rhizosphere Fauna

Observations on the activities of soil animals began long before the landmark studies of Darwin on earthworms during the period 1840–1881, but study of the soil fauna as a recognized discipline has developed only since about 1950 following summarizations of the subject in Kühnelt's (1950) *Bodenbiologie.* Kevan (1965) later reviewed the subject admirably in *The Soil Fauna – Its Nature and Biology.* Books, such as those by Wallwork (1970, 1976) on the ecology of soil animals, as well as many research papers, have emerged since Kevan's review yet, except for a limited number of studies with nematodes, it is relatively rare to find references relating specifically to animal behavior in the rhizosphere.

The major groups of soil-dwelling animals are nematodes and microarthropods, as illustrated in Fig. 4.12. Population estimates of protozoa are less reliable due to rapid daily fluctuations from a few hundred to hundreds of thousands, and are difficult to compare with the other groups because of dissimilar isolation or extraction methods used. Though we often refer to these groups collectively as the "microfauna", a strict categorization based on size would place protozoa and some nematodes in the microfauna and small arthropods of 1–2 mm length (Acarina and Collembola) in the mesofauna. Many developmental stages of insects and other animals with sizes measurable in centimeters, comprise the "macrofauna"; while their biomass may exceed that of smaller animals their numbers are comparatively low. Animals larger than the mesofauna are unlikely to display great sensitivity to the rhizosphere effect, though some may feed directly upon roots or may prey upon certain other organisms in the root zone.

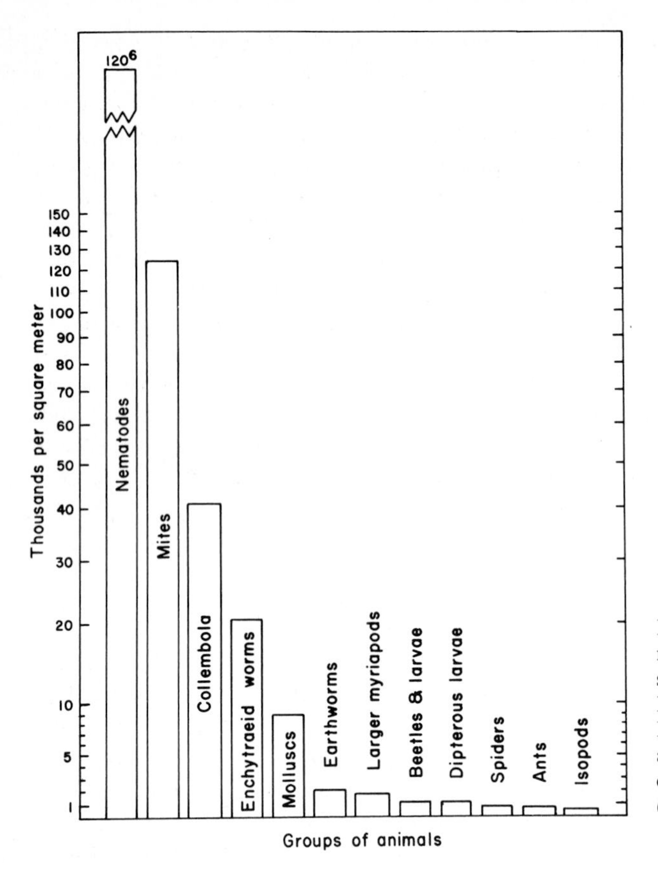

Fig. 4.12. Relative maximum numbers of animals per square meter in soil of a European grassland. Modified from Kevan (1965) as based on the original compilation by Macfadyen (1957)

4.6.1 Protozoa

The Protozoa in soil are mainly rhizopods (such as amoebae) and the flagellates, along with smaller numbers of ciliates. In terms of their percentage of the total microfauna (wet weight) in soil, Overgaard Nielsen (1949) placed them at 86%, well above nematodes (11%). Since bacteria form the major food supply of the protozoan population, one may expect the rhizosphere to harbor large numbers that coincide with microniches harboring large numbers of bacteria. At the same time, because protozoa are somewhat selective in the types of bacteria they ingest, they must exert a selective influence on the composition of the bacterial population (E. W. Russell 1973). One of the earliest efforts to quantitate protozoa in the rhizosphere was that of Katznelson (1946), who found that mangel roots exerted a striking selective action on flagellates, amoebae, and ciliates as well as on numbers of bacteria, actinomycetes, and fungi in both fertilized and unfertilized soil. Later, Katznelson collaborated with Rouatt and Payne (Rouatt et al. 1960) to confirm the rhizosphere effect on protozoa, this time showing highly significant correlations with wheat and barley roots. In wheat rhizosphere, numbers per gram of oven-dry soil were estimated at 2.4×10^3 (R/S ratio of 2:1), and bacteria were estimated at 1.2×10^9 (R/S of 23:1). It was suggested that the protozoan response probably was due to extensive bacterial proliferation. Darbyshire (1966) noted that the protozoan population in ryegrass rhizosphere exceeded that which had been reported previously for mangels, wheat, barley, or cotton. He found no evidence that there were more different species than in root-free soil. Most common were species of *Cercobodo, Cercomonas, Heteromita, Oikomonas, Scytomonas, Tetramitus,* and *Colpoda.* Some of the same genera seem to be common in the rhizosphere of cacti in desert soils, where Bamforth (1975) noted that the proportions of flagellates, amoebae, and ciliates varied in different root samples; there were no trends with regard to location along lateral roots. In contrast to studies in more mesic environments, ciliates comprised 10–95% of cactus rhizosphere populations. Largest populations of protozoa usually occur in warm, moderately wet soils (Nicoljuk 1964; Darbyshire 1966) and can be expected to increase under irrigation. Giant soil amoebae, such as *Arachnula impatiens* of the family Vampyrellidae, are now attracting attention as potential biocontrol agents (Old and Patrick 1979). These mycophagous amoebae perforate fungal spores, allowing the pseudopodia to penetrate and digest the cell contents.

4.6.2 Nematodes

The ecology of soil-inhabiting nematodes is complex. Environmental conditions often determine whether certain species are free-living or parasitic. The majority of terrestrial nematodes are free-living forms but are associated with plants, feeding superficially on underground stems and roots, bacteria, fungal hyphae, or algae (Jones 1959). Plant parasitic forms are mostly endoparasites, as in *Heterodera* spp., or ectoparasites, as in *Tylenchus* and *Dorylaimus* spp., which feed from the outside. Still others are intermediate. Thus, both plant and soil phases in the life cycle are represented. Nematodes generally are most abundant in the upper 30 cm of soil but may extend to considerable depths, as much as 150 cm in porous sandy soils. While nematodes in agricultural soils are affected by factors such as soil texture and structure, temperature, moisture, and aeration, their populations and distribution are closely related to plant cover and the availability of roots. The influence of cropping systems on populations attests to nematode dependence upon plants (Nusbaum and Ferris 1973). Relatively fewer species occur in cultivated lands than in forest soils and grasslands, but monoculture of many crops, or even regular rotations, results in a build-up of specific nematodes. The many ecological phenomena that influence populations and behavior of plant-parasitic nematodes have been discussed by Norton (1978) and others.

Our knowledge of nematodes in relation to the rhizosphere effect is meager compared to that for the microflora. Populations determined by elutriation of large soil samples, or by extraction from root tissue, have little bearing on the subject of the rhizosphere. However, the rhizosphere and root-surface influence on nematodes has been established beyond doubt. Root influence on nematodes has been brought to light largely through egg-hatching reactions to substances in the rhizosphere-rhizoplane zone and the attraction or congregation of larvae around roots. Egg-hatching is stimulated in many nematode species by substances, as yet incompletely characterized, that diffuse from roots. Hatching factors were noted

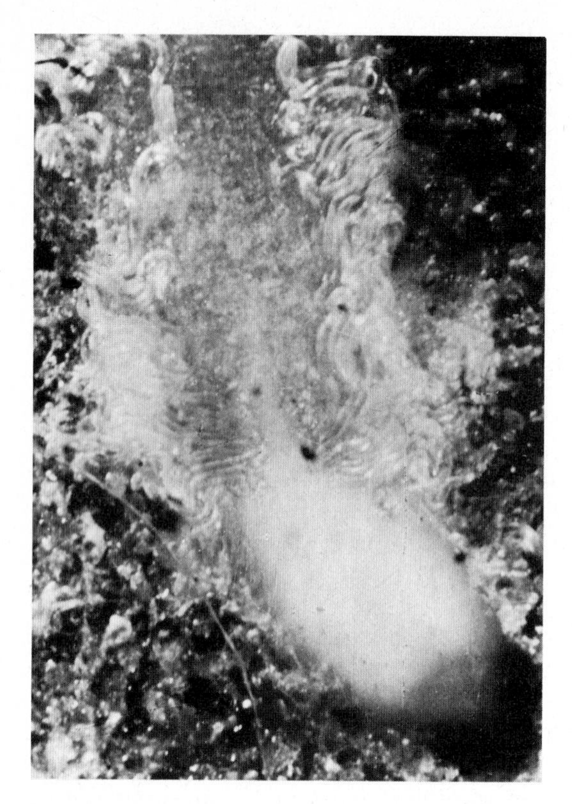

Fig. 4.13. *Trichodorus viruliferus* massed around the zone of elongation of a young apple root. (Pitcher 1967)

55 years ago by O'Brien and Prentice (1930), who demonstrated that potato-root exudates can stimulate the hatching of potato eelworm cysts. Many other examples have been cited (Jones 1959; Rovira 1965a). Tomato, beet, turnip, rape, clover, and a number of other plants have been implicated in induced egg-hatching; some plants are believed to produce more than one kind of hatching factor. Stimulation of egg-hatching has been much studied with *Heterodera*, in which the hatching factor is highly specific for some of the nematode species. Hatching stimulants are not unique for host plants but also may be found in exudates of nonhost plants, leaving the newly emerged larvae without a host if susceptible roots are not available. Some of the Cruciferae and Chenopodiaceae evoke hatching of *H. schachtii* eggs but do not support root invasion by the subsequent larvae.

Since Linford (1939) first observed that larvae of root-knot nematodes were rapidly attracted to the region of cell elongation, numerous other studies have verified such activity at the root–soil interface of both host and nonhost plants. "The observed concentration of nematodes in the region of plant roots is due mainly to the more rapid reproduction of nematodes on the food supply afforded, but also in part to attraction by some of the substances released into the rhizosphere" (Jones 1959). Pitcher (1967) clearly illustrated the congregation of *Trichodorus viruliferus* around the zone of elongation of a developing apple root (Fig. 4.13). The degree of attraction seems to be proportional to the rate of plant growth, as demonstrated for the larvae of *Meloidogyne hapla* (Wieser 1955) around tomato roots. Mobility is closely related to soil pore size and moisture content (H. R. Wallace 1968). Nematodes require at least a water film for movement, but pore saturation is not necessarily advantageous except in soils of fine particle size.

Not only attraction but repellency of nematodes has been observed with bean, eggplant, and soybean roots (Wieser 1956); the distribution of *M. hapla* larvae along the roots probably was attributable to the interactions, perhaps opposing influences, between attractants and repellents.

The attraction mechanisms involved in the movement of nematodes toward roots have been sought for many years through experimentation, often yielding opposing results among different workers (Jones 1959; Klingler 1965; C. D. Green 1971; Norton 1978). Attraction has been attributed to several chemicals under various experimental conditions. These chemicals, including CO_2, gibberellic acid, and tyrosine, may come directly from roots or emanate from areas of high bacterial activity. Attractant compounds in the plant rhizosphere apparently are of low molecular weight, since their effectiveness is not lost after passing through a cellulose membrane of 24 Å pore size, as demonstrated by Viglierchio (1961) for *Meloidogyne hapla, M. incognita acrita,* and *Heterodera schachtii* in association with tomato, oat, rye, or sugar beet, and by Blake (1962) for *Ditylenchus dipsaci* and oat seedlings. Thus, the nematodes appear to move along a concentration gradient of water-soluble, dialyzable chemicals released by roots.

Much study has centered around CO_2 as a primary attractant. Klingler (1961, 1963, 1965), using both artificial and natural sources (germinating seeds) of CO_2, provided some convincing evidence implicating CO_2 as a principle factor in attracting nematodes to the surface of roots. Others (Peacock 1961; A. F. Bird 1962) have presented evidence that the gas may not be the sole factor and that a combination of factors (CO_2, exudates from meristematic tissue, and electric potentials) may be involved. A study of the respiration rates of certain plant parasitic nemtodes in CO_2 and air (Rohde 1960) offered the interesting hypothesis that CO_2 released from respiring roots actually suppresses activity and prevents nematodes from migrating away from the root zone, thus making it appear that they congregate there in response to a stimulus.

Henderson and Katznelson (1961) conducted one of the very few studies designed specifically to determine rhizosphere effect on nematodes. Using a version of the Baermann-funnel method for extraction (see Sect. 4.2) they found that the numbers of nematodes in the rhizospheres of wheat, barley, oat, soybean, and pea were greater than in root-free soil (Table 4.6). *Pratylenchus* sp. was highly favored in the rhizospheres of oat and wheat and *Paratylenchus* sp. in the rhizosphere of pea. This effect may vary according to the level of illumination received by the plants; generally, nematode activity around roots is reduced under low light. A number of examples could be given citing the influence of various crops on the relative abundance of nematodes in field soil. Much of the available information on the activities of nematodes in the rhizosphere comes from experiments with model systems that have little in common with the complexities of field conditions. However, it is evident that species of nematodes and their abundance are closely related to plant growth and that substances in root exudates directly or indirectly influence their behavior.

Table 4.6. Influence of plant roots on nematode populations (numbers per gram oven-dry soil). (Henderson and Katznelson 1961)

Nematodes	Barley	Oats	Wheat	Soybean	Peas	Root-free soil
Pratylenchus	16.1	34	33	3	19	1.0
Paratylenchus	10.1	10	22	20	87	0.7
Aphelenchus avenae	5.0	14	4	3	16	0.1
Tylenchida	37.0	48	66	29	132	2.4
Total nematodes	275.0	135	601	309	708	10.0

4.6.3 The Acari and Collembola

Next in abundance to protoza and nematodes are members of the Arthropoda, commonly referred to as microarthropods but which in fact, based on body size, would be more properly classified among the mesofauna (Wallwork 1970). These are the Acarina (mites) and the Collembola (springtails).

The Acari are the most common but not the most conspicuous of the mesofauna in most soils. The several hundred families of mites represented in soil include predatory species as well as detritus feeders and fungivores. The Orders Mesostigmata and Prostigmata are largely predators and detritivores, the Cryptostigmata are primarily detritivores and fungivores, and the less abundant Astigmata are probably detritus, fungal, and bacterial feeders. Predation in the Acari is most commonly upon nematodes and insect eggs, and sometimes involves other mites abd Colembola. Though studies of mites specifically in the rhizosphere are rare indeed, the feeding habits of these animals and the environmental factors that affect their activities strongly suggest an indirect influence of plant roots. Wallwork (1970), in his book *Ecology of Soil Animals,* stated that fauna of the mineral layer consists mainly of small species and immature mites, which usually are concentrated in the biologically active layer, the rhizosphere, around plant roots. Much is left unexplained about the distribution and movement of mites in soil, but they seem to be favored, along with Collembola, in the fermentation layer, that region between litter and humus. The cryptostigmatid mites are predominantly soil-dwellers, comprising the macrophytophages which feed on woody tissue (xylophages) and species that feed on leaf tissue (phyllophages), and the microphytophages that feed on fungi (mycophages), bacteria (bacteriophages), or algae (phycophages) (Wallwork 1976). Still other groups feed on animal materials (zoophages). The microphytophages are more likely to follow plant roots in their vertical distribution in accordance with the availability of food supply.

Führer (1961) made comparisons of the small arthropod fauna in the rhizospheres of three plant species, *Artemisia campestris, Bromus erectus,* and *Dactylis glomerata,* and found that the mite *Pseudotritia ardua* occurred in large numbers in the root region of *Artemisia* but was absent from the rhizosphere of the other plants. Further tests showed attraction of the mite to areas of high activity of *Pseudomonas* strains.

Table 4.7. Effect of increasing cotton root mass on populations of Collembola in greenhouse pots of Dothan sandy loam, determined 44 days after planting. (Wiggins et al. 1979)

Number of plants/pot	Root weight (g)	Average number of Collembola/liter soil
0	0.0	95.6
1	7.8	253.5
4	31.5	377.4

Root-weight values and numbers of Collembola are average of six pots. All compared values were significantly different, $P = < 0.01$

The Apterygota, referred to by some as "primitive insects", are grouped into the Thysanura, Diplura, Protura, and Collembola. While all are found in soil, the Collembola are most abundantly represented and considered of major importance. Only the Acari are more abundant in certain habitats. The two major groups of Collembola are distinguishable by body form and occupy somewhat different habitats. The Symphypleona with short, globular, indistinctly segmented bodies are more common in the loose upper litter layers of hardwood forests; the Arthropleona, with elongate, visibly six-segmented bodies, are well represented in soil, mostly at depths of 10–15 cm. Like mites, the collembolan fauna shows a vertical stratification of species, the smaller forms of the genera *Onychiurus, Tullbergia,* and *Friesea* sometimes reaching considerable depths in sandy soils of high porosity. Since the Collembola do not burrow, their vertical distribution is restricted, as porosity decreases with increasing depth. It is likely that growing plant roots may afford avenues for vertical movement of small collembolans, which appear to be better adapted to the higher CO_2 concentrations found at greater depths than the predominantly surface dwellers.

The many soil environmental factors which affect both Collembola and the Acari have been discussed by Wallwork (1970, 1976). Aside from the usual effects of soil structure and moisture, agricultural practices and type of crop culture influence populations, species composition, and distribution patterns. Though little study has been made of Collembola and plant-root associations, there is evidently some correspondence between collembolan populations and specific plants. Monoculture of crops, or long-term rotation regimes, may select for certain genera or species. In field plots monocultured in cotton for 40 years, and in other plots rotated with cotton, corn, and soybean for over 10 years, Wiggins et al. (1979) extracted predominantly two species of Collembola from cotton rhizosphere. *Proisotoma minuta* and *Onychiurus encarpatus* were consistently more abundant in the rhizosphere than 20 cm distant from roots in either nutrient-deficient or fertilized soil. A direct relationship was established between increasing cotton-root density and populations of Collembola (Table 4.7); R/S ratios were always higher in a sandy loam than in a clay loam. From the observed close correlation of numbers with root density, it is not surprising that Collembola are much more abundant in grassland soils than in fields of cultivated row crops.

Fig. 4.14. Attraction experiment with Collembola (*Sinella cocca* Schott). Attraction is to the CO_2 source (*left*) but not to the N_2 source (*right*), although both gases reduce the oxygen content near the capillary opening. (Klingler 1965)

Specific reasons for the apparent attraction of Collembola to roots have not been fully determined; they rarely feed upon living plants to the extent that plant growth is affected. However, a strong possibility, which would seem to offer some logic, is that root exudates, sloughed organic matter, microflora, and moisture are all involved. *Proisotoma* and *Onychiurus* are fungal feeders (Wiggins and Curl 1979). Indeed many references have been made to the actual and probable interactions between the soil fauna and microflora (Kevan 1965; Butcher et al. 1971) and specifically to the mycophagous habits of Collembola (Christiansen 1964; Anderson and Healey 1972; Christen 1975). Collembola would be expected to benefit indirectly from root exudates and sloughed root cells, which induce greater fungal and bacterial activity at the root surface. In controlled experiments (Wiggins and Curl 1979), these animals have been observed to move toward cotton roots in slowly drying soil, presumably finding moisture at the root–soil interface.

Studies quite similar to those with nematodes have been applied to Collembola and mites to determine root-related attractants other than food sources and moisture. Moursi (1962) demonstrated that CO_2 introduced by a capillary tube into soil between two glass plates attracted species of Collembola and repelled species of Oribatei. Similar results were obtained with N_2. It was suggested that the animals regulated their position in soil not according to the specific effect of CO_2 or N_2 but in response to O_2 gradients. Other work has not entirely supported this hypothesis, as evidenced in the review and personal studies by Klingler (1965), who found Collembola attracted to a source of CO_2 but not to N_2 (Fig. 4.14). As with nematodes, the attraction of Collembola to the root–soil interface is not yet fully clarified and must surely involve many interrelated factors: moisture and gaseous gradients, amino acids and other substances emanating from roots, pH changes, quantity and nature of the microflora, etc.

It is challenging to reflect upon the possible interacting role of microarthropods, particularly the mycophagus species, in the competitive microbial colonization of roots and rhizosphere and, subsequently, their influence on plant pathogen activity. We shall pursue this thought further in Chapter 7.

Chapter 5 Microbial Interactions

5.1 Introduction

Soil microbial populations are multispecific communities which fluctuate according to the influence of environmental changes induced naturally or through human interference. Such populations rise and fall with time and the availability of nutrients during the succession of substrate colonization and utilization, tending always to return to a state of equilibrium. Microorganisms which share the same ecological niche, e.g., those within a habitat which have similar nutritional requirements and environmental optima, are most likely to be in direct competition for materials that are in limited supply. The complex interactions that occur usually result in suppression of one or more individuals, while growth of others is enhanced.

Logically, interactions are likely to be most intense in habitats with a high organic matter content where populations are abundant, such as the rhizosphere, rhizoplane, and decomposing plant and animal remains. The world ecological literature bulges with information on microbial activities in many habitats, and appropriate terminology, though sometimes confusing and overlapping, has been used to describe the many interactions that may occur in mixed populations. We are concerned in this chapter with interacting rhizosphere organisms and the influence of living roots on their behavior. The primary purpose of studying microbial activities at the root–soil interface is to acquire a better understanding of the phenomena or biological laws governing the colonization of the rhizosphere and roots, which ultimately are related to growth and health of plants.

Much of the general ecological terminology and certain classifications are applicable to the rhizosphere and should be recalled at this point. Growth of two individuals or populations in close proximity without any discernible interaction is referred to as neutralism, but this rarely occurs. Even when two isolates of the same species are brought together a state of mutual aversion may arise between strains. Organisms of widely different nutritional and environmental requirements are likely to be less competitive in the same habitat.

Primarily, two kinds of interaction occur in mixed populations: (a) those which promote growth of individuals within the populations (commensalism and symbiosis), and (b) those in which organisms are inhibited (antagonism: competition, antibiosis, hyperparasitism, and predation). Such interactions occur among both the microflora and fauna, and between the two, since both groups are involved in the food chain within the ecosystem. Other ecological terms for microbial interactions have been used and variations in definition have been expressed (Alexander 1971; Lockwood 1981). However, for our study of the rhizosphere, we shall stay essentially within the bounds of the foregoing terminology and perhaps avoid unnecessary confusion.

We are reminded, as always when dealing with soil microorganisms, that our knowledge about microbial behavior is largely based on results from model experimental systems which may or may not translate to actual behavior in a complex natural soil environment. It is particularly difficult to separate the many microbial interactions that could occur. Brock (1966) expressed the basic potential for population interaction in the form $A \leftrightarrows B$ where A and B represent two populations. The arrows indicate effects of A on B and of B on A. Since the effect of one organism on another can be positive, negative, or zero, nine different qualitative states are possible. Microbial interactions at the root–soil interface are catalyzed by root exudates which either directly affect one or more organisms in the community or indirectly promote interspecific stimulation or inhibition.

5.2 Growth-Promoting Interactions

5.2.1 Commensalistic Relationships

Under the term commensalism, one organism receives benefit from a second organism when the two are growing in close proximity, while the second organism receives no benefit and is not harmed by the beneficiary. The probability for such interactions within the teeming populations of the rhizosphere is understandably great. At least four commensalistic relationships can be expected to occur in the rhizosphere: (a) microbial transformations of compounds or substrates may release previously unavailable nutrient materials for utilization by other microorganisms; (b) one species may synthesize a growth factor which is essential for growth and reproduction by another organism; (c) changes in the physical environment (moisture, pH, CO_2, O_2, osmotic potential, etc.) by one organism may provide a more suitable habitat for another; and (d) some organisms destroy or neutralize antimicrobial factors (toxins), thus allowing uninhibited growth of others. While these activities have not been adequately studied specifically in the rhizosphere, a certain degree of logical extrapolation can be made from what is known about the biochemical processes of soil microorganisms in general.

Microorganisms are not uniformly distributed throughout a soil mass, but tend to proliferate or congregate around two nutrient sources. They either utilize the components of root exudates and sloughed cell materials as a nutrient source or, in the absence of roots, they multiply during the decomposition of fresh organic matter. During the microbial metabolism of such substrates, by-products of respiration are excreted and unassimilated fractions of the food source are left behind. The transformations of both organic and inorganic compounds that occur result in the release of nutrients or the synthesis of substances that can be utilized by yet other microorganisms (commensals) during the succession of substrate colonization. The nature of the ecological niche in the rhizoplane or rhizosphere often determines which of two associates is the commensal. The enormous amount of information that has been compiled on the physiological processes of

Table 5.1. Percentage incidence of bacteria capable of synthesizing various growth factors from rhizosphere and rhizoplane of wheat. (Lochhead 1959)

Growth factor synthesized	Rhizosphere	Rhizoplane
Thiamine	55.3	73.0
Biotin	11.5	22.0
Vitamin B_{12}	21.3	31.0
Pantothenic acid	59.2	77.0
Riboflavin	60.2	85.0
Terregens factor	12.6	13.0
One or more factors	70.8	88.0

bacteria, actinomycetes, and fungi during organic and inorganic matter decomposition (Gray and Williams 1971; E.W. Russel 1973) is also pertinent to aspects of rhizosphere ecology, and relates to our understanding of factors that affect plant growth. In fact, the bulk of such work has focused upon the role microorganisms play in rendering nutrients available to plants. Our inability to measure growth of an organism accurately in a natural soil of mixed populations hampers the clarification of many interactions between microbes. However, higher plants and microorganisms utilize many of the same major and minor nutrient elements and, in fact, compete for these in the rhizosphere. Microbial respiration contributes to a higher CO_2 concentration in the root zone and greater production of carbonic acid which, along with microbially induced organic acids and nitric or sulfuric acid, increases the solubilization of phosphates, ferric oxide, and manganese dioxide. Further, the release of ammonia from amino acids can lead to a rise in pH and precipitation of iron and manganese. D.A. Barber (1968), Alexander (1971), and others have discussed mineralization, solubilization and other processes of microbiologically transformed compounds with respect to numerous elements. Other chemical changes take place in waterlogged soils, or other anaerobic habitats where O_2 has been depleted, creating conditions for increased solubility of calcium, magnesium, and iron and reduced availability of phosphorus. Higher microbial populations and more intense metabolic activity in the rhizosphere should accelerate organic matter decomposition and solubilization of compounds. Katznelson and Rouatt (1957) found that respiration (metabolic activity) of rhizosphere soil from several field crops was consistently higher than for root-free soil. The assumption based on the foregoing discussion is that the activities of some microorganisms in the rhizosphere provide nutrients for commensals in addition to the nutrients that are directly available in root exudates. Reference has already been made in Chap. 4 to the nutritional groups of bacteria established by Lochhead and Chase (1943) and the fact that bacteria with simple requirements (those abundant in the rhizosphere) synthesize amino acids that are required by other bacteria.

B-vitamins and other growth factors are synthesized by microorganisms, and this is accompanied by a proliferation of other microorganisms for which certain of these vitamins are essential. Studies by Lochhead and associates, as reviewed by Lochhead (1959), revealed that 27% of the bacterial isolates from a field soil

Table 5.2. Percentage incidence of bacteria requiring various growth factors from rhizosphere and rhizoplane of wheat. (Lochhead 1959)

Growth factor required	Rhizosphere	Rhizoplane
Thiamine	20.4	17.0
Biotin	9.8	5.0
Vitamin B_{12}	1.9	1.0
Pantothenic acid	8.7	1.0
Folic acid	2.9	7.0
Nicotinic acid	< 1.0	1.0
Riboflavin	3.9	< 1.0
Pyridoxine	< 1.0	< 1.0
p-Aminobenzoic acid	< 1.0	< 1.0
Choline	< 1.0	< 1.0
Inositol	< 1.0	< 1.0
Terregens factor	1.0	< 1.0
One or more factors	36.1	21.0

required one or more vitamins for growth; thiamine and biotin were most commonly required. The percentage incidence of bacteria capable of synthesizing various growth factors is generally somewhat higher in the rhizoplane than in the rhizosphere (Table 5.1), whereas the incidence of vitamin requiring bacteria is higher in the rhizosphere (Table 5.2). Thus, it seems that bacteria capable of synthesizing growth factors are preferentially stimulated at the root surface and this activity creates a favorable environment in which the adjacent rhizosphere bacteria proliferate. Stimulated growth and nitrogen-fixing capacity of *Azotobacter* has been observed in response to the presence of *Pseudomonas, Agrobacterium*, and some actinomycetes (Panosyan 1962). Erwin and Katznelson (1961) demonstrated a stimulating effect of thiamine-synthesizing bacteria on growth of *Phytophthora cryptogea* when cultured together on a synthetic, thiamine-deficient medium. This kind of commensalism also may occur between fungi and even between members of the microflora and fauna.

Though not classed as growth-promoting effects per se, the preferential migration of nematodes toward bacteria, actinomycetes, and fungi has been associated with microbial factors other than the physical availability of the microflora as a food source for predators. Katznelson and Henderson (1962), after observing a striking accumulation of *Rhabditis (Cephaloboides) oxycerca* around actinomycete colonies growing on agar medium, found that shake-culture filtrates of many of the actinomycetes strongly attracted the nematodes. This suggests that nematode attraction to roots may be due in part to probiotic microbial metabolites along with root exudates. The reported migration of *Pelodera chitwoodi* to *Vibrio* sp. from various distances may be linked with attraction of the nematode toward amino acids, vitamins, or other compounds (Joshi et al. 1974). The ultimate result of such nematode activity is predation upon the microflora as a food source. Many other examples could be cited of the unilateral growth-promoting benefits of the microflora and fauna in dual associations (Alexander 1971), but the study of such relationships in the rhizosphere has not advanced much further than speculation. There is ample evidence that growth and function of free-living N_2-

fixing bacteria, such as the aerobic *Azotobacter* spp. and anaerobic *Clostridium* spp., are affected by associations with other bacteria and with fungi (V. Jensen and Holm 1975); specific associations may result in either stimulation or inhibition. Further, nonsymbiotic N_2-fixing bacteria of the genera *Azotobacter, Pseudomonas, Beijerinckia, Arthrobacter,* and others, have been isolated from the rhizosphere of rice plants and their N_2-fixing capacity determined (Balandreau et al. 1975); nitrogenase activity also has been measured in the rhizosphere and rhizoplane of tropical grasses (Dobereiner and Day 1975). Judging from the greater populations and more intense interactions of most of these microorganisms in the rhizosphere than in root-free soil, the magnitude of N_2 fixation near roots under natural conditions should be quite different from such activity in nonrhizosphere soil.

Numerous nonsymbiotic, nonpathogenic, microorganisms, commonly found in the cortex of plants (Davey 1971), surely must interact with resulting influences upon pathogenic and mycorrhizal fungi. The possible role of *Fusarium* has often been mentioned in this regard but little factual information has been provided. A very intriging subject, which has received limited attention, is the interchange of reactions between mycorrhizal fungi and associated microorganisms in what has been termed the mycorrhizosphere (Rambelli 1973). Apparently, metabolites produced by the fungal symbiont, in addition to root exudates, form the basis for stimulation and formation of the mycorrhizospheric association by provoding a selective substrate for the rhizosphere microflora. The qualitative nature of the favored microflora depends on the balance between probiotic and antibiotic substances produced.

Volatile metabolites of microbial origin, while mostly considered to be fungistatic, may have growth-promoting properties. Ethanol vapors produced in culture by the pine root fungal symbiont *Boletus variegatus* are known to stimulate growth of the plant pathogens *Phytophthora cinnamomi* and *Fomes annosus* (Krupa and Nylund 1972); all other volatiles produced by this symbiont inhibited the pathogens. Volatile activity is well known in the case of *Agaricus bisporus*, the cultivated mushroom. Alcohols and aldehydes evolved by the mushroom mycelium cause qualitative changes in the "mycosphere" flora which ultimately select predominantly for *Pseudomonas putida*, and metabolites from this bacterium stimulate sporophore production in the *Agaricus* (Hayes et al. 1969).

Volatile chemicals are released during the microbial decomposition of plant residues and during the metabolism of root exudates and sloughed epidermal cells. These are primarily aldehydes, alcohols, and ammonia. Commensalistic relationships among the associated microorganisms are involved since, in the succession of substrate colonization, certain groups of organisms at a given period are active in the decomposition process while other groups or individuals are subject to the effects of the by-products released. Volatiles in low concentration may enhance the respiration and multiplication of certain species (a commensalistic effect) while higher concentrations are inhibitory. Other examples of stimulation (as well as inhibition) of soil microorganisms by volatile metabolites of microbial origin have been discussed by Linderman and Gilbert (1975).

An interesting example of fungal dependency upon a bacterium is the case in which *Phytophthora cinnamomi* responds to stimuli from species of *Pseudomonas* and *Chromobacterium* by forming sporangia (Zentmyer and Erwin 1970). Bacterial associations with fungi in soil are common, as evidenced by numerous reports of the physical colonization of fungal hyphae by both bacteria and *Streptomyces* spp. and the formation of bacterial colonies along the mycosphere. Such observations were first made using the Cholodny (1930) technique of burying glass slides in natural soil for a time, followed by staining and microscope examination for microbial associations. Many subsequent associations of fungi and bacteria have been reported based largely on modifications of the buried slide method. Other studies showing increased bacterial activity in soil supplemented with fungi can be traced through a review by Lockwood (1968). A nutritional basis for stimulation of bacteria near fungal hyphae is generally assumed. All associations of this type are not commensalistic since lysis or parasitism of the fungus often follows, but in many cases bacteria multiply without harm to the benefactor. Again, the scarcity of information on such associations occurring strictly in the rhizosphere forces the conjecture, supported by logic, that the denser microbial populations at the root–soil interface must increase the opportunities for fungus–bacterium contacts. Whether root exudates actually mask the nutritional benefit offered by the fungal hyphae, and thus lessen the commensalistic effect, is something else to consider.

Along with chemical changes in the root environment, microbially induced physical changes also contribute to the selective growth and development of various groups and kinds of microorganisms. Chemical and physical factors are not always clearly separable, and much less is known about the latter, particularly as they may relate to microbial interactions in the rhizosphere. In the general ecology of soil microorganisms the physical factors that commonly influence microbial activity are temperature, moisture, hydrogen-ion concentration, and gases. These have been discussed in detail by McLaren and Skujins (1968), Raney (1965), D. M. Griffin (1972), and many others. We are concerned here only with microbial alteration of physical factors and the resulting commensalistic effects on other microbes in the root environment.

Temperature regulation is generally not considered to be a microbial function, although some change may occur at sites of high organic matter content and high microbial populations. Water, either in the liquid or vapor phase, is essential for microbial growth. Unicellular organisms, such as bacteria, protozoa, and Oomycete zoospores require higher water potentials for movement and reproduction than do the filamentous fungi and actinomycetes which can traverse pore spaces in drier soils. It is doubtful that microscopic organisms at the root surface or in the rhizosphere under the influence of root exudates can alter the moisture content of a microhabitat sufficiently to affect associated, potentially commensal organisms. F. E. Clark (1965), however, pointed out that microorganisms can be viewed more correctly as commensalistic in their water relations rather than as competitive, since metabolic water may be produced during organic matter decomposition. Very little is actually known about either the production or uptake of water by soil microorganisms in relation to growth of other organisms.

That growing microorganisms can change the pH of unbuffered media is common knowledge. *Sclerotium rolfsii*, for example, grows slowly at first in a liquid medium of high nitrogen content but then, through the production of oxalic acid, it lowers the pH, favoring its own rapid growth and enzyme production. Some *Streptomyces* spp. can raise the pH within their soil microhabitats by the production of ammonia during lysis of fungal mycelia (Williams and Mayfield 1971). Other microorganisms, particularly in fine-textured, poorly drained soils or at considerable depths in the rhizosphere, contribute to CO_2 production and bicarbonate accumulation. Exudates released by roots provide substrate for rhizosphere bacteria which may excrete acids into the soil. Thus, it is conceivable that many organisms in the root zone are commensalistic in their relation to other microbes which favorably modify the soil pH.

Most microorganisms consume oxygen, whether in the rhizosphere or not. Where microbial populations are high and respiration intense, as at the root–soil interface, oxygen uptake will be correspondingly greater and consequently oxygen will be in short supply at specific microsites. This condition would be inhibitory to less competitive aerobic microorganisms but would favor commensalistic facultative anaerobes. The oxygen concentration prevailing is, however, more related to soil structure, moisture, and depth of the inhabited root than directly to microbial comsumption of oxygen.

The soil conditions and microbial activity that lead to oxygen depletion at certain microsites tend to promote relatively high levels of CO_2. Most often CO_2 is considered in terms of microbial tolerance to high concentrations or its inhibition of growth. However, D M. Griffin (1972) provides references to several cases in which CO_2 stimulated growth or reproduction of organisms. Assuming that some of the CO_2 was of microbial origin, this represents a commensalistic situation. The dreaded cotton root-rotting disease caused by *Phymatotrichum omnivorum* in the western United States is most severe in poorly aerated, heavy clay soils where production of sclerotia is favored by high levels of CO_2 (Lyda 1978). Other cases could be cited in which either spore production, spore germination, or hyphal growth of various fungi is enhanced by CO_2.

5.2.2 Mutualistic Relationships

Different species of microorganisms living in close proximity may impart mutually beneficial effects upon each other. This is not to be confused with synergism, a term used to designate the living together of two organisms, which subsequently results in a change that is not effected by either organism alone. The term symbiosis also is quite acceptable for mutualism but usually is applied to the familiar permanent or semi-permanent mutual relationships between microorganisms and plant roots, e.g., the root-fungus mycorrhizal condition and the legume–*Rhizobium* relationship. Some microorganisms also live symbiotically with animals or insect stages in the soil in which the microsymbiont is a necessary component of the gut microflora. Mutualism or symbiosis among soil microorganisms, or between micro- and meso-organisms, has not been investigated with regard to the rhizosphere effect. We can only speculate that if such relationships occur in other habitats they must occur also in the rhizosphere. The most common and well-known microflora symbiosis, the lichen (alga–fungus association), occurs in aquatic habitats (T. W. Johnson and Sparrow 1961) as well as on terrestrial substrates. Both algae–protozoa and bacteria–protozoa mutual relationships have been observed, primarily in aquatic habitats. The necessary components of these symbiotic associations (algae, fungi, bacteria, and protozoa) all occur in the rhizosphere along with many other members of the flora and fauna. Since mutualism is primarily a nutritional relationship (an exchange of metabolites), it should not be surprising to learn that organisms coexisting in the rhizosphere may practice a periodic (nonpermanent) type of mutualism as well as the one-sided commensalism.

Some interesting mutualistic relationships exist between the free-living, nonsymbiotic, nitrogen-fixing bacteria and other soil organisms, or between two nitrogen-fixing bacteria (V. Jensen and Holm 1975). Most of these relationships have been demonstrated in laboratory cultures, which encourages speculation regarding the possibility of similar activities in the rhizosphere or the rhizoplane. Since Beijerinck and Van Delden (1902) first observed that *Azotobacter chroococcum* fixed more nitrogen in the presence of strains of *Agrobacterium, Aerobacter*, and *Clostridium* than in monoculture, similar observations have been made by many other researchers using *Azotobacter* in combination with several other bacterial genera, including *Bacillus, Pseudomonas*, and various actinomycetes. Kalininskaya (1967) applied the term "facultative symbiotrophic nitrogen fixers" for bacteria which fix nitrogen readily in mixed cultures with other microorganisms but have limited fixation ability in pure culture.

Several microorganisms may be involved in interactions that are synergistic in relation to effects on plant growth. For example, vesicular-arbuscular (VA) mycorrhizal fungi (such as *Glomus mosseae*) increase the efficiency of phosphate uptake by plants in low-phosphate soils; this efficiency is further increased in the presence of phosphate-solubilizing bacteria such as *Pseudomonas* spp. Since satisfactory nodulation and nitrogen fixation by *Rhizobium* depend on an adequate phosphorus supply, this constitutes a three-way interaction enhancing plant growth. Azcon-G. De Aguilar and Barea (1978) found that cell-free supernatants of *Rhizobium* and phosphobacteria improved plant growth, nodulation, and mycorrhizal formation in *Medicago sativa*. A similar positive effect was achieved with whole cultures of *Rhizobium*, phosphobacteria, and *G. mosseae*.

5.3 Growth-Inhibiting Relationships (Antagonism)

In ecology the term antagonism is applied to all forms of microbial action in mixed populations that result in the inhibition or unfavorable effect of one organism upon another. Antagonistic effects during the struggle for existence among microorganisms were observed by such eminent microbiologists as Pasteur in 1877 and De Bary in 1879, cited in Waksman's (1947) *Microbial Antagonists and Antibiotic Substances*. In more recent times, the various categories of antagonism have been researched and discussed in considerable depth by soil microbiologists and plant pathologists, largely in relation to plant pathogen survival and biological control of root diseases. The potential importance of this subject is reflected in the relatively large amount of space devoted to the topic since about 1960 in the published proceedings of several international symposia on the ecology of soil microorganisms.

Plant pathologists, more than any other group, have experienced, with considerable anguish, the true meaning of antagonism when attempts have been made to establish potential biological control agents in a natural field soil. Such attempts have often failed due to the greater competitive or other antagonistic action of endemic components of the microflora against the introduced organism. Various meanings, both broad and narrow, have been applied to the categories of antagonism (competition, antibiosis, hyperparasitism, and predation) and, with some manipulation of fact and logic, one might even lump them all together under competition. However, while the reader must realize that these mechanisms are not exclusive, phases of one merging or overlapping with another, for convenience of study they are best treated as separate entities.

5.3.1 Competition in the Rhizosphere

Competition is the active demand in excess of the immediate supply of material or condition on the part of two or more organisms (Clements and Shelford 1936; F.E. Clark 1965). Logically, interactions should be most intense where microbial populations are dense in relation to the amount of an essential resource that is available. Microorganisms initially inhabiting the same ecological niche will require a minimal share of the features of that microenvironment, but one may become dominant due to certain of the cardinal attributes listed by Garrett (1970) as contributing to a high "competitive saprophytic ability" for substrate colonization. In brief summary, these qualities include: (1) rapid germination of propagules and rapid growth from a nutrient base, (2) necessary enzyme capacity for decomposition of organic materials, (3) production of fungistatic and bacteriostatic metabolites, and (4) tolerance to inhibitory substances produced by other microorganisms. Here we see, as mentioned in (3) above, that antibiosis can become a factor in competition; this is sometimes referred to as interference competition (Lockwood 1981). These attributes suggested by Garrett were aimed specifically at measuring the capacity of soil-borne plant pathogens to grow and survive in competition with soil saprophytes. We shall pursue this line further when discussing rhizosphere–pathogen relationships in Chapter 7. However, returning to competition in the narrower sense, the usual assumption is that microorganisms, whether in the rhizosphere or on organic matter substrates, compete with one another for the available nutrients, water, oxygen, and space. Where deprivation of any of these essential resources occurs, the term exploitation competition is sometimes applied. This is not to be confused with exploitation as used in some earlier literature to describe mycoparasitism (hyperparasitism) and predation. Garrett (1956) in one of his earlier books offered as an analogy to saprophytic competition the old game of musical chairs in which a number of people, marching to music around a circle of too few chairs, compete for a seat when the music stops. Any individual may in the early stage of the game obtain a seat merely by position and chance but, as the game progresses, the stronger, more agile players dominate. Microorganisms which possess the necessary attributes for rapid substrate colonization also will win out in the end over competitors. Logically, these remaining, dominant, organisms would be largely of the same species with identical or similar demands for resources. In the ecology of higher plants and animals, a principle introduced by Darwin suggests that the struggle for existence is most intense between individuals of the same species which have equal demands. Applying this principle to the dominant microorganisms in ecological niches of the rhizosphere-rhizoplane, we might expect the occurrence of some intraspecific competition, the intensity of which is determined by nutrient availability.

The struggle among soil microorganisms for occupancy of a place in the rhizosphere is, however, not quite this simple, as was forcefully brought forth by F. E. Clark (1965), whose concept of competition probably most nearly approaches the true situation that exists in nature. F. E. Clark rejected space as a resource for which microorganisms compete, his opinion being that physical space in most soil habitats is sufficient to accommodate the microorganisms present. Direct observation techniques (Gray et al. 1968), as well as respiratory measurements in specific volumes of soil (F. E. Clark 1968), have provided evidence to discount space as an important limiting factor for bacterial activity. In certain specialized circumstances, however, it is conceivable that space may become a limiting factor where particular growth patterns occur, such as the formation of dense mycelial aggregations over the surface of roots, as exemplified by *Phymatotrichum omnivorum* (C. H. Rogers and Watkins 1938). Other dominating types of growth patterns are rhizomorphs produced by *Armillaria mellea* and the mantle formation by ectomycorrhizal fungi on small roots. F. E. Clark also questioned whether microorganisms compete significantly for water, since the availability of water under low-moisture conditions is more likely to be determined by the physical forces that hold it rather than by microbial competition for it. However, available water is essential for microbial growth and movement and it is likely that any absorption by one organism in a microenvironment may affect another to some extent. Fungi, which can function at lower water potentials or in the vapor phase among soil particles, have an advantage over the Oomycetes and bacteria which require higher moisture levels.

The availability of oxygen for microbial consumption is largely determined by other soil physical factors, particularly soil structure, water content, and CO_2/O_2 ratios. Most soil microorganisms require oxygen and some are facultative anaerobes. Thus, consumption of oxygen by aerobes in the micro-spaces among soil particles may inhibit growth of other aerobes unless, with the prevailing soil structure and moisture conditions, the amount of uptake is balanced by adequate oxygen diffusion into the microenvironment. Since microbial activity and plant root respiration both contribute to the carbon dioxide concentration of the rhizosphere, this zone, especially at lower depths, might be expected to maintain a higher CO_2/O_2 ratio. Such a condition results in the inhibition of growth of certain fungi, while many other microorganisms are tolerant to carbon dioxide, or growth may even be enhanced. As F. E. Clark (1965) suggested, oxygen may be more strictly an environmental than a consumptive factor, but in microsites of great oxygen demand and low availability, microbial species with highest rates of oxygen utilization are likely to win a competitive advantage. With the consumption of oxygen by aerobic forms there may be a concomitant rise in the numbers of facultative anaerobes, particularly at deeper soil depths.

This brings us to the apparently logical conclusion that substrate or nutrient supply is the primary resource for which microorganisms actively compete. However, even here, competition is not clearly definable. It is not sufficient to say simply that fierce competition for nutrients occurs among microorganisms where populations are especially dense. As F.E. Clark (1965) stated: "Even in such a specialized and restricted habitat as the rhizoplane of an individual plant, it is probable that most of the microbial species that are present therein are not competitive....... Many of the microbial species that are involved must be in separate ecological niches". This refers to an ecological concept sometimes known as Gauze's principle which contends that two species rarely ever occupy the same niche because one, having some specific advantage, displaces its competitors (Gauze 1934). The security of an individual species within its niche is based largely on its nutritional requirements and enzyme constitution. Gauze's principle, which was aimed primarily at animal life, does not exclude the probability that competition occurs during the initial colonization of a substrate (organic matter or root surface), by organisms such as the so-called sugar fungi. As succession progresses, then microorganisms segregate more into nutritional groups or even to species in ecological niches. Thus, is not competition a necessary prelude to the establishment of a specific niche?

Microorganisms, whether at the root–soil interface or on organic particles distant from roots, are in a changing environment where populations rise and fall quantitatively, and qualitative succession occurs during the selective decomposition of substrate components. One could reason that root exudates provide sufficient nutrients for all types of microbial life, and competition would be minimal until root growth ceased and a whole succession of saprophytes attacked the soluble sugars, hemicelluloses, cellulose, and lignin. Thus, the most intense competition on roots would be delayed, whereas some microorganisms colonizing organic particles distant from roots would be displaced rapidly. Rhizosphere soil perfused with root exudates may be somewhat like a culture medium which is relatively selective for certain species, promoting their rapid growth and reproduction while others are suppressed partially by nutrient deficiency and partially by biostatic metabolites. Basically, in the rhizosphere, organisms genetically endowed with enzyme systems for rapid metabolism and high respiration rates are likely to prevail.

Microbial associates in symbiosis with higher plants also are subject to competition with other soil organisms in the rhizosphere until establishment with the host has been effected. In the rhizoplane of *Pinus radiata*, different bacteria have been found to depress, have no effect, or stimulate growth of mycorrhizal fungi (Bowen and Theodorou 1979). When the mycorrhizal association is established, the fungal symbiont then has the advantage of escaping into living tissue, while saprophytes are excluded. Colonization of the rhizosphere by ectomycorrhizal fungi and initiation of the symbiotic relationship occur more readily in fumigated soils, apparently due in part to reduced antagonism from other organisms. This means, for the most part, reduced competition for nutrients along with partial elimination of antibiosis. Stimulation of the mycorrhizal fungus by metabolites of certain other organisms, which recolonize rapidly after fumigation, also contributes to the advantage gained by the fungal symbiont.

Competition between *Rhizobium* spp. and other microorganisms in the rhizosphere is of great practical significance in the culture of leguminous crops. The rhizobia, like most microbes that are able to coexist with living roots, are considered to be relatively weak competitors with the general rhizosphere flora. Both competition with other soil organisms for essential nutrients and the action of antagonistic fungi, streptomycetes, and bacteria apparently contribute to the low competitiveness of free-living rhizobia (Fåhraeus and Ljunggren 1968). Somewhat contrary to this view are reports that rhizobia are "good rhizosphere organisms" (E. L. Schmidt 1979), meaning that their numbers are frequently found to be higher in rhizosphere soil than in root-free soil. Specific substances in legume root exudates, which are known to inhibit some species of other bacteria, may provide the competitive advantage needed for *Rhizobium* to colonize the rhizoplane prior to infection. Even competition between rhizobial strains is a phenomenon of much concern and importance in legume culture. The true nature of this type of competition often is not defined, but is measured in terms of the percentage of total nodules that are formed by one strain in the presence of another (Caldwell 1969). Vitamin requirements probably influence survival of rhizobia in the rhizosphere as much as any other factor. Though vitamin requirements differ between strains, biotin and thiamine are commonly required and thus provide resources for competition.

The ecology of free-living, nonsymbiotic nitrogen-fixing bacteria has been a subject of great interest since Winogradsky isolated *Clostridium pasteurianum* in 1893 and Beijerinck isolated species of *Azotobacter* in 1901. A number of other bacteria also have been associated with nitrogen fixation in the absence of plants, but the extent to which any of these organisms contribute to the total nitrogen economy of field soils is often questioned. The aerobic *Azotobacter* and anaerobic *Clostridium*, though benefiting from root exudates along with other microorganisms, generally occur in lower numbers in the rhizosphere. There is some doubt that *Azotobacter* can competitively colonize roots (Macura 1968). However, Russian agronomists for many years have claimed to increase wheat yields substantially by inoculation of seed with these bacteria prior to planting. The contention is that the bacteria, when started from infested seed, follow the growing root and provide sufficient coverage to impart benefit to the plant. Some of this benefit, however, may be derived from growth-stimulating substances (gibberellins and indole-3-acetic acid) synthesized by the bacteria rather than from increased nitrogen availability (M. E. Brown et al. 1968). Thus, the free-living nitrogen fixers, when given an initial advantage in colonization, may establish an affinity for the root surface and persist.

Among the requirements for the functioning of free-living, nitrogen-fixing bacteria in a natural environment are (1) the availability of carbon compounds, (2) adequate inorganic nutrients (calcium, magnesium, molybdenum, etc.), (3) optimal pH for growth, and (4) low oxygen tension (Mulder 1975). Any of these may be altered by other microorganisms competing for the same substrate. As already mentioned, some microbial interactions with nitrogen fixers may be mutualistic rather than competitive.

Much is yet to be learned about the specific effects of the rhizosphere flora on both rhizobia and nonsymbiotic bacteria. The stereoscan electron microscope has contributed much to our knowledge of the distribution of microorganisms on root surfaces, but we are still unable to clearly separate competition from other antagonistic mechanisms.

Competition by the soil fauna involves essentially the same resources that determine competition within the microflora, food supply being the most obviously important of these. Both protozoa and saprozoic nematodes ingest bacteria, which may create a competitive situation between the two animal groups or interspecifically within either group, especially at sites on the root surface where bacteria and the fauna congregate. The term competition also has been applied in cases of cohabitation by plant parasitic nematodes (Norton 1978); a stronger competitive ability primarily reflects a greater, more rapid reproductive capacity and a larger quantity of food consumed in a given time. Nematodes in combination may result in reduced numbers of one associate due to altered root morphology and fewer available feeding sites. It has been shown experimentally that a low population of one nematode (50 individuals of *Meloidogyne incognita*) versus a high population of another (200 individuals of *Pratylenchus penetrans*) results in a further reduction of the lower population and consequently reduced invasion of alfalfa (Turner and Chapman 1972). Similar competitive interactions have been observed between ectoparasites and endoparasites, and between species of the two groups.

Some interactions between soil microarthropods are referred to as competition, but the mechanisms responsible for reduced populations of contending species often are not precisely determined and could just as well be due in part to predation of one associate on the other. Van de Bund (1972) made observations on a community of Acari and Collembola among the root systems of potted white clover plants. He found that competition by the collembolan *Onychiurus bicampatus* prevented the establishment of large populations of the mite *Tyrophagus putrescentiae*. Some microarthropods are predacious on nematodes; in addition, they may compete for a common food source such as bacteria and fungi with consequent effects on fecundity and population density of either associate. It is common knowledge, based on the analysis of gut contents and from feeding observations, that the microflora contributes significantly to the food supply of many mites and Collembola (Christiansen 1964; Wallwork 1970; Christen 1975). Few soil animals are restricted to one source of food and, being mobile, they can migrate to new food sources. Thus, while some regulation of populations may occur under stress of competition, death by starvation is unlikely.

Wallwork (1976) reminds us that growing roots and the channels they create form a special kind of microhabitat with its own associated fauna living with a dynamic community of other microorganisms under the influence of the rhizosphere. With the exception of parasitic nematodes, few members of the small soil fauna can be classified as feeders on living plant roots. However, protozoa, saprozoic nematodes, and microarthropods, which feed on the microflora as well as on senescing root tissue and soluble substances, are directly or indirectly affected by root development and decline. It has long been known that bacteria-feeding protozoa and nematodes congregate or proliferate in the rhizosphere and that collembola also are attracted to the root surface. Wiggins et al. (1979) showed that mycophagous species of the collembolan genera *Proisotoma* and *Onychiurus* were more abundant in the rhizosphere of cotton plants in the field than in root-free soil. In subsequent laboratory studies (Wiggins and Curl 1979), the population density of *P. minuta* in sterilized, fungus-infested soil varied with the species of fungus available as a food source. Highest numbers of the insect occurred in the presence of the cotton pathogen *Fusarium oxysporum* f. sp. *vasinfectum* and lowest numbers in soil with *Trichoderma harzianum*. In general, the soil microarthropods are considered to be unspecialized feeders but the apparent "preference" of some Collembola for certain fungi suggests the probability of intra- and interspecific competition among these insects. It would seem to follow therefore, that competition between unrelated groups must occur.

The rhizosphere effect on animal competition and other ecological behavior has not been adequately explored. As suggested by Wiggins and Curl (1979), the activities of mycophagous microarthropods can no longer be ignored in rhizosphere investigations, for the feeding of these organisms may determine the competitive advantage among components of the microflora, including plant pathogens, at the root surface.

5.3.2 Antibiosis and Microbial Stasis

Antibiotics are organic substances that are produced by microorganisms and are deleterious in low concentration to growth or metabolic activities of other microorganisms (Gottlieb and Shaw 1970). Usually, a substance which is effective at 10 ppm or less is considered to be in low enough concentration to qualify as an antibiotic. Inhibitory compounds produced by higher plants and animals are not excluded from the antibiotic group, but we will not dwell on these. We have already seen that root exudates contain inhibitory as well as stimulatory substances which may relate directly to microbial activity in the rhizosphere; the possible role of small animals in this regard is still too vague to be considered at this time.

Antibiosis, a category of antagonism and closely linked with competition, is the inhibition of one organism by a metabolite of another (K. F. Baker and Cook 1982). This broad definition may involve any inorganic inhibitors or organic toxins resulting from microbial metabolism, as described by Alexander (1977) under the all-encompassing term, amesalism. Antibiosis is commonly observed in soil-dilution plates where a fungal colony may be surrounded by a clear zone which other colonies in close proximity cannot penetrate because of an antibiotic substance diffusing outward through the agar medium. The phenomenon is used in standard antibiosis assay (Fig. 5.1) to indicate biological control potential of saprophytic microorganisms against pathogens. Antibiosis is most commonly thought of as a response to antibiotics, such as actinomycin, streptomycin, erythromycin, neomycin, novobiocin, polymixin, penicillin, etc., and many antifungal substances described by Gottlieb and Shaw (1970). The usual methods of testing for antibiosis (L. F. Johnson and Curl 1972) do not distinguish between different antibiotic substances; one organism may produce several such substances or each of several microorganisms may produce the same antibiotic.

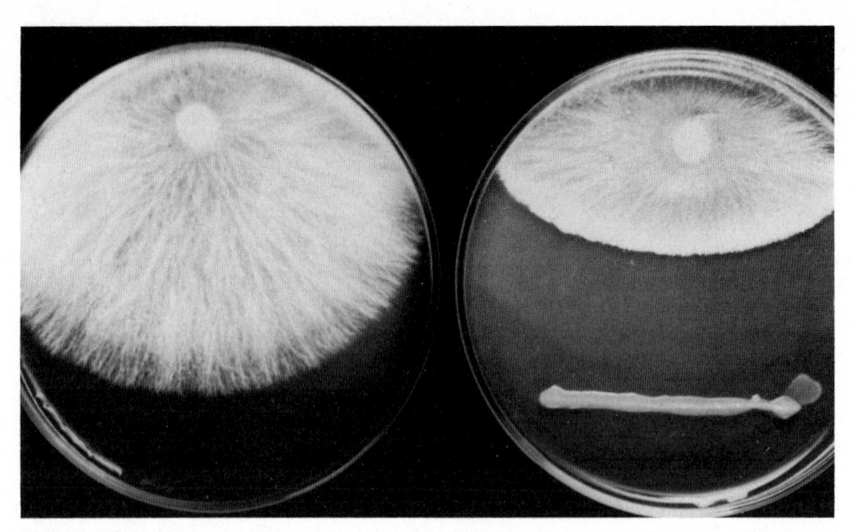

Fig. 5.1. Antagonism (antibiosis) of the plant pathogenic fungus *Sclerotium rolfsii* by a bacterium (*right*)

An interesting, but poorly defined, group of antibiotic-like compounds are the bacteriocins described by Vidaver (1983). These colicins are distinguished from the classical antibiotics largely by their property of restricted biological specificity; specific antagonism occurs between related bacteria or even between strains of the same species. The producer strain is rarely sensitive to its own bacteriocin. These compounds usually kill by affecting protein synthesis, DNA stability, energy flux, or membrane integrity. Species of *Pseudomonas, Corynebacterium, Agrobacterium*, and others produce bacteriocins under the control of extrachromosomal elements called plasmids. Since bacteriocin production is affected by the chemical nature of the bacterial growth medium, as well as by other factors, one might speculate that these compounds could play an important role in the competitive activities of bacteria in the rhizosphere.

The most common antibiotic-synthesizing microorganisms (species of *Streptomyces, Bacillus, Pseudomonas, Penicillium, Trichoderma, Aspergillus,* etc.) are abundant in the rhizosphere and rhizoplane of most cultivated plants and weeds. When a number of microbial isolates from the root zone of the same plant are tested one against the other for antibiosis on a nutrient medium, certain species are inhibited by most of the population while others are principally inhibitors and are themselves little affected. For example, species of *Trichoderma*, which possess all of the attributes required to be strong biological control agents against pathogens, are inhibited by very few other fungi. The difficulty of establishing *Trichoderma* in natural soil as a supplementary biocontrol agent is probably due in large part to interference from bacteria and streptomycetes (Fig. 5.2). Any organism can be inhibited by some other organism, depending upon inoculum density, nutrient availability, and other environmental factors that promote growth and antibiotic production or other forms of antagonism.

Most of our knowledge of antibiosis as a microbial deterrent in natural soil still lies in a cloud of uncertainty, primarily because of the difficulty of detecting

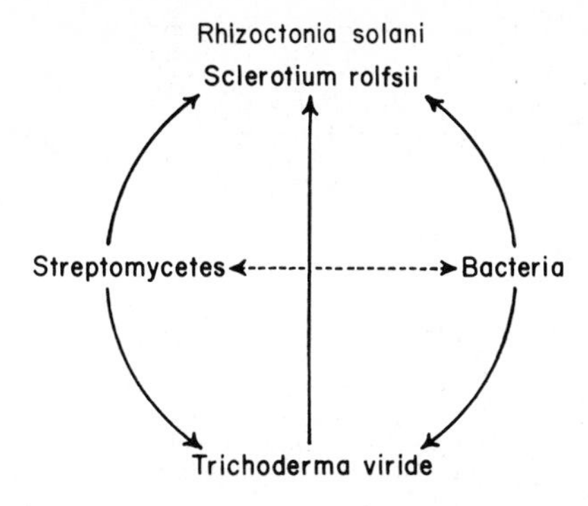

Fig. 5.2. Probable interactions of antagonistic microorganisms in relation to the suppressive potential of *Trichoderma viride* against root-infecting fungi such as *Rhizoctonia solani* and *Sclerotium rolfsii*

the small quantities of antibiotics that must exist at microsites of root exudation and on organic matter particles where the essential nutrients for antibiotic production are available. The present state of our knowledge about antibiotic production, and the probability that these metabolites exert significant biological action in natural environments, have been discussed from various viewpoints (Brock 1966; Gray and Williams 1971; K.F. Baker and Cook 1982; Alexander 1977). There seems to be little doubt that a high organic matter content of soil promotes antibiotic production, but some of these compounds are chemically and biologically unstable, or may be adsorbed to colloidal particles. Logically, the most significant role of antibiosis must occur in microscopic pockets or niches within the rhizosphere influence where nutrients are sufficient for toxin production; as nutrients become depleted the less resistant organisms are suppressed or killed by toxins of others. The fact remains that, while antibiotic production by many bacteria, actinomycetes, and fungi, and the resulting antibiosis, have been demonstrated numerous times in laboratory cultures, the status of antibiosis in a natural environment remains controversial.

Much interest in antibiosis is focused upon its possible relation to phenomena of the root–soil interface affecting plant growth. While probiotic substances of root and microbial origin promote the activities of rhizobia, mycorrhizal fungi, and plant pathogens, antibiotic substances may partially counter such benefits. Bacteria, actinomycetes, and fungi are often said to be antagonistic to rhizobia and to suppress root nodulation, but antibiosis has rarely been specifically implicated as distinct from competition or some other criterion of antagonism.

Difficulty in establishing leguminous crops in many soils cannot always be attributed to deficiencies of major or minor elements or to disease. *Rhizobium* is essentially a rhizosphere or rhizoplane inhabitant, being favored by the presence of living roots and weakly competitive among microorganisms in soil generally. Hely et al. (1957) showed that poor inoculation of subterranean clover in a yellow podzolic (Ultisols) soil in New South Wales, Australia, was due to microbial antagonism which prevented colonization of the rhizosphere by applied *R. trifolii* inoculum; this antagonism was confined to the clover rhizosphere, but the effective principle was not determined. Standard antibiosis tests with bacteria isolated from the rhizoplane of *Trifolium* spp. have indicated that Gram-negative, nonsporing rods, mainly *Pseudomonas*, were the principal antagonists of *R. trifolii*

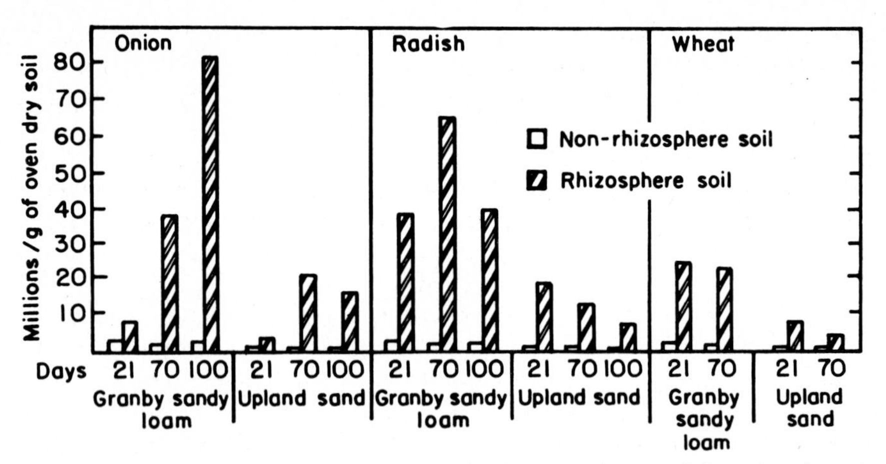

Fig. 5.3. Numbers of microorganisms (bacteria and actinomycetes) antagonistic to *Azotobacter* in the rhizospheres of plants at different stages of growth in two soils. (Strzelczyk 1961)

(Hattingh and Louw 1969). That this problem still remains largely unexplained was indicated by the effort of Patel (1974) to relate the poor nodulation of clovers in New Zealand to antagonism by actinomycetes, which are known for antibiotic production. No significant pattern of antagonism to 12 strains of rhizobia could be established with 279 actinomycete isolates tested. Antagonism by saprophytic soil fungi may be important in the survival and multiplication of rhizobia (Holland and Parker 1966). While the suppressive effects of various soils on rhizobia could be due to any number of interactions with other rhizosphere microorganisms, unidentified toxic factors in soils of problem areas must be involved (Chatel and Parker 1972).

The relationship of antibiosis to nonsymbiotic nitrogen-fixing bacteria in the rhizosphere has not received great attention, probably because some skepticism still prevails regarding the contribution of free-living nitrogen fixers to the overall nitrogen economy of the soil. Attempts have been made (for example Strzelczyk 1961) to correlate the low numbers of *Azotobacter* in both rhizosphere and root-free soils with antagonism by other bacteria and actinomycetes. Wheat, radish, and onion yielded larger numbers of bacteria and actinomycetes that produced toxic zones of inhibition against *Azotobacter* in standard antibiosis tests than were found in root-free soil (Fig. 5.3). Such results suggest that suppression of the nitrogen fixer may be attributable in part to the production of antibiotics in the rhizosphere where these metabolites are most likely to be produced in effective, if not detectable, amounts. Microbial competition, as well as low pH in the rhizosphere, also must create unfavorable conditions for *Azotobacter*. A rhizosphere-mediated effect, whatever the primary principle involved, is evidenced by the fact that different crop plants promote different numbers of antagonists at certain stages of plant growth in the same soil type.

Higher microbial populations are usually found on mycorrhizal roots than on nonmycorrhizal roots (Rambelli 1973). Much of the influence of rhizosphere on mycorrhizae, and hence on plant growth, is derived from the mycorrhizosphere, which is created by the mycorrhiza itself in the form of combined root exudates and metabolites of the responding fungal symbiont or associated microflora. Thus a mycorrhiza can affect its own development, depending on whether the microorganisms it harbors produce inhibitory or stimulatory metabolites.

Soil fungistasis (mycostasis), a phenomenon of virtually all natural soils, is closely related to both competition and antibiosis. Under the influence of this factor, fungal spores fail to germinate even where temperature and moisture are favorable for germination. This imposed state of exogenous dormancy is not limited to spores but may also be associated with sclerotia and mycelia. In fact, it seems that the soil microbial population as a whole is held in a static or semistatic state with periods of activity induced by stimulating factors that release organisms from the grip of fungistasis. Long before Dobbs and Hinson (1953) described a widespread fungistasis in the soils of North Wales, a similar factor was reported limiting the development of bacteria (E. J. Russel and Hutchinson 1909) in untreated soil; subsequently this was termed bacteriostasis (M. E. Brown 1972). However, fungistasis has received greater research attention because of its significant relationship to the survival and germination of pathogen propagules at the root surface of host plants. The variety of methods used to assay soils for fungistasis (L. F. Johnson and Curl 1972) has added somewhat to the confusion and failure among workers to agree on the mechanisms involved.

Clarification of the processes governing the induction, maintenance, and annulment of soil fungistasis has been sought for many years through a continuing controversy about the specific mechanisms involved (Lockwood 1964, 1977; Jackson 1965; Watson and Ford 1972; Griffin and Roth 1979; Balis and Kouyeas 1979; Filonow and Lockwood 1979). A look at the principal opposing hypotheses that have been offered to explain nonconstitutive dormancy of fungal propagules in soil may provide some understanding of the role of a rhizosphere environment in this phenomenon. (1) One hypothesis suggests that inhibitors released by microbial activity are responsible, since the inhibitory effect can be eliminated by soil sterilization and reinstated by reinfestation with microorganisms. This biotic inhibitor hypothesis was favored until about 1964 when Lockwood proposed a second explanation based on a series of studies in model systems. (2) The nutrient-deficiency or nutrient-sink hypothesis states that fungistasis can be caused solely by nutrient depletion in the sporosphere, a condition created by microbial response to spore exudates, resulting in a nutrient diffusion gradient away from the spore. The fungistatic phenomenon affects both carbon-dependent spores, that require exogenous carbon as an energy source for germination, and spores which are carbon-independent (G. J. Griffin and Roth 1979). In the latter case any exogenous nutrient sink created by microbial activity must also drain endogenous nutrient reserves of the spore.

While individual researchers have focused upon one hypothesis or the other, it is generally agreed that neither explanation is adequate alone. Watson and Ford (1972) reappraised the data that have been offered to explain the phenomenon, then proposed as a third explanation, that soil fungistasis is caused by complex inhibitors of biotic and abiotic origin which are most effective in microenvironments of low stimulator concentration (low nutrients). This combines factors of the first two hypotheses, suggesting that nutrient depletion at the soil–spore interface, which may directly inhibit germination, also predisposes the spore to greater suppression by inhibitors. According to this view, both the onset of fungistasis and its annulment would depend upon the specific balance of inhibitor and stimulator concentrations.

The compromise hypothesis of Watson and Ford (1972) has served to refocus some attention upon the role of inhibitors without loss of emphasis on the nutrient deprivation hypothesis. Volatiles of plant origin may either inhibit or stimulate fungal growth, depending on the nature and source of the volatile substance and the test organism involved. Such volatiles are known to emanate from undecomposed plant materials, decomposing plant residues, mycorrhizal roots, and root exudates (Linderman and Gilbert 1975). Volatile inhibitors of microbial origin are now believed to be implicated in the fungistatic phenomenon. Many microorganisms produce volatile or gaseous metabolites such as ethanol, ethyl acetate, acetone, acetaldehyde, ethylene, hydrogen cyanide, ammonia, formaldehyde, and CO_2. *Trichoderma* species, widely known for production of nonvolatile antibiotics, also produce volatile substances (Bilai 1956; Dennis and Webster 1971; Pavlica et al. 1978) which may contribute to soil fungistasis. Along with *Trichoderma*, actinomycetes are among the most active components of the soil microflora that produce volatiles inhibitory to spore germination in soil. This can be demonstrated by applying various assay techniques to sterilized soil which has been recolonized with monocultures of the volatile-forming organisms (Hora and R. Baker 1972). Characterization of volatile inhibitors is incomplete, since several chemicals apparently are involved and their fungistatic activity may vary with the test fungus, concentration of the volatile factor, and certain soil conditions. Ammonia has been found most consistently, and is believed to be a principal contributor to soil fungistasis. Intermediate products released through the metabolism of ethylene, and perhaps other alkenes, by microorganisms may be responsible for fungistatic effects (Balis and Kouyeas 1979). Volatile inhibitory factors are usually soil-pH related, being most evident in alkaline soils. Nullification of their activity on spore germination is often correlated with increasing nutrient concentration, supporting the possibility that the inhibitor hypothesis and the nutrient deficiency explanation for soil fungistasis may be inseparable.

The relationship of rhizosphere to soil fungistasis would seem evident, though this specific subject has been somewhat neglected during the frantic search for mechanisms to explain the phenomenon. Fungal propagules lie dormant in field soil, usually embedded in organic matter, until the fungistatic principle is annulled by nutrients from decomposing organic matter, microbial synthesis, or root exudates. Stimulation has been observed with common components of exudates, including various sugars and amino acids (Lockwood 1964) but opposing results can be found, depending upon the test fungi and soils used. It is generally agreed that spores of various plant pathogenic fungi germinate readily when adjacent to germinating seeds and in the rhizosphere. Long-chain unsaturated fatty acids in rhizosphere soil of bean and certain other plants may contribute to induced germination of endoconidia and chlamydospores of the pathogen *Thielaviopsis basicola*. Fatty acids that can be extracted from rhizosphere soil exceed quantities in extracts from nonrhizosphere soil, and when a mixture of palmitic, stearic, oleic, linoleic, and palmitoleic acids (representative of those found in the

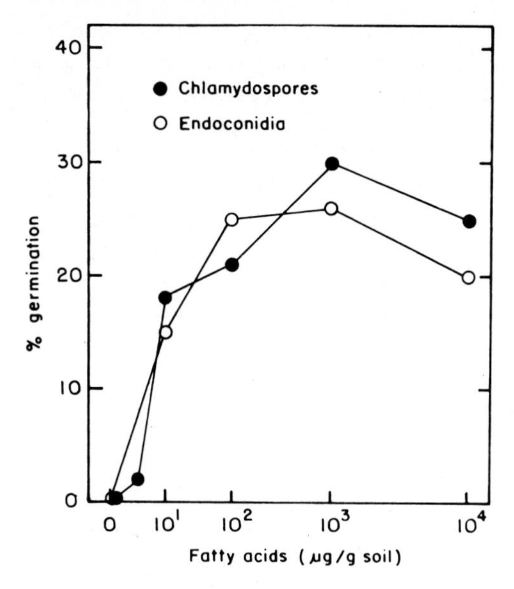

Fig. 5.4. Percent germination of endoconidia and chlamydospores of *Thielaviopsis basicola* in fungistatic soil fortified with hexane solutions of fatty acids added in concentrations from 0 to 10^4 μg g^{-1} soil. (Papavizas and Kovacs 1972)

rhizosphere of bean) were added to fungistatic soil containing spores of the fungus, germination increased with increasing concentrations of fatty acids up to 10^3 μg g^{-1} soil (Fig. 5.4). No spore germination occurred in fungistatic soil without fatty acid amendment (Papavizas and Kovacs 1972). Both root exudation and microbial processes probably serve as contributing sources of these substances in the rhizosphere.

While the volatile compounds produced by root exudates or those from products of microbial metabolism may serve as germination inhibitors, such compounds can also act as stimulators releasing some fungal propagules from the restraint of fungistasis. Volatiles that diffuse primarily from the root tips of *Allium* spp. stimulate sclerotial germination of *Sclerotium cepivorum* (Coley-Smith and Cooke 1971). This action seems derived from the metabolism of nonvolatile components of root exudates by soil bacteria resulting in the production of volatile alkyl sulfides (King and Coley-Smith 1969). Volatile compounds emanating from the ectomycorrhizal association between *Pinus sylvestris* and the Basidiomycete, *Boletus variegatus*, are of great interest because of the inhibitory effect they have upon *Phytophthora cinnamomi* and *Fomes annosus*, pathogens of pine. However, one volatile component (ethanol) produced by *B. variegatus* has been shown to stimulate both pathogens (Krupa and Nylund 1972).

The study of volatiles is in an early stage of development. The potential significance of these compounds would seem greatest in the rhizosphere where root exudates and microbial populations combined are most likely to produce them in quantities that are biologically active. Such substances of high vapor pressure can move some distance from their source and make contact in either a liquid or gaseous phase with fungal propagules in specific soil microhabitats. They may become an important link between the inhibitor and nutrient deficiency hypotheses for fungistasis.

Bacteriostasis inhibits the development of many bacteria in soils of low nutrient availability, and its persistence is governed by some of the same factors that influence soil fungistasis. Most bacteria isolated from wheat rhizosphere (98%) were found to be sensitive to bacteriostasis, whereas bacteria that were not stimulated by roots were not sensitive (M. E. Brown 1973). The effect could be nullified by wheat root exudate and sometimes by the addition of glucose or mineral salts. The phenomenon is less active in acid soils and can be removed partially by air-drying the soil and totally by sterilization. Soil bacteriostasis interferes with the inoculation of *Azotobacter* (Barea et al. 1978) and phosphobacteria (Ocampo et al. 1978) into the rhizosphere, but this effect can be overcome by soil amendment with NPK fertilizers.

5.3.3 Parasitism and Predation

The parasitic action of one microorganism upon another, and predator–prey relationships, may be considered as ultimate forms of antagonism since the host or prey can be totally destroyed. However, this usually does not occur in the case of parasitism where the host is exploited to the extent that its metabolic activities are adversely affected and growth and development are suppressed. Predators may consume entire individual organisms but rarely eradicate whole populations upon which the predator depends for continued existence. In either case, populations in a microhabitat, around organic particles or at the root–soil interface, are reduced in density or activity. Much of our knowledge about these interactions is derived from studies in artificial systems (synthetic media and sterilized soil) and can be related to a natural environment only by conjecture. Least is known about these antagonistic phenomena in the rhizosphere; nevertheless, they must be taken into account when considering factors in rhizosphere ecology capable of altering population densities or microbial vigor leading to a shift in the balance of competition between individuals or physiological groups of microorganisms.

Parasitism of one microorganism upon another is common, but the highly specific nature of parasite toward host casts some doubt upon the possibility that whole populations in the rhizosphere, or elsewhere in the soil, could be significantly changed. The most commonly recognized parasites of bacteria are phages, which either live inside the organism (endoparasites) in a stable association or they cause lysis of the cell. Bacteria are also parasitized by another bacterium, a very small vibrioid to rod-shaped Gram-negative organism of the genus *Bdellovibrio* (Stolp 1973). This unique bacterium (Fig. 5.5), measuring only 1–2 μm in length and 0.25–0.40 μm in width, was discovered accidentally and isolated in Berlin (Stolp and Petzold 1962) during a study of bacteriophage active against a phytopathogenic pseudomonad. Populations of *Bdellovibrio*, now believed to be widely distributed in soils, are made up of both parasitic and saprophytic strains; the parasites are extremely motile and attach themselves to the host by a projection from one end of the cell. This at first gave the appearance that they were ectoparasites, but in fact they are endoparasites as they penetrate and multiply intracellularly, eventually causing lysis of the host cell. Bdellovibrios are in turn subject to destruction by the lytic action of phages.

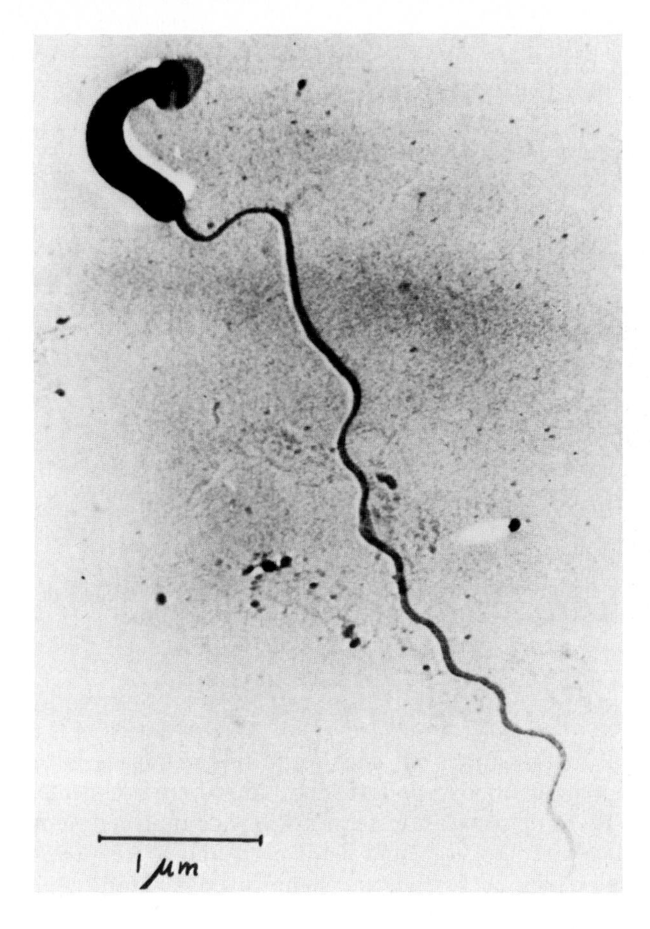

Fig. 5.5. Electron micrograph of *Bdellovibrio*. (Courtesy of H. Stolp, Lehrstuhl für Mikrobiologie, Universität Bayreuth)

While relatively little is known about the role of phages or bdellovibrios in rhizosphere ecology, their potential influence must be considered in relation to colonization of roots by *Rhizobium, Azotobacter*, phosphobacteria, plant pathogenic bacteria, and species used in seed bacterization to control pathogens or stimulate plant growth. Phages in particular may be significant in the proliferation and relative nodulating success of rhizobia. For example, a bacteriophage reduced the population of a susceptible strain of *Rhizobium trifolii* in the rhizoplane of clover and resulted in the appearance of variant substrains which were less susceptible to the phage but also less effective in symbiotic nitrogen fixation (Evans et al. 1979).

The root zone, with its relatively dense populations and varied species of the microflora, should provide many opportunities for hyperparasitism to occur. This refers to the action of one parasite on another. Examples already mentioned are phages and bdellovibrios which attack plant pathogenic bacteria, but perhaps the greatest interest has been focused upon mycoparasitism in which one fungus is parasitic upon another (Boosalis 1964; Boosalis and Mankau 1965; Lumsden 1981). Most of our knowledge of the host range, mode of parasitism, and factors

Fig. 5.6. Parasitism of *Didymella exitialis A* on *Ophiobolus* (Gaeumannomyces) *graminis; B* in the rhizosphere; *C* root fragment. × 570. (Siegle 1961)

affecting parasitism has been obtained from tests in dual culture systems in the laboratory, while ecological investigations on soil–borne mycoparasites have been delayed due to inadequate methods applicable to natural environments. However, the proven capacity of mycoparasites to attack a wide range of plant pathogenic fungi common to both above-ground plant parts and root systems provides reason to speculate on their occurrence and significance in nature.

Mycoparasitism primarily involves two groups, the destructive parasites which kill the host and the balanced parasites which inflict little damage to the mycohost. Soil–borne fungi are most frequently victims of the destructive parasites. The host range of a specific mycoparasite may be confined to a single species, a few species, or fungi of several classes; mycoparasites isolated from soil or roots are known to attack a wide range of phytopathogenic species. Lumsden (1981) has provided a list of mycoparasite–mycohost combinations that have been described from natural ecosystems. Examples are *Penicillium vermiculatum* on *Rhizoctonia solani* (Boosalis 1956), *Coniothyrium minitans* on sclerotia of *Sclerotinia trifoliorum* (Tribe 1957), and *Sporidesmium sclerotivorum* on sclerotia of *Sclerotinia minor* (Adams and Ayers 1980). Greenhouse experiments have shown that the severity of take-all disease caused by *Gaeumannomyces graminis* can be reduced about 50% by simultaneous inoculation of wheat seedlings with the pathogen and *Didymella exitialis* under sterile conditions (Siegle 1961). The mycoparasite penetrates and destroys the hyphae of *G. graminis* in the rhizosphere (Fig. 5.6). Amino acids produced by *D. exitialis* apparently are responsible in part for suppression of pathogenicity, not by inhibition of hyphal growth but by interference with nitrogen metabolism and cellulase production of the pathogen. Destructive parasites grown in dual culture on agar media or in sterilized soil can destroy all of the reproductive structures of Phycomycetes, e.g., sporangia, oogonia, oospores, and zoospores. Some plant pathogens which serve as hosts for certain mycoparasites may themselves act in a parasitic role against other fungi. For example, certain *Pythium* species can parasitize *Rhizoctonia solani*, but the latter also can parasitize several species of *Pythium* (Butler 1957).

Little direct study of this phenomenon has been made in the rhizosphere or other natural soil environments, but our knowledge of the factors that induce or influence mycoparasitism leaves little doubt that nutrients from root exudates must be important in the interaction. Both the quality and quantity of nutrients, particularly nitrogen and carbohydrate, may determine the susceptibility of the host fungus to parasitism. This can be demonstrated by varying the kinds or amounts of nutrient supplements on agar media supporting dual cultures of host and parasite.

Other forms of parasitism among soil organisms are certain to occur in the rhizosphere, but their specific activities and significance there are not known. Among these are bacterial parasites of soil nematodes and the minute sporozoan protozoa which are wide-spread parasites of plant pathogenic and other soil nematodes. Certain fungi of the Blastocladiales and Lagenidiales, as well as endozoic Hyphomycetes (species of *Verticillium, Acrostolagmus, Cephalosporium, Harposporium,* etc.) also parasitize nematodes. These and others have been described (Barron 1977; Tribe 1977; Kerry 1981 b).

Predation among soil organisms involves predator–prey interactions between the microflora and fauna and between members of the soil fauna. Whether such interactions are more intense in the vicinity of roots than in soil generally is largely undetermined, though certain indirect relationships to rhizosphere and plant growth are evident.

General usage of the terms predator and predation in the biological sciences does not always conform to the standard dictionary definition, which declares that a predator is an animal that preys upon another animal for food, and predation is the killing and eating of one species by another species. Nematode-trapping fungi are merely parasites which have been elevated to the status of predators because of a unique capicity to capture their "prey" in a network of constricting or nonconstricting mycelial rings or on short, sticky lateral branches. Most nematode-trapping fungi are Hyphomycetes with species in the genera *Arthrobotrys, Dactylaria, Dactylella,* and *Monacrosporium* (Drechsler 1941; Duddington 1956; Duddington and Wyborn 1972). The fact that predacious fungi need a carbohydrate energy source other than that provided by the nematode host (Cooke 1962) suggests a rhizosphere benefit. *Arthrobotrys oligospora* is a common species in the rhizosphere of various plants (Peterson and Katznelson 1965), but all plants do not show a rhizosphere effect in terms of higher populations than in soil generally. For example, development of the fungus is favored in soybean but not wheat rhizosphere. This difference is believed attributable to an antagonistic bacterial flora in the wheat rhizosphere and a favorable effect on growth of *A. oligospora* by bacteria in the soybean rhizosphere. Predacious fungi other than the nematophagous Hyphomycetes are found in the Zygomycetes (Order Zoopagales); these primarily attack protozoa. Further information on parasitism and predation among soil microorganisms has been provided by Cook and Baker (1983).

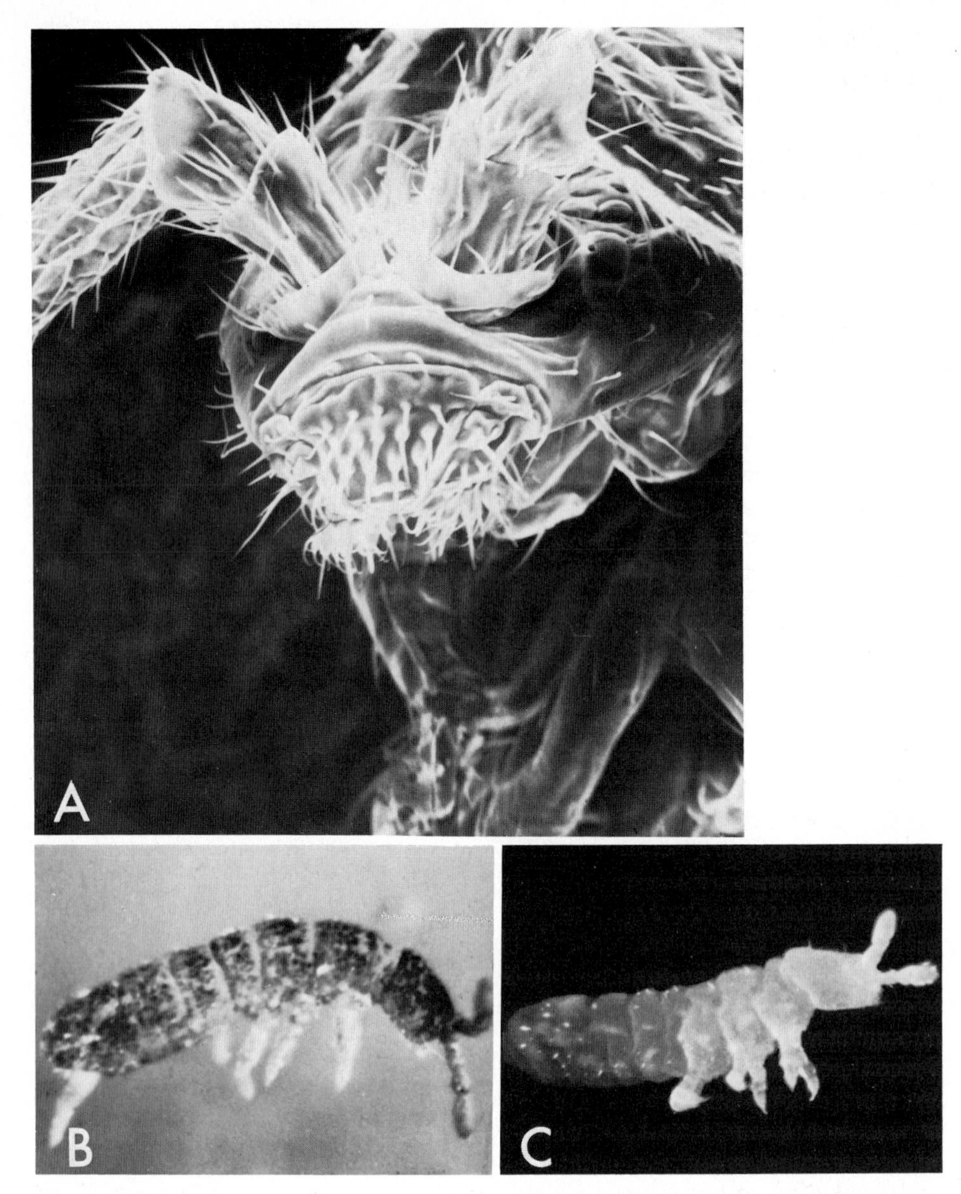

Fig. 5.7 A–C. Mycophagous *Collembola* from the rhizosphere of cotton seedlings. (A, B) *Proisotoma minuta,* average length 1 mm. (C) *Onychiurus encarpatus,* average length 1.5 mm. (Scanning electron micrograph (A) courtesy of Emilio C. Mora, Department of Poultry Science, Auburn University, AL)

Predators of soil nematodes include other nematodes, protozoa, and microarthropods. Among the small arthropods, the Acari of the Order Mesostigmata include a number of predaceous species which feed on Collembola. Instances have been cited in which the application of pesticides (such as the organochlorine insecticides BHC and DDT) to agricultural soils reduced populations of the more

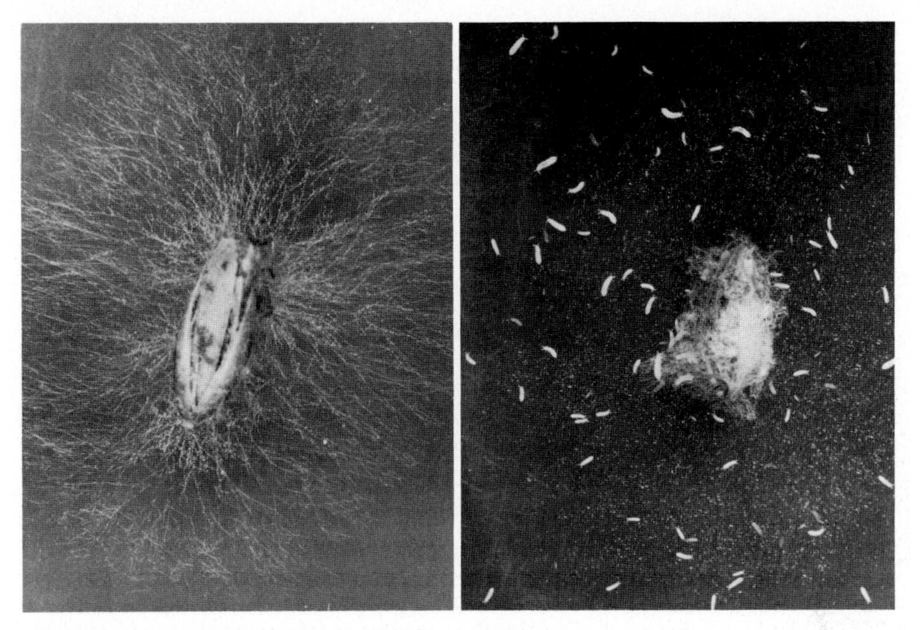

Fig. 5.8. Feeding of *Collembola* species on *Rhizoctonia solani* on oat grain (*right*) compared with young oat-fungus culture without *Collembola* (*left*). (Curl 1979)

susceptible mites and increased numbers of resistant Collembola (Edwards and Thompson 1973; Wallwork 1976). The nematicide Nemagon-20 (19% dibromochloropropane), which has strong acaricidal properties, reduced the population density of predatory mites in the rhizosphere of azaleas, whereas Collembola numbers, as a consequence of reduced predation, were 84% higher than in untreated soil (Heungens 1968).

While certain fungi trap and consume nematodes, species of several genera of soil nematodes are mycophagous and depend on fungi as a significant energy source (Norton 1978). Saprozoic nematodes also feed on bacteria and may serve as vectors of plant pathogens. Protozoa are primarely bacterial feeders and, along with bacteriophagous nematodes, tend to congregate in the rhizosphere where bacterial populations are high. Mycophagous amoebae, such as the giant amoeba *Arachnula impatiens*, are common in soil. These can perforate both hyphal and conidial walls and subsequently absorb the fungal protoplasm (Old and Patrick 1979; Alabouvette et al. 1979). Estimations have indicated that populations of amoebae are low generally but can be induced to increase by the enrichment of soil with fungal spores (T. R. Anderson and Patrick 1978).

Many Collembola and mites are also mycophagous, as evidenced by the variety of fungal spores and mycelial fragments that have been observed in their gut contents. A series of studies in Alabama has provided information suggesting that mycophagous Collembola in the rhizosphere of cotton plants may be instrumental in reducing the inoculum density of *Rhizoctonia solani* and subsequently lessening seedling-disease severity (Wiggins et al. 1979; Wiggins and Curl 1979; Curl 1979). Populations of Collembola in the genera *Proisotoma* and *Onychiurus*

Table 5.3. Effect of mycophagous Collembola on populations of fungal spores in the rhizosphere of cotton seedlings

Fungus (inoculum)	Initial spore concentration on roots (colonies plate^{-1})[a]	Colony-forming units g^{-1} soil (× 1000)[b]	
		(−) Insects	(+) Insects
Fusarium oxysporum f. sp. *vasinfectum* chlamydospores	106	104.1	39.6
Trichoderma harzianum conidia	130	318.6	89.2

[a] Seedlings in duplicate were dipped in spore suspensions and immediately washed in sterile water, which was plated on Ohio medium

[b] Seedlings in duplicate were dipped in spore suspensions, planted in sterile soil, later recovered, and rhizosphere soil dilution-plated

(Fig. 5.7) were consistently more abundant in cotton rhizosphere than in root-free soil, and species of these tested in vitro rapidly destroyed colonies of *R. solani* (Fig. 5.8). A number of other common fungi, both saprophytic and pathogenic, have been observed in or isolated from the gut contents of Collembola (Christen 1975). While these animals are generally considered to be unspecialized feeders, a certain degree of selectivity is evident. In addition to *Rhizoctonia*, they readily feed on mycelia of *Pythium* spp., *Phytophthora* spp., *Fusarium* spp., *Cylindrocladium crotalariae, Gaeumannomyces graminis,* and others. The mycelia of common saprophytes such as *Aspergillus, Penicillium*, and *Trichoderma* are not favored food sources, yet washed spores are readily consumed. Our laboratory tests with Collembola have demonstrated that populations of *T. harzianum* conidia, as well as *Fusarium oxysporum* f. sp. *vasinfectum* chlamydospores, can be greatly reduced by insect feeding in the rhizosphere of seedling cotton plants (Table 5.3). The ingested spores are redeposited in fecal pellets where some continue to germinate, but at a lower rate. Collembola may also be repelled by certain fungi. We have observed a distinct aversion of the collembolan *Onychiurus encarpatus* to the plant pathogen *Sclerotium rolfsii* in Petri-dish cultures (Curl and Snell 1981). This reaction is presently unexplained, but probably is related to an unfavorably low pH of the medium created by oxalic acid production by the fungus.

Chapter 6 Rhizosphere in Relation to Plant Nutrition and Growth

6.1 Introduction

Just as microbial populations respond to plant growth through the influences of root exudates, plants in turn may either derive benefit from or suffer the consequences of microbial activities in the rhizosphere. Thus, from seed germination to maturity the growing plant creates a root environment which affects its own development. Microorganisms, in the rhizosphere or in soil generally, can affect plant growth by influencing the following:

1. nutrient availability and uptake: mineralization, phosphates, minor elements, competition, root morphology, fauna activity;
2. nonsymbiotic nitrogen fixation;
3. symbiotic relationships: rhizobia and mycorrhizae;
4. plant responses to microbial metabolites;
5. plant pathogen activity and disease.

Effects of plants on the growth of other, contiguous, plants as a consequence of intermingling root systems also must be considered as rhizosphere effects, inasmuch as such effects can be due to root exudates, microbial activity and competition between plants for nutrients. Weeds growing among crop plants provide the most common opportunity for such interactions.

It would not serve the purpose of this chapter to review the nutrient requirements for plant growth and the fundamental processes of plant nutrition. The essential major and minor elements and their common sources are detailed in E. Walter Russell's *Soil Conditions and Plant Growth* (1973), as well as in numerous physiology texts. Also, we need not dwell on the nature and functions of the well-known symbiotic relations of nitrogen-fixing bacteria and mycorrhizal fungi, for these too have been dealt with in great detail in published symposia and texts in the past 10 years. However, in a natural environment, these growth-related processes in higher plants must proceed in the face of intense microbial activity at the root–soil interface and therefore are subject to changes induced by rhizosphere phenomena. We shall examine some of these rhizosphere-related changes.

Plant growth also is unfavorably affected by the action of plant pathogens and favorably affected by the natural biocontrol properties of pathogen-inhibiting saprophytes, mycoparasites, and predators. The relationship of the rhizosphere to plant disease will be discussed in detail in Chap. 7.

6.2 Nutrient Availability and Uptake

The rhizosphere microflora, nourished by root exudates and root debris, can affect plant growth indirectly by influencing the availability and uptake of nutrients, resulting in either benefit or detriment to the plant. According to logic, and Liebig's *Law of the Minimum*, growth of plants will be limited by any essential nutrient which is in short supply, i.e., when the ratio of supply to demand reaches a minimum. Thus, deficiences of soil nitrogen, phosphorus, or potassium are most often involved in the limitation of plant growth and frequently must be supplemented by application of commercial fertilizers. Other major and minor elements may become the limiting factors for specific plants under certain environmental conditions.

6.2.1 Mineralization from Organic Sources

The most obvious contribution of microorganisms to plant nutrition is in the decomposition of organic matter leading to the release or subsequent formation of ammonia, nitrates, sulfates, phosphates, etc. along with CO_2 and water. These activities are more intense in the rhizosphere of crop plants where the metabolic activity of organisms, as shown by measured respiration, may be as much as four times higher than in nonrhizosphere soil (Rovira and McDougall 1967).

Nitrogen is a constituent of protein and is understandably essential for plant growth. It is made available to plants through mineralization of soil organic compounds with the liberation of ammonia which is then oxidized to hydroxylamine, nitrite, and nitrate by nitrifying bacteria. The nitrifying bacteria are largely chemoautotrophs represented most commonly by the genera *Nitrosomonas*, which oxidizes ammonia to nitrite, and *Nitrobacter* which completes the oxidation process to nitrate. Many heterotrophic microorganisms, including fungi and actinomycetes, also have been shown to oxidize nitrogen compounds (Alexander 1965), but these nonobligate activities probably do not contribute significantly to the total nitrogen economy of the soil.

Soil conditions that favor nitrifying bacteria also favor plant growth, e.g., good aeration, neutral pH, and an adequate nitrogen supply. The quantity of mineral nitrogen (especially nitrate ions) in soil depends on the rate of mineralization from organic matter by microorganisms and the rate of removal by leaching or utilization by crop plants and microbial populations. The rhizosphere has a definite effect on mineralization and ammonification. Just as these processes can be accelerated by the addition of organic matter to soil, root exudates and sloughed root tissues also provide fresh organic substances that stimulate activities of the rhizosphere flora, resulting in accelerated turnover of nitrogen (Harmsen and Kolenbrander 1965; Rovira and McDougall 1967). The R/S ratio of ammonifying bacteria may exceed 50 to 1 (Rouatt et al. 1960), suggesting a high potential for release of ammonia in the rhizosphere; indeed, Starkey (1929) demonstrated this potential more than 50 years ago. With the development of more precise techniques, the relative seasonal contributions of bacteria and fungi to the mineralization processes are becoming better understood. Use of the antibiotics streptomycin and actidione for the selective inhibition of bacterial or fungal activities in combination with a rapid radioactive assay showed that bacteria played a major role in glucose mineralization in both the rhizosphere and rhizoplane of plants from a shortgrass prairie ecosystem (Nakas and Klein 1980). Glucose, together with other sugars, water-soluble nitrogenous compounds, and organic acids, represent substrates readily available for initiation of the decomposition process.

The rhizosphere effect on the soil nitrification process varies with different plant species. This might be expected, since nitrifying bacteria are very sensitive to microbial toxins and to pH changes that occur with qualitative differences in root exudates and the responding microbial activity. In some instances, numbers of *Nitrosomonas* and *Nitrobacter* have been found to increase in response to root exudates, but in other cases populations and nitrification have been inhibited or nitrogen immobilized. The finding by Goring and Clark (1948) that less nitrogen was available to plants in the rhizosphere than the amount that would have been transformed to nitrate in unplanted soil was attributed to immobilization of nitrogen by microorganisms. After removal of the roots, and thus a reduction in microbial activity, the nitrate content of the soil increased. Other work has suggested that nitrification is inhibited by grass roots, resulting in low levels of nitrate in permanent-grassland soils. Generally, the nitrifiers are not strong competitors with heterotrophs for ammonified nitrogen, yet numbers of *Nitrosomonas* and *Nitrobacter* often are higher in the rhizosphere of grain and grass crops than in nonrhizosphere soil. Smit and Woldendorp (1981) showed that nitrifying bacteria in the rhizosphere of *Plantago* spp. were not inhibited and nitrification was not suppressed.

Inorganic nitrogen compounds, in addition to being taken up by growing plants and microorganisms, or lost by leaching, can be reduced through the process of denitrification. In the presence of the required reductases and associated electron transport compounds, nitrate can be converted to gaseous nitrogen and nitrous oxide which are then lost to the atmosphere. This process is carried out by many facultatively anaerobic bacteria, most commonly in the genera *Pseudomonas, Micrococcus*, and *Bacillus*, under conditions of poor aeration such as in waterlogged soils. These organisms grow well in the oxygen atmosphere of laboratory cultures, but under limited oxygen supply they utilize nitrate as a hydrogen acceptor. While only a low level of microbiological activity is required for denitrification under low oxygen, the process also occurs readily in aerated soils when large amounts of decomposable organic matter (such as farmyard manure) are applied. In such cases, the soil oxygen is being used up by the highly intensified microbial activity at a greater rate than it is replaced by diffusion from the atmosphere (E. W. Russell 1973).

Similar activity may occur in the rhizosphere where a relatively large amount of decomposable organic matter is available. In permanent grassland, Woldendorp (1963) attributed a high rate of nitrate loss from applied fertilizer to high numbers of denitrifying bacteria in the rhizosphere, greater oxygen consumption, and the availability of hydrogen donors from amino acids in root exudates. A greater amount of gaseous nitrification in the presence of pea roots than of ryegrass roots was related to more amino acids released from pea. Experiments with intact oat plants have shown that denitrification decreased rapidly with increasing distance away from the roots (M. S. Smith and Tiedje 1979). When soil NO_3 concentration is high, the rate of denitrifaction in the rhizosphere tends to increase.

6.2.2 Phosphate Availability and Absorption by Plants

Microorganisms are involved in the availability of phosphorus, potassium, sulfur, calcium, iron, and other elements through the decomposition of organic compounds and the oxidation or reduction of inorganic compounds in soil. The more important aspects of microbial solubilization and mineralization, with emphasis on phosphates in the rhizosphere, are discussed in this section.

Different species of plants grown in similar environments may differ in chemical composition, due in part to differences in nutrient availability at the root–soil interface and the capacity of various species for nutrient absorption and utilization. The role of microbial mineralization in the release of phosphates, sulfates, and other nutrient products from organic sources is well documented. Depending upon the species of plant involved, the qualitative and quantitative nature of the microbial population present, and the experimental techniques employed, either enhanced or restricted availability and uptake of nutrient ions in nonsterile systems has been demonstrated when compared with sterile systems. Most of our present knowledge is derived from extensive work with phosphates.

Phosphorus occurs as a constituent of both organic and inorganic compounds in soil, plants, and microorganisms, performing an essential function in plant growth and the biology of soil. Microbial communities in either the rhizosphere or bulk soil regulate the phosphorus cycle through the processes of: (a) heterotrophic mineralization of organic phosphorus compounds and the regeneration of orthophosphate; (b) immobilization of inorganic phosphorus by autotrophic and heterotrophic microorganisms, resulting in a lower available phosphorus supply; and (c) solubilization of aluminum, iron, and calcium phosphates (Alexander 1977).

As extracellular mineralization by soil phosphatase of either plant or microbial origin proceeds, the regenerated phosphate is rapidly immobilized under conditions that are highly favorable for microbial activity, such as occur in the presence of root exudates. Not only the microflora, but the microfauna also may play a major role in nutrient cycling (Cole et al. 1978). Bacteria assimilate and retain labile inorganic phosphorus as carbon substrates in the rhizosphere are metabolized. This bacterial phosphorus is then mineralized and returned to the inorganic phosphorus pool by bacteriophagous amoebae. This process was suggested by Cole et al. from the results of an experiment designed to simulate biological activities in the rhizosphere by using glucose amendments to represent input of root exudates in microcosms with and without amoebae. Nematodes also may participate in the process in a similar manner but apparently less effectively. Thus, the contributing role of microorganisms in determining the availability of phosphate through mineralization or immobilization is evident.

Table 6.1. The phosphate-dissolving ability of 317 bacterial isolates obtained from phosphate-treated pots. (Louw 1970)

Source of isolates	Number tested	Percentage of the isolates dissolving		
		Dicalcium phosphate	Tricalcium phosphate	Hydroxyapatite
Uncropped soil	67	32.83	32.83	31.34
Wheat rhizosphere	66	54.54	45.45	40.91
Wheat rhizoplane	40	50.00	47.50	35.00
Lupin rhizosphere	65	58.46	58.46	43.08
Lupin rhizoplane	79	58.23	58.23	56.96

Phosphorus, and certain other essential elements, must be in solution in the immediate vicinity of roots if they are to be adequately absorbed, since the rate of diffusion of ions through soil to the roots is extremely slow. A significant role for rhizosphere microorganisms in determining the solubility of phosphates and their uptake by plants is not frequently proposed by soil scientists. Indeed, phosphate solubilization is largely a function of soil pH, the cation exchange capacity of roots, the adsorption or absorption of Ca from calcium phosphate, and the complexing of aluminum and iron by organic anions to solubilize Al and Fe phosphates (Drake and Steckel 1955; A. W. Johnston and Miller 1959; W. B. Johnston and Olsen 1972; Gardner et al. 1983).

Nevertheless, these processes may be induced by or related to the action of root exudates and the activities of microorganisms at the root–mineral interface. The role of microorganisms in the solubilization of phosphate in soil and its uptake by plants has attracted increasing attention since Gerretsen (1948) showed that phosphate uptake and the growth of oat plants in sterilized sand could be increased by adding a small amount of fertile nonsterilized soil to the sand. Subsequently, the solubilization of phosphates in soil has been attributed to a number of microorganisms isolated from the rhizosphere of plants. Sperber (1958a, b) found that organisms dissolving apatite were most commonly rhizosphere inhabitants that produced acids in liquid culture, although the degree of solubilization was not always proportional to the decline in pH. Solubilization of apatite by these microorganisms occurred under conditions of high carbohydrate level accompanied by the accumulation of end-products of metabolism, suggesting a relation to the higher carbohydrate availability in root exudates in microhabitats of the rhizosphere. Common fungi, bacteria, and actinomycetes have been implicated in the dissolution of phosphates.

A high percentage of bacterial isolates from the rhizosphere and rhizoplane of wheat and lupine indicated phosphate-solubilizing ability (Table 6.1; Louw 1970). Many of these "phosphobacteria" are Gram-negative, nonsporing, short rods as well as pleomorphic types, some of which are known to produce 2-ketogluconic acid that may contribute to the solubilization process. The root exudate of young tea plants contains appreciable amounts of malic acid which may be effective in releasing phosphates that have been immobilized as iron and aluminum salts, thus making such phosphates available for utilization by plants (Jayman and Sivasubramaniam 1975).

Moghimi et al. (1978 a) declared that under natural conditions the phosphate-dissolving power of plants depends on the presence of both root exudates and associated microbial products, together referred to as "rhizosphere products". Fractions of these rhizosphere products from wheat plants released phosphate from calcium phosphate. This dissolution of phosphate was associated with negatively charged and UV-absorbing fractions of the rhizosphere products, and also with a drop in pH of the ^{32}P-labeled hydroxyapatite suspension in which the roots were growing.

Microorganisms in the root environment not only affect nutrient availability through mineralization processes and the dissolution of relatively insoluble materials, but also contribute to the process of nutrient absorption. Nutrient uptake is closely linked with the rate of diffusion of ions through soil and the rate of their arrival at the root–soil interface (R. S. Russell 1977). Whether root exudates or microorganisms can influence the mobility of ions in the rhizosphere has not been clearly resolved. However, in root zones of exceedingly high microbial activity, it is unlikely that these factors could be of no consequence. While slow diffusion of ions is largely due to their interaction with clay mineral surfaces, these surfaces may be coated with microorganisms which affect the integrity of the mineral particles. The rate of nutrient diffusion toward the root also depends in part on the rate of uptake by the root and lowering of the concentration at the root surface. Microbial activity may be involved in this process if competition with the plant for nutrients in the rhizoplane is sufficient to accelerate the formation of a nutrient void. Interactions between intercropped plants (wheat and *Lupinus albus*) also occur below ground, resulting in wheat having access to a larger pool of available P, Mn, and N than when wheat is monocultured (Gardner and Boundy 1983). This suggests that wheat is able to take up nutrients produced or made available by the intercropped lupins, but the mechanisms involved are not yet clear. Some evidence has been presented suggesting that a chelating agent released from roots of *L. albus* acts in conjunction with reductants and hydrogen ions resulting in phosphorus mobilization (Gardner et al. 1983).

Since phosphorus uptake is related to pH across the rhizosphere-rhizoplane, knowing the factors which determine pH of the soil solution can provide clues to the mechanisms involved in nutrient absorption. A general assumption that roots make the rhizosphere soil more acid, thus increasing the availability of nutrients, does not hold true in all situations. Changes in pH and HCO_3 accumulation are largely governed by the relative uptake of cations and anions, and therefore by the source of nitrogen (NH_4 or NO_3) utilized by the plant (Riley and Barber 1969; Nye 1981). Under field conditions, plants generally utilize more NO_3-N, which would tend to increase the pH at the root surface. This is not to minimize the probability of direct effects of materials released by plant roots into the rhizosphere as uncharged compounds. These include soluble exudates, mucilage, and sloughed cells and lysates, consisting of carbohydrates, carboxylates, and amino acids and amides. Along with any undissociated acids from these compounds, other acids may be released by rhizosphere bacteria that utilize the root materials.

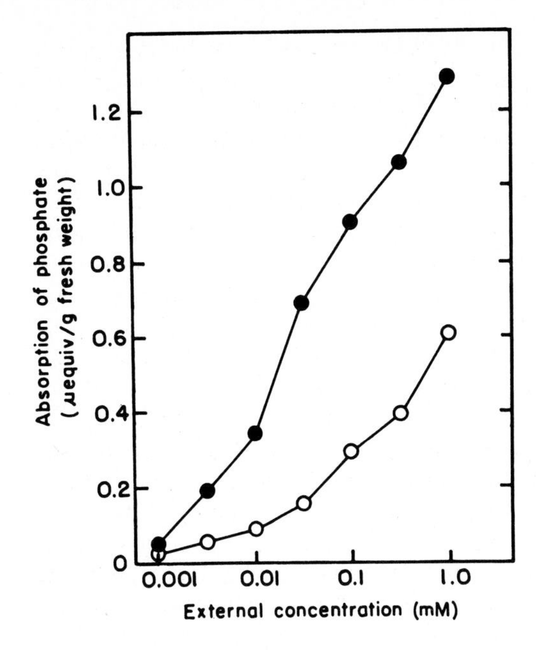

Fig. 6.1. Absorption during 1 h of phosphate from solutions of KH_2PO_4 by excised roots grown under sterile (○) and nonsterile (●) conditions. (D. A. Barber and Frankenburg 1971)

Microorganisms often are not considered when interpreting results of experimentation on nutrient uptake by plants, even though such plants may be cultured in nonsterile environments. Some specific evidence of a microbial role in nutrient absorption has been obtained with plants grown in highly artificial systems. D. A. Barber and Frankenburg (1971) cultured excised roots of barley in sterile and nonsterile solutions of KH_2PO_4 and measured the absorption of phosphate ions. Roots growing under nonsterile conditions had a greater capacity for ion uptake than roots growing in the absence of microorganisms (Fig. 6.1); furthermore, greater incorporation of phosphate into nucleic acids occurred in the presence of microorganisms. Increased phosphate uptake by plants also is likely to be associated with increased phosphatase activity by the root itself (Estermann and McLaren 1961). Soil phosphatase can be enriched up to 40% by corn roots (Boero and Thien 1979). There is some doubt whether the presence of a rhizosphere microflora can increase phosphatase activity above that normally produced by roots (Ridge and Rovira 1971; J. K. Martin 1973).

Whether microorganisms significantly affect phosphate uptake and distribution depends to a large extent on the existing concentration of phosphate in soil or in the experimental growth medium (D. A. Barber 1966; Benians and D. A. Barber 1974). When the supply of phosphate is adequate for the metabolic requirements of both the plant and microorganisms, any effect of microbial activity becomes masked and probably negligible. At low concentrations of soil phosphate, competition occurs between plant and microorganisms with a resultant restriction of phosphate uptake by the plant.

Thus, the dissolution of phosphate and its absorption by plants are related to various chemical, physical, and biotic factors of the soil environment. Although the specific mechanisms involved are still largely undetermined, the practice of inoculating seed or soil with isolated "phosphobacteria" to increase crop yields is receiving renewed attention following many years of such efforts in the Soviet Union, where considerable faith is evident in the economic value of bacterial fertilizers. Since Barea et al. (1976) found that phosphate-dissolving bacteria from rhizosphere soil produce indole-3-acetic acid, gibberellins, and cytokinin, the question also has been raised regarding the relationship between vitamin production and phosphate solubilization in the benefits to plant growth. Phosphate-dissolving isolates of bacteria from the rhizosphere and rhizoplane may be more active producers of vitamin B_{12}, riboflavin, and niacin than solubilizing isolates from nonrhizosphere soil or nonsolubilizing isolates (Baya et al. 1981).

6.2.3 Availability and Uptake of Other Elements

Microorganisms on the root surface and root hairs can affect the availability and uptake of ions other than phosphate. Both enhanced and reduced uptake of either cations or anions have been demonstrated with nonsterile plants as compared with sterile plants growing in nutrient solution or in nonsterile versus sterile soil. Along with the action of microorganisms it is to be expected that the chelating compounds in root exudates might increase the availability and uptake of minor elements such as zinc. Some observed differences between plant species in the solubility and absorption of calcium in the root zone have been attributed to root exudate effects (S. A. Barber and Ozanne 1970), probably mediated by a change in pH. Comparisons of the uptake and translocation of ^{86}Rb and ^{45}Ca in sterile and nonsterile red clover plants have shown higher contents in the tops of sterile plants (Trolldenier and Marckwordt 1962); this suggested a more efficient utilization of the elements by nonsterile plants rather than greater uptake by sterile plants. The absorption of rubidium, as well as phosphorus, is said to be greater in barley roots infested with microorganisms than in plants grown under sterile conditions (D. A. Barber and Frankenburg 1971). Thallium, which is phytotoxic at concentrations above 0.2 mM, also may be absorbed more readily by plant roots in nonsterile soil (D. A. Barber 1974). Obviously, it is essential that such information be obtained through the use of techniques which can distinguish between uptake by roots and absorption by the associated microorganisms.

More often the effect of microorganisms has been one of reduced nutrient availability or uptake by plants, reflecting the capacity of microbes to concentrate and tie up elements on the root surface, particularly at sites of increased exudation and intense microbial activity. That microorganisms influence the availability of trace elements to plants has been known for many years; Timonin (1946) first suggested that the rhizosphere microflora might be involved as an explanation for manganese deficiency in oats. He showed that a variety susceptible to manganese deficiency harbored a higher population of Mn-oxidizing bacteria in the rhizosphere than did a nonsusceptible variety.

In the Hanke's Bay area of New Zealand, vegetables grown in Napier soils contain higher concentrations of molybdenum, aluminum, and titanium and lower concentrations of barium, copper, manganese, and strontium than vegetables grown in Hastings soils (Healy et al. 1961). The marked differences in molybdenum content of harvested crops were of particular interest because of the pos-

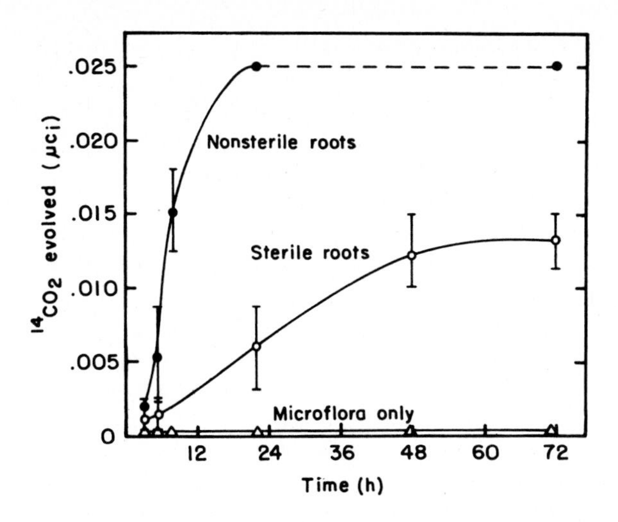

Fig. 6.2. Production of $^{14}CO_2$ from ^{14}C-glucose by sterile and nonsterile roots. (Zwarun 1972)

sible relationship to nutrition and the incidence of dental caries in these areas. Subsequently, differences were shown in molybdenum uptake by microorganisms isolated from nonrhizosphere soil and from the rhizosphere of radish grown in Napier and Hastings soils (Loutit et al. 1967). Organisms from the rhizosphere of plants in Hastings soil concentrated more molybdenum than those from the rhizosphere of Napier soil and more than organisms from root-free soil. Thus, molybdenum concentration may be different for the same plant species grown under similar conditions in two different soils. Differences in kinds rather than numbers of bacteria seem to be responsible. The molybdenum may be bound in the extracellular slime of rhizosphere bacteria, thus preventing free entry of the element into roots (E. L. Tan and Loutit 1976). There is evidence to show that microbial cells enveloped in a polysaccharide slime can absorb 20 times the amount of ions as cells without the slime (Rorem 1955). This kind of immobilization of nutrients may be accompanied by competition for elements between microbe and plant; a more rapid assimilation of nutrients by microorganisms with their short generation times can create a deficiency for the plant. Upon the death of microbial cells, the nutrients are released and again become available for absorption by either microorganisms or plant roots. Absorption by plants is most likely to be restricted by microbial competition where an element is in short supply, except where mycorrhizal associations enhance the absorptive capacity of the plant.

Our knowledge that microorganisms on roots and in solutions can limit the uptake of materials by plants through competitive action has been based largely upon experiments with inorganic salts (D. A. Barber 1968) which dissociate to give charged particles subject to ion exchange. Thus, some microbial competition is due to adsorption of nutrient ions on microbial cell walls without absorption into the cytoplasm. Less information is available on competition for organic molecules. Competition for glucose between excised soybean roots and root surface microorganisms was readily demonstrated where the criterion for comparison was the amount of $^{14}CO_2$ evolved from the metabolism of ^{14}C-glucose by sterile and nonsterile soybean roots (Zwarun 1972). The nonsterile roots produced

$^{14}CO_2$ rapidly and in much higher quantity than either sterile roots or the microflora alone (Fig. 6.2).

6.2.4 Microorganism Effects on Root Morphology

The absorptive capacity of roots is related to the density of the root system, the total surface area, and the volume of soil occupied by roots and root hairs. These features are governed by the kind and age of plant, soil type, moisture, fertilization, and other factors. Root surface microorganisms also affect root morphology and subsequently enhance or reduce nutrient absorption. Root stunting and retarded root-hair development have been observed in several crop plants following exposure of root systems to soil-water suspensions (Bowen and Rovira 1961); the effects did not occur when dilute suspensions (reduced numbers of microorganisms) were used as inocula. Pea plants inoculated with a species of *Pseudomonas* which was isolated from the rhizosphere of timothy grass, showed a lack of root-hair development and some damage to epidermal and cortical cells (Darbyshire and Greaves 1970). Studies with corn have shown that only certain soil organisms induce marked morphological and metabolic changes in seedlings; root development was inhibited by soil microorganisms but unaffected by aerial contaminants (Williamson and Wyn Jones 1973). In this case, even though root development was suppressed in the presence of soil microbes (compared with sterile soil), ^{86}Rb uptake and translocation were enhanced. This seems to relate nutrient absorption to some action of specific microorganisms rather than strictly to modified root morphology.

In addition to the well-known plant associations with free-living nitrogen-fixing bacteria, legume rhizobia, and the mycorrhizal fungi, noninfecting rhizosphere microorganisms also can increase the absorptive capacity of plants by stimulating the development of "proteoid roots" (Malajczuk and Bowen 1974; Lamont and McComb 1974). These are dense clusters of roots which are not formed in sterile soil. They apparently are induced by stimuli from specific microorganisms which inhabit the rhizosphere of such root systems, or which proliferate in the upper soil horizon rich in organic matter.

6.2.5 Fauna Influence on Nutrient Uptake

Small fauna in the rhizosphere can influence nutrient availability and uptake by plants indirectly through their predatory action upon the microflora. Our first thought might be that bacteriophagous protozoa and nematodes could consume enough bacteria to interfere with the normal mineralization of nutrients. However, it is more likely that such feeding will liberate nutrients immobilized in bacterial cells and accelerate the mineralization process. This has been demonstrated in gnotobiotic microcosms where soils containing both amoebae and bacteria, or nematodes and bacteria, mineralized significantly more NH_4-N and inorganic P than soils with bacteria alone (Coleman et al. 1977; Elliott et al. 1979). We noted in Chap. 4 that the rhizosphere is a highly favorable habitat for the rapid build-up

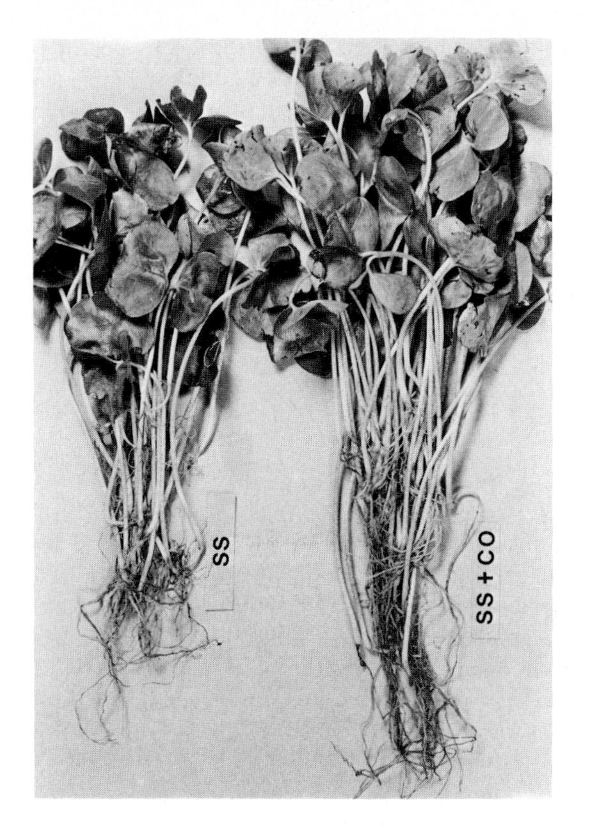

Fig. 6.3. Cotton seedlings grown in sterilized soil (*SS*) without mycophagous *Collembola,* compared with those grown in sterilized soil with *Collembola* (*SS*+*CO*). (Wiggins and Curl 1979)

of protozoan populations and for the congregation of saprophagous nematodes. A more in depth review of animal–microbial interactions relating to soil-nutrient transformations in the rhizosphere is provided by Coleman et al. (1984). A number of studies conducted in controlled microcosms leave little doubt that bacteriophagous amoebae and nematodes play important roles in determining the availability of phosphorus, nitrogen, and perhaps other major nutrients, as well as growth factors, for plant utilization.

Populations of microphagous small arthropods (Acari and Collembola) are especially abundant in habitats of dense, fibrous root systems, suggesting a close relation to roots for feeding and reproduction (Chap. 4). The common occurrence of bacteria, fungal spores, and mycelial fragments among their gut contents is evidence that they consume a portion of the soil microflora. The Collembola are attracted to living roots, and they can transport bacteria and fungal spores on their bristled bodies into the rhizosphere (Wiggins and Curl 1979). These activities suggest a potential for altering the quantitative and qualitative nature of the microflora around roots, but we can only speculate at this time that nutrient availability and plant growth could be indirectly affected. In controlled experiments, cotton seedlings initiated from surface-disinfested seed and grown in sterilized (autoclaved) soil grew 3 cm taller when field-collected Collembola were added than in soil maintained sterile with uncontaminated plants (Fig. 6.3). One or more of the following activities could explain the stimulated plant growth: (a) insect-transported bacteria proliferating at the root–soil interface released additional nitrogen or phosphorus for plant absorption, (b) the bacteria synthesized plant-growth stimulating factors, or (c) microbial degradation of toxins formed during heat sterilization of the soil released the plants from inhibition and promoted their growth. In either case the insects probably served only as vehicles for the microflora. It seems likely that such transport could be significant in a natural soil environment where specific microorganisms are shifted from one microhabitat to another habitat (the root surface) which is more favorable for rapid growth.

6.3 Nonsymbiotic Nitrogen Fixation

Since the isolation of *Clostridium pasteurianum* by Winogradski in 1893 and the subsequent isolation of *Azotobacter chroococcum* and *Azomonas agilis* by Beijerinck in 1901, other free-living microorganisms, including a number of blue-green algae and yeasts, have been shown to convert atmospheric nitrogen to ammonia which is then available to plants. Bacteria of the genus *Beijerinckia* were initially described as belonging to *Azotobacter* but were later separated on the basis of morphological and nutritional differences. These two genera, along with *Clostridium*, now comprise the major groups of free-living nitrogen-fixing bacteria, though species of *Bacillus, Azospirillum, Pseudomonas, Klebsiella,* and other bacteria are known to fix nitrogen in the rhizosphere of plants. The nitrogen-fixing process is similar for all species, and involves an ATP-driven nitrogenase. This reaction can also catalyze the conversion of acetylene to ethylene which is easily measured, hence the rate of acetylene reduction has become much used to assess the activity of nitrogen-fixing enzyme systems (Mulder 1975; Dart and Day 1975). The product of nitrogen fixation is ammonia which is then incorporated into amino acids and proteins. When this activity occurs in the rhizosphere the fixed nitrogen is immediately available to plants for the synthesis of organic compounds.

Azotobacter species are the most commonly known of the free-living nitrogen-fixing bacteria and have been credited by some as playing a major role in the nitrogen economy of soil. The soil species are Gram-negative, obligate aerobes which grow well on nitrogen-deficient media and produce a capsular slime when grown in the presence of glucose as a carbon source. These bacteria are widespread but usually low in numbers except in the rhizosphere where a two- to three-fold increase may occur. Even these numbers, however, are relatively small for rhizosphere populations of bacteria, and their occurrence on the root surface (rhizoplane) has been observed infrequently.

Mulder's (1975) discussion of the physiology and ecology of free-living, nitrogen-fixing bacteria provides several examples of actual and potential influences of the rhizosphere on activities of organisms. Among their requirements for functioning in soil are: (a) available carbon compounds; (b) adequate amounts of inorganic nutrients such as phosphorus, calcium, magnesium, molybdenum, etc.: (c) optimal pH for growth; and (d) a relatively low oxygen supply. From what we already know about root exudates and the rhizosphere, it is apparent that roots of specific plants may determine the suitability of the environment for nitrogen fixers. High amounts of available carbon compounds and low amounts of combined nitrogen (high C/N ratio) favor free-living nitrogen fixers. Growth of most *Azotobacter* species is favored by a neutral or alkaline medium. Oxygen affects all free-living nitrogen-fixing bacteria, the anaerobic, facultatively anaerobic, and aerobic types. Though *Azotobacter* is obligately aerobic, nitrogen fixation occurs most readily in lower soil layers or specific environments of relatively low oxygen content.

Azospirillum, also aerobic, can inhabit the cortical layers of tissue and utilize the root exudate energy source while benefiting the plant by substantial nitrogen-fixing activity (Atlas and Bartha 1981). Species of *Beijerinckia* grow aerobically and are most common in tropical acid soils. *Clostridium* species are intermediate in pH requirement and grow under anaerobic conditions in many soils. The nitrogen-fixing clostridia are primarily saccharolytic bacteria which utilize carbohydrates and organic acids as carbon sources.

Since nitrogen-fixing bacteria are stimulated and nitrogenase activity increases in the presence of high soil organic matter, it is not surprising that such activity is prominently associated with the rhizosphere of many plants. Though the significance of nonsymbiotic nitrogen fixation in natural environments is often questioned, it seems certain that this process contributes considerably to the maintenance of soil fertility in tropical agriculture. Rice can be grown continuously for many years in the same fields in Asia without supplemental nitrogen fertilizer, because the nitrogen removed by the crop is replaced in part by nonsymbiotic nitrogen fixation. Both aerobic and anaerobic nitrogen-fixing bacteria usually are more abundant in the rhizosphere of rice than distant from the roots (Balandreau et al. 1975). In wild rice (*Zizania aquatica*), fixation of nitrogen in the upper rhizosphere soil appears to be primarily of algal origin involving the blue-green algae, *Aphanizomenon* and *Oscillatoria*, whereas *Azotobacter* and *Clostridium* are major contributors on root surfaces at lower soil dephts (Ogan 1979). Nitrogenase activity occurs in association with many monocotyledonous plants (Dart and Day 1975) which stimulates the multiplication of nitrogen-fixing bacteria in the rhizosphere. Counts of *Beijerinckia* in soil samples from the rhizosphere and rhizoplane of sugarcane have indicated populations 20 and 50 times higher than in root-free soil. *Azotobacter* may require several weeks or months to establish the association with plant roots, and populations are generally lower than those of the other major nitrogen fixers. Yet well-known associations such as that of *A. paspali* with *Paspalum notatum* (Dobereiner et al. 1972; Dobereiner and Day 1975), which has been reported as fixing 250 g N $ha^{-1}day^{-1}$, suggest that significant nitrogen fixation can occur under specific environmental conditions. Data from the studies of Macrae (1975) show a clear rhizosphere effect with respect to nonsymbiotic nitrogen-fixing microorganisms associated with rice seedlings grown in either flooded or nonflooded soil (Table 6.2). Supplemental nitrogen applied as either NH_4 or NO_3 inhibited nonsymbiotic nitrogen fixation in the rhizosphere.

Nitrogen fixation by free-living organisms also occurs in the rhizospheres of other plants growing in aquatic environments. The freshwater angiosperms *Gly-*

Table 6.2. Effect of applied nitrogen upon nitrogenase activity in the rhizophere soil of rice seedlings grown in flooded and nonflooded soil. (Macrae 1975)

Form of nitrogen	Application rate	C_2H_4 production[a] (nmol g^{-1} soil d^{-1})	
	(parts/10^6)	Flooded	Nonflooded
"Native"	–	59.00 (0.03)	22.40 (0.04)
NH_4–N	50	6.20 (0.03)	0.93 (0.03)
NH_4–N	200	0.35 (0.03)	0.65 (0.03)
NO_3–N	50	0.47 (0.03)	0.73 (0.03)
NO_3–N	200	2.92 (0.18)	0.48 (0.03)
		LSD 49.75	

[a] Figures in parentheses represent C_2H_2 reduction activity in non-rhizosphere soil

ceria borealis and a *Typha* species collected from aquatic habitats and cultured in a controlled system showed a marked stimulation of acetylene reduction as compared with sediments without roots or rhizomes (Bristow 1974).

Along with accrued knowledge of the activities of nonsymbiotic nitrogen-fixing bacteria in the rhizosphere of plants, much interest has developed in the potential for application and management of these organisms to promote growth and vigor in nonleguminous crops. *Azotobacter* in particular has been studied for many years in the Soviet Union as a soil or seed inoculant with resultant claims of significant increases in grain-crop yields. Attempts by some other scientists have failed to reproduce these results, but several studies have provided sufficient evidence that free-living nitrogen-fixing bacteria can be established in the rhizosphere by seed inoculation, subsequently benefiting plant growth. Recent advances related to this will be discussed in Chap. 8.

6.4 Symbiotic Relationships

The most common symbiotic alliances between plant roots and microorganisms involve *Rhizobium* in the nodulation of legume roots, and various fungi which form either ectomycorrhizae or vesicular-arbuscular endomycorrhizae in many plants.

6.4.1 Legume–Rhizobium Interactions

Rhizobium spp. are essentially rhizosphere and root inhabitants which, in the absence of a suitable host plant, decline over a period of several years due to their lower competitive ability against less specialized soil saprophytes. Opportunites for the biotic inhibition of rhizobia in the rhizosphere are numerous, as evidenced by the antibiotic action of soil extracts and the antagonism or lysis of *Rhizobium* by other rhizosphere bacteria, actinomycetes, and fungi (Hely et al. 1957; Nutman 1975). However, it is unlikely that these bacteria, under normal agricultural systems, ever decline to a point of extinction since they are stimulated to multiply by root exudates of both susceptible and nonsuceptible plants. In soil deficient in available nitrogen, but otherwise favorable for plant growth, factors which oppose the establishment and multiplication of *Rhizobium* in the rhizosphere, and subsequent formation or function of nodules, can be expected to influence plant growth merely by denial of an adequate nitrogen supply.

The occurrence of *Rhizobium* bacteriophages is believed to be as widespread as the host bacteria, and apparently they are most abundant in the rhizosphere of leguminous plants (E.K. Allen and O.N. Allen 1958). Thus, their prevalence in the rhizosphere would be expected to determine to some extent the nature of *Rhizobium* populations, under some conditions reducing population levels below what are required for adequate infection (Evans et al. 1979). Bdellovibrios are minute bacterial parasites of other bacteria including *Rhizobium* (Stolp 1973). Their culture filtrates are active in the lysis of *R. meliloti* and *R. trifolii*, as well as the plant pathogen *Agrobacterium tumefaciens* (Parker and Grove 1970). Some strains of *Rhizobium* also are known to be lysed by soil myxobacteria. To these adversities must be added the predacious activities of animal agents such as bacteriophagous amoebae and nematodes, which probably play some role in the disfunction of rhizobia at the root surface.

Nodulation and effective root-nodule function are sensitive to low pH, calcium deficiency, aluminum and manganese toxicity, and many other factors of the soil environment that affect host physiology (Vincent 1965; Munns 1978). Some rhizosphere effects are induced by agricultural practices. Fertilizers may be toxic to the rhizobia because of acidity from superphosphate, contact with heavy-metal trace elements, or because of local osmotic effects (Vincent 1965). Spraying urea on leaves can affect nodulation adversely. Foliar sprays of 1% aqueous urea on *Phaseolus vulgaris, Vicia sativa,* and *Pisum sativum* applied three times weekly either prevented or reduced nodule development; nodulation in other legumes also was reduced or delayed (Cartwright and Snow 1962). Rhizobia are sensitive to certain concentrations of fungicides, herbicides, and insecticides, but the specific mechanisms by which these effects are mediated through altered rhizosphere activity have not been determined.

The foregoing account represents largely the gross effects of natural or induced environmental changes on rhizobia and nodule formation. As observed for many other microbial activities in natural soil, the precise effects of specific microbial, chemical, and physical factors on the function of *Rhizobium* have not been determined. Nodulation in legumes often fails to occur despite evidence of rhizosphere competence of rhizobia, i.e., the establishment of adequate populations in the root vicinity. Following the "arrival" of *Rhizobium* at the root surface, whether by specific stimulatory root exudates (Nutman 1965) or by chemotaxis (Currier and Strobel 1976; Kush and Dadarwal 1981), the most intriguing questions regarding rhizosphere-rhizoplane influence on establishment of the *Rhizobium*–host partnership remain unanswered. After arrival, the next step required in the process is cellular recognition, the cell-to-cell communication resulting in specific biochemical, physiological, or morphological responses between bacterium and host (Clarke and Knox 1978). It is currently believed that host plant lectins (carbohydrate-binding proteins) interact with microbial cell surface carbohydrates to determine recognition, host specificity, and attachment. Root-hair curling and infection thread formation then follow, as *Rhizobium* cells are transported from the root surface into the cortex. All of these phenomena relating to rhizosphere competence, recognition, and attachment of rhizobia to the root have been discussed in great detail (E. L. Schmidt 1979; Bauer 1981), revealing numerous gaps in our knowledge of the specific roles of root–soil components of the microenvironment in the pre-nodulation process and their consequences to plant growth.

6.4.2 Rhizosphere Effect on Mycorrhizae

Many soil fungi that form symbiotic associations with plant roots enhance the growth of plants by increasing their nutrient absorption capacity. Species in several genera of Basidiomycetes form ectomycorrhizae primarily with trees in the families Pinaceae, Betulaceae, Myrtaceae, and Fagaceae. In this association, short lateral roots become swollen with increased branching (Fig. 6.4). A fungal sheath or mantle forms over the roots (Fig. 6.5) and hyphae penetrate between cells of the outer cortex, forming the Hartig net which establishes a close cellular association in which nutrient exchange can occur. The bulk of the mycelium is in the rhizoplane and rhizosphere but extends beyond these zones of intense competition for nutrients, drawing upon additional sources of nitrogen, phosphorus, and potassium. The ectomycorrhiza-forming fungi liberate growth hormones and growth regulators which induce morphological changes (branched and swollen rootlets) resulting in an increase in surface area and absorptive capacity of the root system. In addition, some fungal symbionts, such as *Suillus luteus* and *Amanita rubescens*, apparently produce growth-promoting metabolites that directly affect the development of tree seedlings (Slankis 1973).

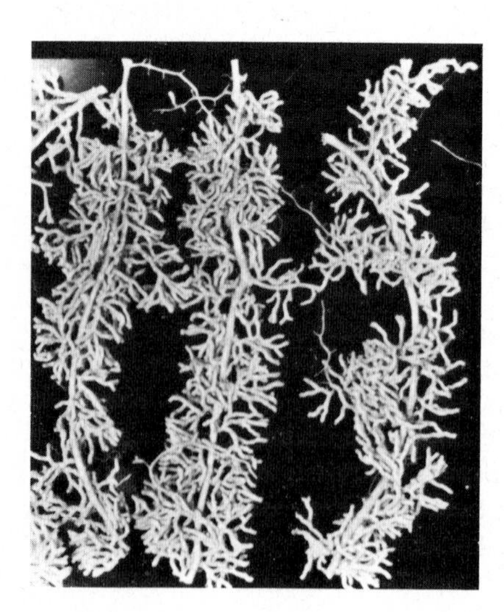

Fig. 6.4. Typical ectomycorrhizal roots of loblolly pine with *Pisolithus tinctorius*. (Photograph by Craig Bryan, USDA Forestry Sciences Laboratory, Athens, GA)

Endomycorrhizae, particularly of the vesicular-arbuscular (VA) type, formed largely by Zygomycete species in the genera *Acaulospora, Glomus*, and *Gigaspora*, can be found associated with virtually all crop plants. The hyphal network extends from the root into the rhizosphere and beyond (Fig. 6.6) but does not form a mantle around the roots. The fungal symbiont forms vesicles (terminal hyphal

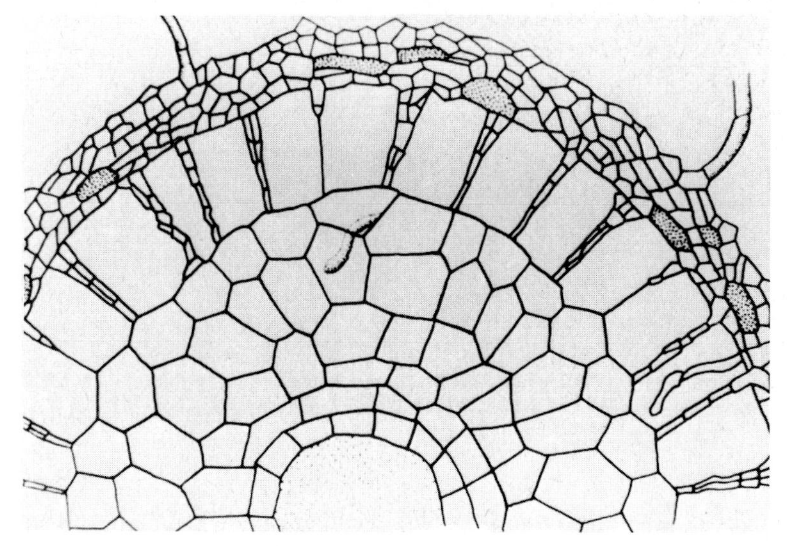

Fig. 6.5. Cross-section of ectomycorrhizal rootlet showing the exterior fungal sheath or mantle and intercellular penetration. (Harley 1965)

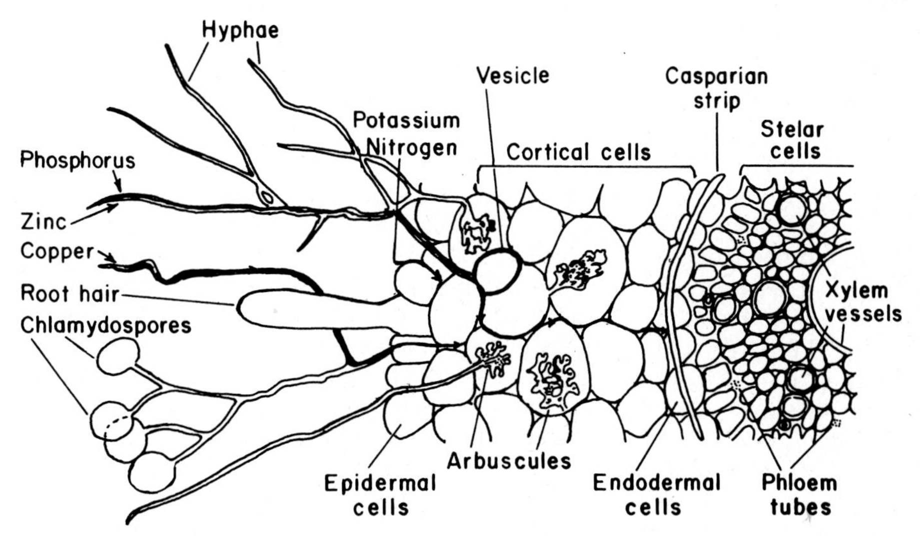

Fig. 6.6. A typical endomycorrhiza showing hyphae extending beyond the root epidermis into the rhizosphere. Intracellular arbuscules are haustoria-like structures. (Redrawn, permission of J. A. Menge, University of California, Riverside, CA)

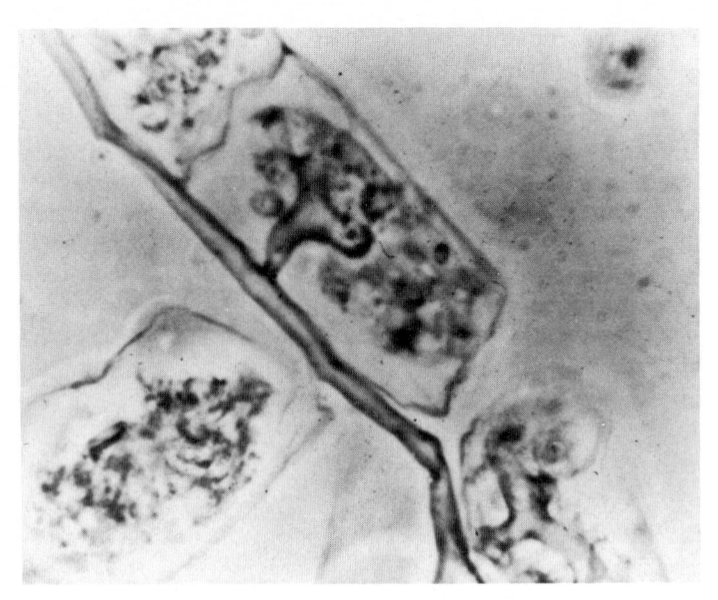

Fig. 6.7. Intercellular hypha with intracellular arbuscules formed by the endomycorrhizal fungus *Gigaspora margarita* on a cotton root. (Courtesy of R. W. Roncadori, University of Georgia, Athens, GA)

swellings thought to be storage organs) and intracellular haustoria-like structures called arbuscules (Fig. 6.7). Like the ectomycorrhizae, the VA mycorrhizae increase the nutrient absorptive capacity of plants, particularly in phosphorus-deficient soils or where phosphorus is not readily available to the plants. Where either type of mycorrhiza is present seedlings are usually better developed and produce a greater total dry weight.

Since it is well established that mycorrhizal fungi affect plant growth, then any rhizosphere effect on the mycorrhizal association may be expected to alter plant development. The major rhizosphere-related factors or processes affecting the initiation, development, and function of mycorrhizae are root exudates and interactions of the fungal symbiont with other microorganisms (antagonism or stimulation). Light, temperature, soil moisture, and soil fertility indirectly affect mycorrhizal morphogenesis largely because they affect root physiology and morphology, the quality and quantity of root exudates, and microbial activity (Marks and Foster 1973; Mosse 1973; Moawad 1979). An interesting relationship is one often observed between the extent of root-hair development and the formation of endomycorrhizae. Plants with few or no root hairs show a low capacity for nutrient absorption and tend to form mycorrhizae more readily than plants with abundant root hairs (Baylis 1972). For example, many trees and shrubs of the moist podocarp forests of New Zealand seem to be deficient in root hairs and dependent on the fungal symbiont for phosphorus.

A poorly understood relationship, but one probably of great significance, is the role of ectomycorrhizae as biological deterrents to feeder root infection by pathogenic species of *Phytophthora, Pythium, Fusarium*, and other fungi which reduce plant growth (Marx 1973). If seedling infection by a pathogen occurs before the mycorrhizal fungus invades the root, the symbiotic association may not form. But the formation of a mycorrhizal mantle over roots before contact with a pathogen may serve as a protective barrier and lessen the chance of fungal or nematode-induced disease. On the other hand, mycophagous nematodes and perhaps other micro- or mesofauna may destroy the fungal symbiont.

There is evidence indicating that certain pesticides used to control plant parasitic nematodes or fungi can create a rhizosphere environment favorable for infection and sporulation by VA mycorrhizal fungi. This is probably in response to increased root exudation or possibly due to the reduction of competitors and antagonists of the mycorrhizal fungus (Menge et al. 1979).

Aside from the influences of root exudates and the general rhizosphere population on mycorrhizae and plant growth, other work suggests an interesting relationship between VA mycorrhizae and nitrogen-fixing bacteria. In rubber plantations, where legume plant covers are rarely fertilized, levels of available phosphorus in the soil are usually suboptimal for plant growth. In such areas, it appears that VA mycorrhizal inoculation is an essential pre-condition for satisfactory growth of legumes (Waidyanatha et al. 1979). Growth and nodulation of *Pueraria phaseoloides* and *Stylosanthes guyanensis* in methyl bromide-treated soil were severely retarded unless the plants were inoculated with mycorrhizal fungi along with *Rhizobium*, or given large amounts of rock phosphate. Dual inoculation with *Glomus fasciculatum* and *Rhizobium japonicum* can significantly increase the number, dry weight, and nitrogen content of root nodules in soybean and enhance plant growth over that in plants with either symbiont alone (Bagyaraj et al. 1979). A synergistic interaction between *G. fasciculatum* and *Azotobacter chroococcum* also has been observed in inoculated tomato plants (Bagyaraj and Menge 1978).

6.5 Microbial Metabolites

Plants respond to specific microorganisms applied to seeds or roots. The response most often reflects a growth-promoting effect, but growth inhibition also occurs in the presence of nonparasitic bacteria or fungi. Such effects have been observed many times with various organisms and crop plants (Mishustin and Naumova 1962; M. E. Brown 1974; Merriman et al. 1974; Kloepper et al. 1980; Harper and Lynch 1980; Suslow and Schroth 1982a, b). Though the mechanisms responsible for the observed effects are not entirely clear, it is likely that they are related to a combination of factors: increased availability and absorption of nutrients, biological activity against pathogens (see Chap. 7), and the production of growth-promoting or growth-inhibiting metabolites by rhizosphere microorganisms.

6.5.1 Growth-Promoting Factors

Plant responses to bacterial inoculation of seed usually appear as: (a) increased vegetative growth, (b) early flowering, (c) change in root-to-shoot weight ratio, and (d) increased yields. While some of these effects on plant growth can be attributed largely to nitrogen-fixing activities where *Azotobacter* is used, this organism, along with a wide range of other microorganisms, can also produce growth-regulating substances in the root zone (M. E. Brown 1975). Microorganisms in the rhizosphere and rhizoplane of wheat release growth regulating substances with the properties of indole-3-acetic acid (IAA) and the gibberellins (Rivière 1963; M. E. Brown 1972), which can be readily absorbed in the region of root-hair development. Tryptophan (an IAA precursor), which can be supplied by root exudates and probably as a microbial metabolite or by microbial autolysis, promotes maximum IAA production. Vitamins also are synthesized by microorganisms in the rhizosphere and have a definite role in the growth of plants. Bacteria-inoculated plants contain more B-group vitamins than sterile plants (Rempe 1972).

Plant-growth-promoting factors are probably involved in many instances where the mechanisms have not been determined. *Bacillus subtilis* and *Streptomyces griseus*, when applied to seeds of barley, oat, wheat, or carrot, can induce increases in marketable yields of the crops (Merriman et al. 1974). These organisms are antagonistic to *Rhizoctonia solani*, but since seed bacterization does not control the pathogen and disease incidence, the observed benefits to plant development must be due to other factors, probably including growth factors synthesized by the applied microorganisms. In other cases, growth-stimulating bacteria in the rhizosphere are known to inhibit weakly pathogenic bacteria and fungi (Suslow and Schroth 1982a, b). Thus, in a natural soil environment, it seems likely that plant growth is affected both by microbially synthesized growth factors and the competitive interactions of growth-promoting versus deleterious microorganisms at the root surface.

Though bacteria are most frequently implicated in the production of substances affecting plant growth, a number of fungi isolated from plant rhizospheres also synthesize auxin and gibberellins in culture (Youssef and Mankarios 1975; Kampert et al. 1975). Thus, the phenomenon seems to be quite common and offers interesting possibilities for increasing plant growth and yields. However, many mechanisms of action upon plants remain to be explained, and practical methods for promoting the multiplication of growth-benefiting microorganisms on the root surface while excluding the growth-inhibiting organisms are in a developmental state. Even among plant-growth-promoting microorganisms some species induce undesirable changes in root morphology. For example, some species of *Pseudomonas* stimulate outgrowths on storage organs such as sweet potato (Goto and Makino 1976), somewhat resembling the gall-forming action of *Agrobacterium tumefaciens*.

6.5.2 Growth Inhibition

Some nonparasitic microorganisms in the rhizosphere can inhibit growth of young plants, as evidenced in bioassays with pea seedlings and lettuce hypocotyls (M. E. Brown 1972) and in tests with corn seedlings (Rempe 1972). Where such plants are grown in water-culture solutions with microorganisms, high concentrations of auxins and vitamins occur in the solutions and in the plants. These and other compounds, generally viewed as growth-promoting factors, may inhibit or catalyze the inhibition of plants when such compounds are present in certain concentrations. A relatively low concentration of IAA applied to pea roots can induce the formation of ethylene with subsequent inhibition of root extension (Chadwick and Burg 1967). Ethylene is a known inhibitor of tissue growth in plants, though it also has been known to stimulate growth, particularly root-hair development. It may occur in anaerobic or partially anaerobic soils at concentrations injurious to plants. Split-root experiments with barley plants in solution culture have revealed that the growth of roots exposed to ethylene becomes modified, whereas growth in an ethylene-free environment is unaffected (Crossett and Campbell 1975). Species vary considerably in sensitivity of their roots to ethylene, usually depending on their tolerance to waterlogged (anaerobic) environments.

Among other products of microbial origin that adversely affect root development are nitrites and antibiotic compounds. Rice plants are sensitive to nitrite, which is produced from nitrates by bacterial action and accumulates in the rhizoplane in quantities sufficient to suppress plant growth in paddy soils (Asanuma et al. 1980). Trichodermin, the sesquiterpene antibiotic isolated from *Trichoderma viride*, inhibits the growth of many fungi, and also is a potent inhibitor of plant growth (Cutler and Lefiles 1978). *Trichoderma* spp. are common inhabitants of plant rhizospheres.

While some microorganisms in the rhizosphere produce plant-growth inhibitors, other species may counter this action by degrading the inhibitory compound. Anaerobic organisms such as *Desulfovibrio* spp. are abundant in the submerged soils of rice paddies where they produce high concentrations of hydrogen sulfide by reduction of sulfate; hydrogen sulfide is toxic to rice. However, a catalase-like enzyme activity around the root tips of rice plants favors the growth of *Beggiatoa*, a filamentous bacterium capable of oxidizing the hydrogen sulfide to sulfur (Pitts et al. 1972). Any toxic peroxides produced by *Beggiatoa* are believed to be decomposed by rice-root catalase. Further evidence of the ability of soil microorganisms to detoxify plant-inhibiting substances is suggested by the apparent removal of phytotoxins from heat-sterilized soil following reinfestation of the soil with dilute nonsterile soil suspensions or with pure cultures of bacteria (Rovira and Bowen 1966). Such detoxication by microorganisms could be misinterpreted in experiments as being a direct stimulation of plant growth by microbial metabolites.

6.6 Plant Effects on Other Plants

Since De Candolle (1832) suggested that certain species of plants excrete from their roots growth inhibitors that are injurious to other plants (toxic theory), a great deal of study has been made of the interactions of plants growing in close proximity. Substances released by one plant may directly or indirectly influence the growth of another plant either by stimulation or inhibition. Molisch (1937) used the term allelopathy to refer to both beneficial and detrimental biochemical interactions between all kinds of plants including microorganisms, but comprehensive reviews on the subject (Tukey 1969; Rice 1974, 1979) generally reflect adoption of a more restricted definition involving only harmful effects by one plant on another through the production of chemical compounds released into the environment. Thus, any effects due to competition for space, water, nutrient, etc. are excluded. In consideration of the rhizosphere effect of one plant on growth of another, we must take into account both stimulation of growth and allelopathy, keeping in mind that these effects can be mediated through the direct action of root-exudate components or synthesis of secondary compounds by microorganisms. These then can be taken up by the plant.

6.6.1 Growth Stimulation

The most celebrated examples of the stimulation of one plant by the roots of another are provided by the action of host-root exudates on the germination of seeds of plant-parasitic phanerogams. Seeds of broomrape (*Orobanche ramosa*), a root parasite of solanaceous and leguminous crops, germinate when root exudates from host plants are released near the seeds (Koch 1887). Gibberellins which have been isolated from both higher plants and fungi, and which are also released into the rhizosphere, may be some of the primary stimulatory agents for *Orobanche* seeds (Nash and Wilhelm 1960). The exudate effect varies with plant species (Hameed et al. 1973); for example, flax exudate induces a high percentage of *Orobanche* seed germination, whereas marigold has little or no effect.

Striga lutea, an angiospermous parasite belonging to the Rhinanthoideae, attacks the roots of maize, sorghum, sugarcane, and other economically important species. Seed of the parasite usually germinate only when in close proximity to a host root (R. Brown and Edwards 1944). That the frequency of seed germination declines with increasing distance from the root suggests a gradient in the concentration of stimulant from the root. Germination along a host root is enhanced by the production of laterals with the formation of additional root tips. We noted earlier (Chap. 3) that maximum exudation usually occurs at the point of lateral root emergence and in the meristematic region behind the root cap.

The foregoing examples assume that a direct effect of exudates from one plant on seed germination of another plant also occurs under natural field conditions. Field observations and experiments in nonsterile systems show these stimulatory effects but do not assess the contribution of rhizosphere microorganisms to the effect. Microorganisms rapidly attack the amino acids and sugars in exudates and produce secondary products, some of which may stimulate growth of a contiguous plant. The microflora of one plant can be altered by the exudation and microflora of an adjacent plant. There is a general tendency for fungi and, to some extent, bacteria to be more abundant where grassland plants are grown in mixture than when monocultured (Christie et al. 1978). This effect is no doubt related in part to competition between plants, resulting in altered physiology of some plants in the mixture and increased root exudation. One might also expect that endomycorrhizal infections would be affected by deficiencies in N, P, or K brought about by plant competition or microbial utilization of nutrients.

6.6.2 Allelopathy

Field observations frequently show unthrifty plants growing near healthy dominant species. In many instances toxic materials have been extracted or leached from tissues of the dominant plants and found to inhibit seed germination and growth of other species. An obvious question then arises regarding the potential for allelopathic effects of root exudates on contiguous plants, particularly in weed–crop interactions where minimal tillage is practiced.

Although there are reported instances in which weed-root excretions stimulated crop plants, or crop plants stimulated weed growth, weeds are generally considered to be harmful and undesirable in crop culture. Much of this effect is attributed to competition for nutrients or water, but many experiments designed to exclude competition as a limiting factor have demonstrated toxic effects between plants (Rice 1974, 1979). Employing a "stairstep-pot" apparatus in which pots were positioned in series from high to lower levels, Bell and Koeppe (1972) passed a nutrient solution through the rhizosphere of giant foxtail into the rhizosphere of corn, thus excluding competition, and found that mature giant foxtail inhibited the growth of corn approximately 35% through an allelopathic mechanism. Newman and Rovira (1975) tested the effects of root leachates from a number of plant species on growth of each of the same species and showed that all species did not respond to exudates in the same way, even though all of the "donor" species inhibited each "receiver" species. Some species were less inhibited by leachate from their own species than by leachate from any other species, whereas others were more suppressed by leachate from their own species.

The principal chemical compounds that have been identified as growth inhibitors have been described by Rice (1974). Many of these are known to be common components in root exudates. Since Bonner and Galston (1944) identified *trans*-cinnamic acid as the toxin produced by guayule roots and known to inhibit other guayule plants, many other organic compounds that are released by roots have been assigned to groups of allelopathic chemicals. These include organic acids, alcohols, acetaldehyde, coumarins, alkaloids, sulfides, mustard oil glycosides, and others.

Although components of root exudates can be isolated and determined to be inhibitory to test plants in vitro, few instances can be cited in which the direct toxic action of a substance from roots has been shown to affect growth of another plant in natural (nonsterilized) soil. Many compounds are unstable in the presence of microorganisms; therefore, the questions of origin and chemical nature of the effective agent often remain unanswered. Phytotoxic substances are released into soil as a result of microbial decomposition of plant residues, including sloughed root-epidermal cells in the rhizosphere. These may affect plant growth by chemical injury to young roots, thus inducing pathogenesis without parasitism; or such injury may predispose plants to infectious disease. Therefore, under the broad definition usually applied to allelopathy, these effects represent at least an indirect adversity to plant growth due to chemical substances originating from the decomposition of other plants.

Common among the growth inhibitors produced by living plants or released from decaying plants by microbial action and leaching are the phenolic acids and associated compounds. Patterson (1981) tested ten such compounds for effects on soybean seedlings grown in nutrient solution. At concentrations of $10^{-3}M$, caffeic, t-cinnamic, p-coumaric, ferulic, gallic, and vanillic acids significantly affected plant growth, as evidenced by reductions in total dry weight, leaf area, and plant height. Lower concentrations did not affect growth.

Even the foliage of some plants may affect the root function and growth of the same or different species of contiguous plants. While it is common knowledge that plants such as potato, tomato, alfalfa, and apple can be injured when their roots grow in close proximity to walnut roots, the leaves of walnut also contain the inhibitory juglone (5-hydroxy-napthoquinone). The leachates, washings, and extracts from leaves of many plants can inhibit seed germination and growth of other plants. The relation of substances from these sources to rhizosphere phenomena and subsequent plant growth has attracted far too little attention.

Chapter 7 Rhizosphere Relation to Plant Disease

7.1 Introduction

Root diseases may be induced by soil-borne fungi, bacteria, actinomycetes, nematodes, or viruses, but most investigations of rhizosphere influence have focused upon fungal pathogens and this will be reflected in the present discussion. Some highlights of rhizosphere influence on pathogen behavior and root disease have been previewed (Curl 1982).

7.1.1 General Ecology

First, we should recall some of the key ecological factors and phenomena that determine pathogen populations, growth and survival in soil, and host–parasite relationships, for these are the phases of the pathogen life cycle subject to alteration by root exudates and rhizosphere activities. Pathogens, together with the associated nonpathogenic microbial population, function under the influence of two distinctly different but inseparable environments: the subsurface root-soil environment, and that environment which is external to the soil profile but affects plant growth (Fig. 7.1). Within the subsurface environment, the major root-soil factors which control the behavior and fate of pathogens also are interlinked and inseparable under natural conditions. These interacting factors are: root exudates, the general microflora and fauna under the influence of exudates, and the chemical-physical nature of the root–soil interface. A pathogen can be affected by exudates and the rhizosphere biota of either a host (suscept) or nonhost plant, and where the root systems of two contiguous plants intermingle, the resulting allelopathic effects of nonhost on the host plant also may translate into influences on pathogen behavior and root infection.

Park (1963) visualized a triangle of interactions to illustrate the ecological relationships between host (H) and the soil microbial populations (S), between host

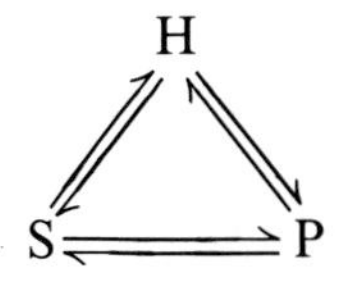

and pathogen (P), and between pathogen and the soil microbial populations. The principal effects resulting from these interactions may be summarized as follows:

H→S Host-root exudates and sloughed root cells influence the activities of soil saprophytes, including microorganisms which inhibit pathogens.

S→H Nonpathogenic soil microorganisms unfavorably affect growth of the host plant resulting in predisposition to disease, or favorably affect the host by enhancing nutrient availability and uptake (Chap. 6).

H→P Effect similar to H→S; root exudates stimulate or inhibit plant pathogens directly or indirectly, major role being the negation of soil fungistasis, thus inducing the germination of pathogen propagules.

P→H Pathogenesis occurs either with or without parasitism.

P→S Pathogen may inhibit competing microorganisms; or synergistic relationships may occur between primary pathogen and facultative parasites during host infection.

S→P Soil saprophytes impose some form of antagonism upon the pathogen, providing a basis for natural biological control; in some instances the pathogen is stimulated by growth factors synthesized by saprophytes.

Factors of the environment external to the soil profile may affect root disease indirectly. Any biotic, physical, chemical, or mechanical agent that affects the physiological functions of a plant through either stimulation or stress can be expected to alter root exudation and microbial activity in the rhizosphere. These in-

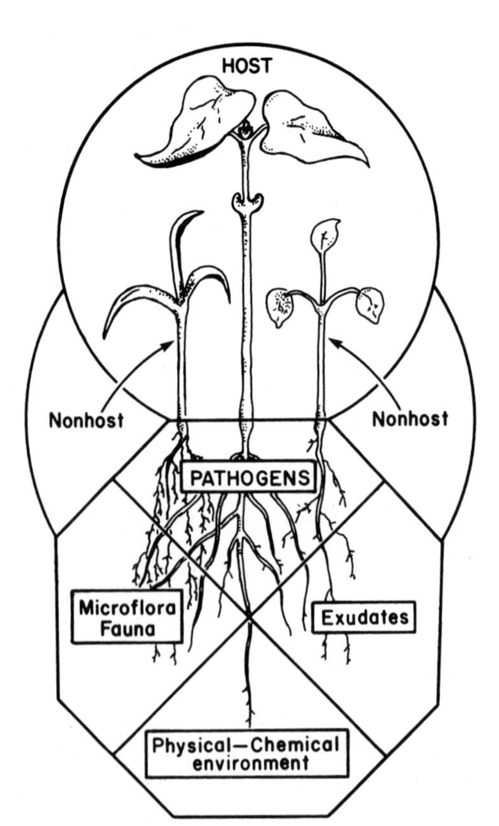

Fig. 7.1. Interacting environments above and below the soil surface affecting pathogen activity and plant growth

Table 7.1. Some common root-infecting fungi and the kinds of propagules subject to rhizosphere effects. (After Rodríguez-Kábana et al. 1977)

Pathogen	Host range	Disease	Mode of survival	Primary inoculum
Rhizoctonia solani	Many plants	Seed decay, damping-off, root rot	Sclerotia, colonized organic matter	Sclerotia or mycelia
Pythium spp.	Many plants	Seed decay, damping-off, crown rot, feeder root disease	Oospores, sporangia	Sporangia zoospores
Phytophthora spp.	Many species of conifers, orchards, ornamentals	Damping-off, feeder root disease, root rot	Chlamydospores in crop residues	Chlamydospores, zoospores
Fusarium solani f. sp. *phaseoli*	Bean; weed plants	Root rot	Chlamydospores	Chlamydospores
Sclerotium rolfsii	Many plants, most in Leguminoseae and Compositae	Sclerotial blight or stem rot	Sclerotia and colonized organic matter	Sclerotia, mycelium in organic matter
Aphanomyces euteiches	Pea, sugarbeet, and many others	Root rot	Oospores	Zoospores
Thielaviopsis basicola	Tobacco, cotton, bean, many other hosts	Black root rot	Chlamydospores	Chlamydospores or endoconidia
Cylindrocladium scoparium	Over 66 genera of 31 families	Root and stem rot (black rot)	Microsclerotia	Microsclerotia
Cephalosporium gregatum	Soybean	Brown stem rot	Mycelium in crop debris	Conidia
Macrophomina phaseolina	Soybean, corn, cotton, and many others	Root and stem rot, charcoal rot	Sclerotia	Microsclerotia
Sclerotinia sclerotiorum	Many plants	Stem rot, white mold, wilt	Sclerotia	Ascospores, mycelia from sclerotia, seedborne
Phymatotrichum omnivorum	Cotton and 2,000 other broadleaf plants	Root rot	Sclerotia, colonized roots	Sclerotia, mycelial strands
Gaeumannomyces graminis	Cereals and grasses	Take all	Mycelium in crop residue	Mycelium in organic matter
Cochliobolus sativus	Barley, wheat, other cereals, and grasses	Crown and root rot	Dormant conidia, mycelium in old stubble	Conidia
Armillaria mellea	Forest, fruit trees, and others	Root rot, damping-off	Colonized roots	Basidiospores, rhizomorphs
Heterobasidion annosum	Conifers and other forest trees	Root rot	Colonized stumps and roots	Basidiospores, diseased-to-healthy root grafts
Fusarium oxysporum f. sp. *vasinfectum*	Cotton	Vascular wilt	Chlamydospores, colonized organic matter	Chlamydospores or mycelia
Verticillium dahliae; *V. albo-atrum*	Cotton and many other plants	Wilt	Microsclerotia	Microsclerotia or conidia

clude all of the natural factors that contribute to or reduce photosynthesis, along with various cultural practices and pesticide injury. Presumably, foliage leachates washed by rain down the plant stem into the upper regions of the rhizosphere also could add either nutrients or toxins to the root zone and create a phyllosphere effect upon the rhizosphere.

Representative species of major soilborne fungal pathogens are listed in Table 7.1 along with their principal means of survival and the propagules that serve as primary inocula for infection. Each species may respond differently to rhizosphere components according to its physiological requirements for growth and reproduction and its tolerance to biological and chemical stress factors. Most of these fungal pathogens are found in the upper 15 cm of the soil profile, where oxygen, organic matter, and the root systems of cultivated plants are most concentrated; vertical distribution, however, can vary considerably with soil type and the specific pathogen.

Using Garrett's (1956, 1970) ecological classification as a guide, we can group soil fungi into *soil saprophytes* and *root-infecting fungi*, then further subgroup these according to their host relationships and their competitive or survival characteristics. Common organic matter inhabitants, such as many species of *Penicillium, Aspergillus, Trichoderma, Fusarium,* the Mucorales, and others, are generally referred to as *obligate saprophytes*, but in fact may be pathogenic as seed-rotters or may attack roots of plants weakend by environmental stress factors.

The root-infecting fungi can be further divided into *specialized parasites*, which are highly efficient but host-dependent, and *unspecialized parasites* which are largely host-independent and possess some of the survival characteristics of true saprophytes. The most efficient of the specialized group are the mycorrhizal fungi which, in a symbiotic alliance with higher plants, obtain nutrients from host cells without pathogenesis, and at the same time enhance nutrient uptake by the plant symbiont.

Specialized root-infecting fungi which cause root disease and are characterized by a declining saprophytic phase after death of the host plant are referred to as *root-inhabiting fungi*. Examples are *Plasmodiophora brassicae, Phymatotrichum omnivorum, Gaeumannomyces graminis, Armillaria mellea, Fomes annosus, Verticillium dahliae*, the vascular wilt fusaria, and others. They survive in soil largely in a passive state as resistant spores, sclerotia, or rhizomorphs. The unspecialized root parasites are said to be *soil-inhabiting fungi*, inferring that they can survive and even grow and reproduce on soil organic matter in the absence of a host plant; thus, they have an expanding saprophytic phase following death of the host. This group, exemplified by *Pythium* spp., *Rhizoctonia solani, Aphanomyces euteiches, Fusarium solani* f. sp. *phaseoli, Cochliobolus sativus*, and others, also may produce resistant survival structures in adverse environments.

The capacity of root-infecting fungi to survive in the face of fierce competition from the general saprophytic population has been termed *competitive saprophytic ability* (CSA). The main characteristics which provide for a high CSA are (from Garrett 1970):

a) rapid germination of fungal propagules and rapid growth of hyphae in response to nutrients from the substrate;

b) essential enzyme systems for colonization and decomposition of plant tissues;
c) production of fungistatic and bacteriostatic substances, including antibiotics;
d) tolerance to fungistatic substances produced by other soil microorganisms.

The CSA of a pathogen is a genetic characteristic, just as pathogenicity is controlled by the gene composition, but other factors, such as population density and the quantity of soil organic matter, may determine the probability of one organism gaining an initial "advantage of position" for substrate (organic matter) colonization. Root exudates and sloughed root debris in the rhizosphere contribute to the organic nutrient base required for competitive growth and for infection.

Bacterial plant pathogens also have been grouped into ecological categories (Buddenhagen 1965) according to their dependency upon a host or their ability to survive in soil after death of the host. However, the positions of different species are less well defined than for fungi. *Corynebacterium sepedonicum* (cause of potato ring rot) and *Pseudomonas solanacearum* (cause of wilts in the Solanaceae) are likely to persist for a relatively short time in soil following death of the host plant, whereas *Agrobacterium tumefaciens* (crown gall pathogen) is more likely to benefit saprophytically in the rhizosphere. Admittedly, with our meager knowledge of bacterial ecology in soil, statements of this kind are hazardous.

7.1.2 Inoculum Potential

The major influence of the rhizosphere on both the saprophytic and parasitic activities of root-infecting organisms is mediated through the action of root exudates (Fig. 7.2). Direct effects of exudates are reflected in pathogen population changes, effects on growth and survival, and the germination of infective propagules. Indirect effects are imposed by the general microbial population responding to root exudates, this activity contributing to nutrient availability and uptake by plants, synthesis of growth factors that affect both host plants and pathogens, and the initiation of antagonistic phenomena as described in Chapter 5. All of these activities influence the inoculum potential of a pathogen.

The inoculum potential of pathogenic fungi was originally defined by Garrett (1970) as "... the energy of growth of a fungal parasite available for infection of a host at the surface of the host organ to be infected, per unit area of the host surface". "The absolute inoculum potential is a measure of the maximum capacity of a pathogen population to infect fully susceptible plant tissue under optimum conditions for infection" (Mitchell 1979). This attribute is controlled by the gene complement of the pathogen, which determines how the pathogen population will respond to environmental factors of the microecosystem. It follows then that the inoculum potential will vary with the inherent nature of different pathogens to produce propagules, to survive in soil, and to infect host tissue. Disease potential (the susceptibility of the host as influenced by disease proneness) can be considered as distinct from inoculum potential. Therefore, disease = in-

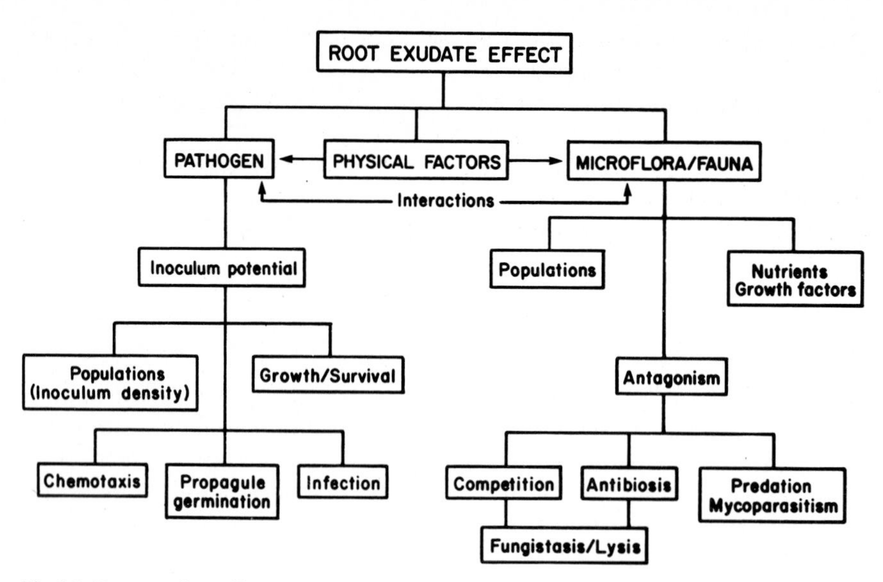

Fig. 7.2. Root exudate effects on pathogen activities imposed either directly or mediated indirectly by interactions with saprophytic microbial populations

oculum potential × disease potential (R. Baker 1978). In summary, the minimal requirements for disease to occur are:

a) a susceptible plant (host or suscept);
b) a sufficient pathogen population or inoculum density at the root surface;
c) a nutrient energy source for rapid propagule germination and host infection;
d) a biotic and physicochemical environment favorable to pathogen activity.

With this introductory background, and with pertinent facts learned in previous chapters about root exudates, general microbial populations, microbial interactions, and plant growth, we can now examine the specific rhizosphere influences on pathogen behavior and the potentiality for root infection. Logically, four major topics should be considered: pathogen populations (inoculum density), growth and survival, infection-pathogenesis, and disease control.

7.2 Pathogen Populations

Pathogen population refers to the number of propagative units of bacteria, fungi, or nematodes per unit of soil contributing to the inoculum potential or chance that disease will occur. Little is known about the populations of other pathogens in the rhizosphere. Flagellate protozoa of the class Mastigophora are associated with roots and may cause disease in coffee and coconut plants as they migrate through root grafts from tree to tree and vertically in the phloem (Agrios 1978), but the rhizosphere effects have not been determined. Soil-borne viruses are vectored by nematodes or by certain fungi of the Chytridiales and Plasmodiophorales; thus, their concentrations in the rhizosphere and their potential for disease must be studied in relation to activities of the vectors.

Since the inoculum density of a pathogen contributes to inoculum potential, the assessment of viable populations in field soil is often used to predict disease incidence and severity. Such assessments have no immediate relation to the rhizosphere, because the estimates are usually made prior to planting a crop. However, the rhizosphere effects of various crop plants used in a rotation system may determine the concentration and nature of inoculum available for infection from season to season. If a pathogen is seed-borne, broad field assessments of populations may hold little relevance, as sufficient inoculum can develop from the initial colonization of the rhizosphere of the germinating seed.

7.2.1 Bacteria and Fungi

A discussion of rhizosphere effect on populations of bacterial and fungal pathogens can be superimposed upon our broader discussion in Chapter 4 of the root–soil factors that determine the proliferation or decline of the general microflora. The responsiveness of bacteria to root exudates is reflected to a large extent by their short generation time and rapid generation rate when provided a suitable nutrient source. Also contributing to the population increase of pathogens in the root zone is the oozing of bacteria from root tissue during the decline of diseased plants. Great differences in numbers of a single species may occur between plots or fields subjected to different cropping systems. Numbers of *Agrobacterium tumefaciens*, estimated by soil-plating on a selective medium, were found to be extremely low (42 propagules g^{-1} soil) in a pasture as compared with 200 in a stone-fruit nursery and 316 in another fruit-tree nursery (Weinhold 1970). These differences suggest a plant-root effect on population build-up. Populations of *Streptomyces scabies* (potato scab) in soil samples from rotation plots in California ranged from 6,000 to 7,000 propagules g^{-1} of soil, and these values showed a direct relationship to scab severity.

Common rhizosphere bacteria can become minor pathogens under certain environmental conditions. Species of *Pseudomonas* in particular, along with some species of the Enterobacteriaceae and the Corynebacteriaceae, produce substances that either inhibit plant growth or stimulate fungal pathogens such as *Pythium* spp. to colonize roots, thus predisposing plants to disease. These organisms are sometimes referred to as "deleterious rhizobacteria", as opposed to "plant-growth-promoting rhizobacteria" which are also predominantly *Pseudomonas* spp. (Suslow and Schroth 1982a). The high frequency of isolation of these bacteria from the rhizosphere and rhizoplane suggests that their activities might seriously affect crop yields. Indeed, some diseases attributed to specific fungi might be due in part to deleterious rhizosphere bacteria. For example, root and crown deterioration of the perennial legume sainfoin (*Onobrychis viciaefolia*) in irrigated fields in Montana appears to be caused by a complex of rhizosphere bacteria (Gaudet et al. 1980) rather than by *Fusarium solani* as previously reported.

The inoculum density of fungal pathogens required to produce disease varies widely among different pathogens and according to the type of inoculum unit. Inocula may consist of conidia, ascospores, basidiospores, zoospores, oospores, sporangia, chlamydospores, sclerotia, rhizomorphs, or mycelia embedded in organic matter particles. Some pathogens produce two or more of these infective propagules. The inoculum density required for induction of disease (K. F. Baker and Cook 1982) ranges from less than 1 unit g^{-1} of soil for pathogens such as *Sclerotium rolfsii, Phymatotrichum omnivorum*, and *Rhizoctonia solani* that form multicellular structures (sclerotia) to more than 1,000 g^{-1} for *Fusarium solani* f. sp. *phaseoli* and *Thielaviopsis basicola* that persist as thick-walled resting spores. The inoculum densities of *Gaeumannomyces graminis*, which survives as mycelium in host-tissue fragments, and *Aphanomyces euteiches*, which produces zoospores from resting oospores, are more difficult to assess.

There seems to be a negative correlation between propagule size and population density of many pathogens and other soil microorganisms (Chuang and Ko 1979). Also, an inverse relationship may be expected between propagule size and the reproductive capacity, or numbers produced per unit area or volume of nutrient medium. In the rhizosphere, root exudates provide the energy source for vegetative growth and the production of new propagules, which may vary in size according to the quantity and nutrient quality of the exudates. The size as well as numbers of inoculum units contributes to the inoculum density and the potential for infection and disease to occur. For example, no symptoms occurred in bean seedlings when grown in soil infested with *R. solani* propagules (pieces of mycelial mat plus sclerotia) smaller than 250 μm in diameter at a concentration of 4 propagules g^{-1} soil, whereas symptoms were observed in all plants after 7 days when propagules of 250–1,000 μm in size were used (Henis and Ben-Yephet 1970). *Rhizoctonia* can cause 50% damping-off in beet at a concentration of 1 sclerotium/10 g soil (Ko and Hora 1971). The minimum numbers of microsclerotia and conidia of *Verticillium dahliae* required for 100% infection of tomato were 100 and 50,000 g^{-1} soil, respectively (R. J. Green 1969); microsclerotia are many times larger than conidia.

Stanghellini et al. (1983) determined the spatial pattern of distribution of *Pythium aphanidermatum* in the rhizospheres of individual mature sugar beet tap roots, and the environmental factors affecting expression of the absolute inoculum potential of the pathogen. Bioassay with potato tuber tissue showed that an average of 36% of the surface area of the root–soil interface of sugar beet contained an infective population of *P. aphanidermatum*. Inoculum densities ranged from one to five oospores 0.1 cm^3 of root–soil interface. Inoculum effective for infection was located primarily within 1 mm of the host surface. The surface area of sugar beet tap roots usually exceeds 500 cm^2. Assuming a 100% efficiency of inoculum, along with knowledge of the percentage of root–soil interface infested, one may expect about 170 lesions per tap root. However, Stanghellini et al. observed a mean of 1.6 lesions per root, indicating that distribution of inoculum was not the limiting factor restricting the number of lesions on roots. The limiting factors to expression of the absolute inoculum potential were suboptimal soil moisture and/or temperature.

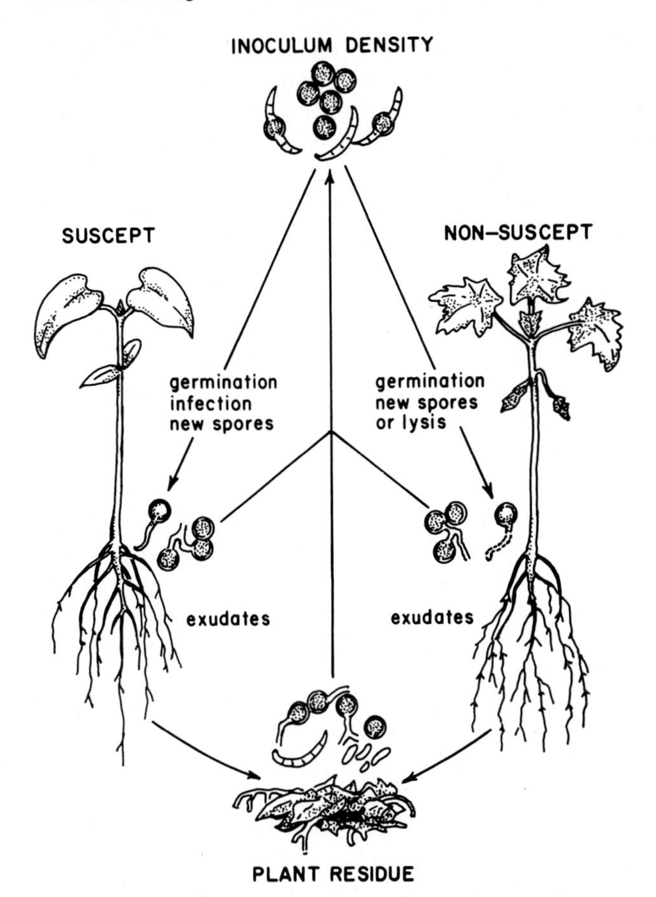

Fig. 7.3. Representation of a portion of the life cycle of *Fusarium solani* f. sp. *phaseoli* featuring inoculum production

Both susceptible and nonsusceptible plants contribute to the inoculum density. The activity of *Fusarium solani* f. sp. *phaseoli* is primarily associated with susceptible bean plants, but populations of propagules (largely chlamydospores) g^{-1} of soil also may be greater in rhizosphere soil of the nonsuscepts corn, lettuce, and tomato than in nonrhizosphere soil (Schroth and Hendrix 1962). Onion rhizosphere, however, does not favor reproduction of the fungus.

The extensive work of Schroth, Snyder, and associates with *F. solani* f. sp. *phaseoli* in California has provided clear insight into the role of root exudates, rhizosphere, and organic residues during the saprophytic and pathogenic phases of the pathogen life cycle (Schroth and Hildebrand 1964). A simplified version of the life cycle presented in Fig. 7.3 features the rhizosphere influence on inoculum production. The fungus survives for many months in the form of thick-walled chlamydospores embedded in decomposing diseased plants. These spores persist in a state of exogenous dormancy imposed by soil fungistatic factors. Fungistasis is broken when plant roots grow into the inoculum site and release in their exudates the amino acids, sugars, or other energy-rich compounds required for spore germination. Spores may germinate readily in either the bean rhizosphere or the rhizosphere of a nonsusceptible variety of crop or weed plant.

Infection occurs only in bean, but new populations of spores can be produced in either rhizosphere as well as in the organic debris left when these plants decline. Though chlamydospores serve as primary inoculum, the fungus also forms macroconidia and microconidia in the soil organic matter, and these may germinate to produce a new thallus and more spores or the macroconidia may convert to chlamydospores. Thus, there is a high potential for reproduction and build-up of inoculum density as influenced by both living and dead plants. At the same time, stimulation of other microbial activities in the rhizosphere and the induction of antagonistic deterrents, including lysis of spores and germ tubes, determine the final concentration of inoculum available for infection.

Different form species of *F. solani* are host-specific in parasitism, but their spores may germinate in response to root exudates of nonsusceptible plants, subsequently leading to new spore production. Such is the case for *F. solani* f. sp. *phaseoli* which attacks bean and *F. solani* f. sp. *pisi* which infects pea (Kraft and Burke 1974). Only the population of the pea pathogen, however, is significantly increased in the rhizosphere of both pea and bean. The fact that monoculture of pea or bean for 3 years in a field resulted in a build-up of the specific pathogen suggestes a long-range root or rhizosphere effect. Crop refuse left in the field also contributes to increased inoculum density.

A large number of form species of *Fusarium oxysporum* cause wilt diseases of many crop plants. These fungi, though classed ecologically as root-inhabitating, specialized parasites of low competitive saprophytic ability, can live for many months in soil in a passive state as chlamydospores or as mycelia in organic matter. It is not surprising that populations of the specialized wilt fusaria, such as *F. oxysporum* f. sp. *melonis* (Banihashemi and deZeeuw 1975) and *F. oxysporum* f. sp. *spinaciae* (A. A. Reyes 1979), increase in the presence of living susceptible plants and decrease in their absence. Populations are greater in the rhizosphere of wilted host plants than in the rhizosphere of healthy susceptible plants or of nonhost plants, though roots of the latter may be externally colonized by the pathogens.

A high frequency of cropping with susceptible plants in a rotation system is frequently accompanied by increased populations of a pathogen and increased disease severity. Sugar beet yield decreased considerably where sugar beet comprised 33% of the crops in a rotation plan (Schäufele and Winner 1979); reduced yield was attributed primarily to increased inocula of *Aphanomyces cochlioides* and *Pythium* spp. Populations of a pathogen are not always reduced by rotations with nonsusceptible crops, even though general recommendations for disease control may include such practices. *Verticillium dahliae* and *Rhizoctonia solani*, which have the advantage of very wide host ranges, were not reduced in population in potato fields when the potato crop was rotated with barley (Davis and McDole 1979). A higher infestation of *R. solani* occurred during a potato-grain rotation sequence than in continuous potato culture. Both of these pathogens produce sclerotia in dead tissue of diseased plants; thus, a rhizosphere effect from living roots is of primary significance only when sclerotia are induced to germinate by root exudates, followed by production of secondary sclerotia. Volatile chemicals emanating from root exudates and the volatile metabolites from microbial activity in the rhizosphere also can affect growth of a fungal pathogen, either by inhibition or stimulation, and thereby influence the potential for reproduction of spores or sclerotia. Extensive research by Coley-Smith and coworkers in England (reviewed by Linderman and Gilbert 1975) has shown that water-insoluble alkyl cysteine sulfoxides that diffuse into the soil from *Allium* spp. are metabolized by soil bacteria resulting in the production of volatile alkyl sulfides which stimulate sclerotial germination of *Sclerotium cepivorum*; subsequently, new sclerotia are produced.

The extreme complexity of the soil ecosystem poses great difficulties in pinpointing specific factors that stimulate or suppress the reproduction of pathogens under field conditions. The rhizosphere effect surely must play a prominent role. The very nature of modern agriculture, growing plants in pure stands, offers the pathogen a favorable and abundant substrate for growth and reproduction in the rhizoplane, and for further reproduction as the host declines to a state of dead refuse.

7.2.2 Nematode Populations

The majority of nematodes in field soil are free-living, feeding superficially on fungal hyphae, algae, and bacteria that occur on underground stems, roots, and organic debris. Since the availability of food source for these animals may be greatly influenced by root exudates and related factors, their reproductive capacity is obviously subject to rhizosphere effect. Plant parasitic nematodes, though having a soil phase in their life cycle, feed directly upon living plant tissue and in this respect their populations are influenced by susceptible host roots. Eggs constitute a part of the population but, except for the easily extractable eggs of *Heterodera* and *Meloidogyne*, inadequate extraction methods make their study difficult. Galls on host plants also reflect population size since they contain nematodes. Populations may increase suddenly, as when eggs hatch, or decrease suddenly due to a drastic environmental change. Soil environmental factors that determine nematode numbers and distribution have been reviewed by Yeates (1981).

Population densities of nematodes may be as high as 500 eggs g^{-1} soil, the equivalent of 100 million eggs m^2 and capable of liberating 80 million larvae m^2 (Jones 1959). Because nematodes tend to congregate around roots of growing plants, populations there are usually greater than the average per unit weight or volume of soil. The amount of food available to a nematode is affected by the number feeding at the same site; this could be construed as competition in the rhizoplane resulting in an increase or a decrease in populations of species. For the most part, however, the actual rhizosphere effect is limited to the stimulation of egg-hatch by root exudates or the inhibition of nematode activity by toxic substances released by roots of some plants. Examples of substances from living roots that inhibit plant parasitic nematodes have been cited by K. F. Baker and Cook (1982). Lower numbers of *Pratylenchus penetrans* have been observed near marigold (*Tagetes* spp.) roots than near other plants, whereas numbers of cysts of *Heterodera rostochiensis* were unaffected. Polythienyls, highly nematicidal to *Pratylenchus* spp., have been isolated from roots of *T. erecta*. The stubby-root nematode, *Trichodorus christiei* multiplies rapidly on tomato roots, but does not feed on asparagus roots which produce a toxic glycoside. Hatching of the cysts of the golden nematode (*H. rostochiensis*) of potato can be prevented by growing mustard with potato; the effective agent is phenyl isothiocyanate released by mustard plants.

Conceivably, populations may be affected by parasites and predators which have been suggested as potential biocontrol agents (Mankau 1980; Kerry 1981 b; K.F. Baker and Cook 1982) and which tend to be concentrated in the rhizosphere. These are discussed under Biological Control of Nematodes in Section 7.6.

Some evidence of rhizosphere effect on nematode populations has been observed through the systemic activity of foliar applied chemicals such as oxamyl (methyl *N'N'*-dimethyl-*N*-[(methyl carbamoyl)oxy]-1-thiooxamimidate), which reduces the hatching of eggs (as in *Meloidogyne javanica*) and the numbers of extractable microbial feeding nematodes (Atilano and Van Gundy 1979). Oxamyl applications to tomato after soil infestation with *M. javanica* also significantly reduced female size and the production of eggs in roots.

7.3 Growth and Survival in the Rhizosphere

Relatively few microorganisms in the rhizosphere or within the general soil population have evolved the capacity to induce disease in plants. Considering the overwhelming populations of saprophytes with which the few pathogens must compete for a niche in the microecosystem, growth and survival of the pathogens seems remarkable indeed.

7.3.1 Bacteria

Through the evolutionary process no clear and consistent physiological, morphological, or cytological differences have been established between bacterial plant pathogens and closely related saprophytes (Buddenhagen 1965). Yet, given these close similarities, most of the major root-infecting and vascular-disease bacteria are relatively weak saprophytes, their survival time being closely linked with the degeneration time of the diseased host and competition with the general microflora. Plant pathogenic bacteria are non-spore-forming rods and, except for *Corynebacterium*, are Gram negative. These forms, along with various pleomorphic forms, are most active physiologically in the metabolism of amino acid and sugar substrates common in root exudates; therefore, they are generally faster growing and more responsive to exudates than Gram-positive types and the cocci. Growth of pathogenic bacteria in the rhizosphere-rhizoplane is subject to either inhibition by antibiotic-producing microroganisms or stimulation by vitamin-synthesizing bacteria. Much is unknown, however, about the survival capacity of plant pathogenic bacteria outside of the living host. Different pathogens, even within a single genus such as *Pseudomonas*, vary greatly in their ability to survive in soil. *Erwinia carotovora*, associated with soft rots of many field and vegetable crops, can be recovered from the rhizosphere of both cultivated and noncultivated plants. A biotype of *E. carotovora* var. *atroseptica* appears to be endemic to Arizona on native *Lupinus blumerii* and may persist indefinitely in agricultural field soils as a rhizosphere inhabitant, and for at least 5 months in fallow soil (de Mendonca and Stanghellini 1979).

7.3.2 Host-Dependent Fungal Pathogens

Natural populations of root-infecting fungi consist of many clones and races constituting a variable genome for the process of selection in a variable environment. Physiologically variable hosts, along with changes in the micro- and macro-environments of soil, exert selective pressures upon root-infecting fungi (Stover 1959) and determine their adaptability or fitness to the environment. Thus, different species or genotypes of pathogenic fungi may be expected to display differences in nutritional responses to root exudates which affect growth, reproduction, and survival.

Whereas nutrients in the rhizosphere benefit both pathogens and antagonistic saprophytes, a pathogen has the advantage of escaping into living tissue. The more host-dependent the pathogen, the greater is its affinity for living roots and the probability of eluding the suppressiveness of the general microflora and the predacious activity of the soil fauna. The root, in effect, acts as a selective medium. The amino acids, sugars, and growth factors in root exudates induce germination of dormant chlamydospores, oospores, and sclerotia, or they activate quiescent mycelia embedded in organic matter. The host-dependent fungi, labeled root-inhibitants by Garrett, survive in an active state only as long as the host plant is living or for a limited time after death of the host. Thereafter, survival again becomes the function of dormant, resistant propagules which, being immobile, must await activation by chance contact with roots of other susceptible plants. The propagules of some pathogenic fungi, when stimulated to activity by root exudates, may lyse upon encountering temporary host resistance. But those endowed with the ectotrophic growth habit (Garrett 1956) can grow over the surface of roots under the stimulus of exudates until the aggregation of mycelium is sufficient to overcome the protective mechanisms of the host. *Ophiobolus graminis (Gaeumannomyces graminis)*, the destructive pathogen of cereal crops, exemplifies this survival habit. However, antibiosis and competition imposed by specific saprophytes may deter or prevent establishment of the ectotrophic growth phase of a pathogen before infection. *Pseudomonas* spp. are well adapted antagonists of this pathogen and are selectively stimulated near wheat roots (Sands and Rovira 1970). Their rapid growth rate and high tolerance to antibiotics and acidic environments give them a competitive advantage in colonizing the same ecological niche occupied by the pathogen; consequently, it is believed that these bacteria contribute to the suppression of *O. graminis* in soils (Smiley 1979).

Phymatotrichum omnivorum, the cotton root-rot pathogen, was once thought to have strong competitive saprophytic capabilities until it was learned that the fungus could survive for many months on deep, yet living, taproots of cotton. The mysterious mycelial strands can grow slowly and produce sclerotia in abundance at depths to 90 cm in clay soils of low O_2 and relatively high CO_2 concentration commonly found in the southwestern USA (Lyda 1974; Alderman and Hine 1982). The pathogen is not prevalent in acids soils where CO_2 is not retained. Apparently, microbial antagonists are suppressed more by relatively high levels of CO_2 than is the pathogen.

Several Basidiomycetes cause destructive root and butt-rot diseases of trees the world over. *Fomes annosus (Heterobasidion annosum)*, most prevalent on conifers, can spread extensively through forest plantations following cutting operations. Basidiospores from the air initiate the colonization of freshly cut stumps from which the fungus grows ectotrophically over the roots in alkaline soils, or through the inner wood in acid soils, making contact with adjacent healthy trees through natural root grafts. The extensive work of Braun (1958) in Germany showed the weak competitive nature of *F. annosus*, and its inability to become established and grow extensively free of living roots. *Fomes* is naturally suppressed by another Basidiomycete, *Peniophora gigantea*, a common component of forest organic matter.

Soil fungi which cause vascular wilt diseases in plants are said to be weak saprophytes, unable to make more than limited hyphal growth through natural soil. Saprophytic activity of *Verticillium dahliae*, cause of wilt in many plants, is limited to the rhizosphere where nutrients in root exudates nullify fungistasis, releasing conidia or microsclerotia from exogenous dormancy and allowing hyphal growth toward roots (Schreiber and Green 1963). Amending soil with various organic materials may adversely affect propagule germination and germ-tube growth in the rhizosphere, resulting in a lower inoculum potential; chitin and laminarin amendments were effective against *V. dahliae* in strawberry (Jordan et al. 1972). The mechanism of action against the pathogen, following organic amendments, is not always clear, but it is usually assumed to be due to an increase in production of antifungal substances by the enriched microflora of the rhizosphere. Probably more important is competition between the pathogen and the increased microflora for nutrients at the root surface, since there is still doubt that antibiotic substances produced in natural soil can endure microbial degradation long enough to act as effective pathogen inhibitors. It has been shown experimentally that competition between *Fusarium oxysporum* f. sp. *pisi* (cause of pea wilt) and other rhizosphere microorganisms for nutrients in root exudates is important to growth of the pathogen near host-root surfaces (Buxton 1960). Thus, the direct effects of pea-root exudates on this fungus can be altered by other members of the microflora. This and related interactions can be expressed diagrammatically:

→ *F. oxysporum* f. sp *pisi* ←

Root exudates ⇄ Other microflora

Buxton verified in laboratory studies that the ability of prevalent rhizosphere fungi to inhibit the pathogenic *Fusarium* increased when the contending organisms were grown in media containing rhizosphere soil extract, root exudate, or both. Therefore, aside from the probability that antibiotic substances may break down rapidly in natural soil, the potential for exudate-mediated antibiosis exists.

7.3.3 Fungal Pathogens with High Saprophytic Ability

Plant pathogens described by Garrett (1970) as soil-inhabiting, unspecialized parasites attack mostly juvenile tissues of seedlings or the feeder roots of older plants. Species of *Rhizoctonia, Pythium*, and nonvascular fusaria are exceedingly important. These, as well as pathogenic species of *Phytophthora, Aphanomyces, Sclerotium,* and *Cochliobolus* are highly competitive with obligate saprophytes at the root surface prior to infection. They can colonize dead organic matter competitively with other microorganisms and survive for varying periods in the absence of a living host. However, long-term survival of these fungi is enhanced by periodic contact with host plants which provide new vigor and support reproduction for perpetuation. *Rhizoctonia solani, Sclerotium rolfsii,* and *Gaeumannomyces graminis* can survive many months as mycelia embedded in organic matter if the fungi have been preconditioned by natural environmental conditions to withstand desiccation. With some specific exceptions, both unspecialized fungal pathogens and the specialized species survive mostly in the form of resistant structures (chlamydospores and sclerotia), the essential rhizosphere influence being primarily the exudate-induced stimulation of propagule germination.

Rhizoctonia solani is perhaps the best-known representative of the soil-inhabiting unspecialized pathogens of high saprophytic capability. Strains occurring in both cultivated and uncultivated soils worldwide have evolved the ability to survive competitively with other microorganisms under a wide range of conditions (Papavizas 1970). Certain strains are relatively host-dependent, but most can be isolated from natural soils by trapping methods, such as the colonization of buried seeds or stem segments, thus demonstrating that the fungus exists as active mycelium in the sparophytic state. It would seem, therefore, that growth and survival of pathogens such as *Rhizoctonia* may be less affected by rhizosphere phenomena than are the more host-dependent types. However, growth of *R. solani* at the infection site (the formation of infection cushions) is very much affected by root exudates. Though relatively little study has been made of the saprophytic growth of root pathogens specifically in the rhizosphere, except for those with ectotrophic growth habits, the saprophytic phase of unspecialized parasites should be no less affected by the increased nutrient pool of the rhizosphere than are the obligate saprophytes. Since different plant species or cultivars vary in the quantitative and qualitative nature of root exudates, the growth response and sclerotial production by *Rhizoctonia* and similar fungi also can be expected to vary, with an ultimate influence on virulence beyond the intrinsic nature of the pathogen.

A review of the physiology (R. T. Sherwood 1970) and metabolism (Tolmsoff 1970) of *R. solani,* far beyond the scope of this chapter, would provide many clues to the actual and potential roles of root exudates in the government of *Rhizoctonia* behavior. Different isolates or strains of the pathogen vary greatly in the utilization of the carbon and nitrogen compounds released in exudates; these sources also determine numbers and size of sclerotia produced. Any competitive advantage that *Rhizoctonia* might gain at the root surface may be largely due to a more efficient use of these nutrient sources as compared with the associated microflora. Not only the kinds or carbon and nitrogen sources but also the C/N ratio has a definite influence on the competitive behavior and the inoculum potential of *R. solani* (Papavizas 1970). Growth factors, such as thiamine and biotin, also are supplied in root exudates and synthesized by bacteria in the rhizophere; many isolates of *R. solani* have absolute requirements for these vitamins.

Reference was made earlier in this chapter to the important bean pathogen, *Fusarium solani,* f. sp. *phaseoli,* in which a high reproductive capacity and survival ability are promoted by exudates of both susceptible and nonsusceptible plants. Production of resistant chlamydospores in response to exudates, and reproduction on dead-plant refuse, provide a high potential for long survival. Let us consider in summary the fate of a chlamydospore or other typical fungal propagule lying dormant in the soil. Depending upon a combination of biotic, chemical, and physical factors of soil and host origin several things may happen to that propagule (R. Baker 1965). The propagule may:

a) die of senility in the absence of a stimulant;
b) germinate in response to nutrients from organic matter or from nonhost plant exudates, then either lyse or produce new spores;
c) germinate in response to root exudates of a host plant and establish infection;
d) germinate at the suscept root surface but die or lyse due to inadequate energy source for immediate infection.

7.4 Pathogenesis

Pathogenesis is the sequence of events that occurs during disease development, and this sequence begins when a pathogen contacts an infectible site on a plant (Wheeler 1975). The primary factor determining infectibility is the genetic constitution of both pathogen and the host. Thereafter, as stated by Wheeler, the outcome of the struggle between pathogen and plant is determined by the nature of environmental factors at the time of contact; this relates back to inoculum potential described earlier in this chapter.

The attributes of a successful parasite (K. F. Baker and Cook 1982) involve some of the properties suggested by Garrett as requirements for a strong competitive saprophytic ability and also correspond to the conditions essential for root disease to occur. For example, the successful parasite must maintain an adequate population density and high survival capacity, and possess the necessary enzymes to invade host tissue rapidly when conditions are favorable. The ability to parasitize more than one plant species also is a favorable attribute.

7.4.1 Fungal Pathogens

Soil-borne pathogenic fungi which lie in a dormant or quiescent state in the absence of plant roots must be activated before they can enter the phase of pathogenesis. Assuming plant susceptibility and the availability of an adequate pathogen population, the major factor controlling infection is a suitable energy source for propagule germination and host-cell penetration. For unspecialized facultative parasites, such as *Rhizoctonia solani* and *Sclerotium rolfsii,* this source may be provided largely by dead organic matter, although root exudates and substances in the laimosphere contribute to the energy for penetration. Host seed and hypocotyl exudates (e. g., in cotton), in addition to providing nutrients for growth and formation of infection structures, also may promote the infection process by inducing the pathogen (*R. solani*) to produce endo-polygalacturonase (Brookhouser and Weinhold 1979). Spores of different species of fungi vary in the requirement for exogenous nutrients and, therefore, are variously affected by soil fungistasis (Griffin and Roth 1979) which prevents the germination of fungal propagules in natural soil (see Chap. 5). A relationship between root exudate and the annulment of soil fungistasis was suggested (Jackson 1957) long before wide acceptance of the nutrient deficiency hypothesis to explain the fungistatic phenomenon (Lockwood 1977). Exudates can promote germination of oospores, zoospores, conidia, chlamydospores, and sclerotia, and activate the quiescent mycelia of fungi embedded in organic matter. Exudate effects can be demonstrated by simple techniques. The sexual stage spores (oospores) of *Pythium mamillatum,* produced in the interstices of glass-fiber tapes on agar-plate cultures of the fungus, were buried beneath turnip seedlings in garden soil and in soil without plants. When the tapes with spores were recovered, stained, and examined, it was discovered that significant germination occurred only in the proximity of roots (Barton 1957). Oospores of *Aphanomyces euteiches,* when buried in cellulose casings in the rhizosphere, germinated in higher numbers next to roots of pea plants than next to roots of soybean, bean, and sweet corn (Scharen 1960). L. F. Johnson and Arroyo (1983), using an agar-slide technique for burial and recovery of oospores of *Pythium ultimum,* observed that germination of oospores occurred mostly in the root-hair zone of cotton seedlings at distances not greater than 1.5 mm from the root surface. Oospores, sporangia, zoospores, and zoospore cysts of other pathogenic species of the Oomycetes are similarly affected by root exudates.

Pathogenesis by species of *Phytophthora, Pythium,* and *Aphanomyces* in many instances is preceded by a pre-penetration phase in which zoospores from germinated oospores or sporangia accumulate around roots along the region of elongation in chemotactic response to root exudates. Zoospores of *Phytophthora cinnamomi* have been observed to encyst at various distances from avocado roots, indicating response to a concentration gradient of stimulatory chemicals released by the roots (Zentmyer 1980). Germ tubes also exhibit chemotropism as they become oriented toward the site of exudation. This response appears to be specific to living roots, and there is some evidence of differences in chemotaxis of zoospores to resistant and susceptible plants.

The specific root-exudate components or other mechanisms involved in the chemotactic response of zoospores have not been entirely clarified. Amino acids, sugars, and possibly growth factors moving from foliage to roots seem to be implicated, since the removal of leaves or girdling of large roots of avocado seedlings results in reduced attraction of zoospores of *P. cinnamomi* and a lower rate of infection (Zentmyer 1980). Ethanol produced by plant roots may promote chemotaxis (R. N. Allen and Newhook 1973). Weak bioelectric currents that occur around plant roots also may be involved in drawing negatively charged zoospores to positively charged zones of the rhizoplane, thus contributing to eventual pathogenesis (Khew and Zentmyer 1974).

Chlamydospores of *Fusarium solani* f. sp. *phaseoli* germinate readily in soil in close proximity to germinating bean seeds and near root tips of seedlings where exudation of amino acids and sugars is most abundant (Schroth and Snyder 1961). Germination can be demonstrated in solutions of various common chemical constituents of bean exudate. However, germinating bean seeds and hypocotyls differ in susceptibility to attack by the pathogen and in the amount and quality of exudate released (Cook and Snyder 1965). Bean seed exudate stimulates high germination of chlamydospores, but ectotrophic growth of the fungus is not established and germ tubes undergo a high degree of lysis. Bean seedling hypocotyls stimulate less germination of spores but less lysis occurs and hyphae develop the necessary ectotrophic growth for eventual infection. Thus, the development of *F. solani* f. sp. *phaseoli*, and perhaps other pathogens in the host rhizosphere, is determined largely by the kind of nutrients in exudates, particularly the availability of exogenous carbon and nitrogen sources.

Fatty acids of high molecular weight have been implicated in the germination of endoconidia and chlamydospores of *Thielaviopsis basicola*, the pathogen that causes black rot of many crop plants (Papavizas and Kovacs 1972). Total fatty acid was one to two times higher in rhizosphere soil of bean plants than in nonrhizosphere soil, and rhizosphere soil of susceptible plants (bean, cotton and tobacco) had more fatty acid than soil near roots of nonsusceptible plants (corn, wheat, kale). Cytokinins, which are produced by many soil fungi and bacteria around roots, may promote higher levels of free fatty acids and sterols in plant tissues and root exudates. This was demonstrated with 0.1 mM kinetin (6-furfurylamino purine) in a nutrient solution supporting the growth of peanut plants (Thompson and Hale 1983). The implications relating to root disease are evident, since some pathogenic fungi which colonize plant roots require an external source of free fatty acids and sterols. Sterols stimulate the growth of a number of fungi and are required for sexual and asexual reproduction in species of *Pythium* and *Phytophthora* (Hendrix 1970). Conceivably, sterol precursors may inhibit the action of sterols in reproduction of pythiaceous fungi and thus play a role in plant disease resistance.

Sclerotia of pathogenic fungi also respond to exudates from roots. The macrosclerotia of *Sclerotium cepivorum* germinate readily in the presence of host plants (species of *Allium*) but are only slightly affected by nonhost plants; sclerotia are most affected near the root-tip region. Certain volatile organic sulfides, which are common constituents of onion and garlic, are important in the stimulatory process. Advanced studies with *S. cepivorum* have shown that stimulation of sclerotium germination is not a simple direct effect of volatiles, but the process is mediated by microbial activity which may release inhibitors as well as stimulants (Linderman and Gilbert 1975). Nutrient materials released by the sclerotia contribute to this microbial activity.

Microsclerotia are produced by several soil-borne pathogens. These structures are formed by septation of hyphae into chains of chlamydospore-like cells which are thick-walled and pigmented. The microsclerotia of *Verticillium albo-atrum*, cause of vascular disease and wilt of numerous crop and garden plants, are freed from the grip of soil fungistasis when in contact with the root exudate from a susceptible tomato plant but are less affected by exudate of wheat, a nonhost crop (Schreiber and Green 1963). *Macrophomina phaseolina,* causal agent of root and stem decay in over 300 species of plants, persists in fields and forest-tree nurseries as microsclerotia. Natural fungistasis holds germination of sclerotia to less than 5% in the absence of host roots but germination can be increased to 50% by addition of a pine-root exudate (W. H. Smith 1969 a). In most instances where specific exudate constituents have been tested, the amino-acid fraction has been more effective than the sugars in inducing spore and sclerotium germination. However, the important relationship of carbon-to-nitrogen balance to pathogen survival and activity is well known and no doubt has a vital role in the annulment of fungistasis.

7.4.2 Nematode Activity

Root exudates are important in the activation of quiescent nematode stages and in the movement of plant parasitic forms to infection sites (Shepherd 1970). Although nematodes do not move great distances, even in moist soil (0.1 to 1.0 cm day^{-1}), some movement increases the probability of contact with roots and permits aggregation at specific infection sites in response to either a nutritional or sensory stimulus. This was essentially the thesis of H. R. Wallace (1978) in his discussion of plant pathogen dispersal in time and space relative to surviving environmental fluctuations.

The eggs of cyst-forming nematodes, *Globodera* and *Heterodera* spp., are protected inside the cuticle remains of the dead female which forms the envelope-like cyst. Each of 200–600 eggs within the cyst develops to the first-molt larval stage, then remains quiescent until stimulated to hatch by exudates from either host or nonhost plant roots. Under the root-exudate stimulus, the activated larva thrusts its mouth spear through the eggshell and eventually emerges. The specific constituents of root exudates responsible for the egg-hatch stimulus have not been determined with certainty because of the difficulty of distinguishing between exudate influence and the effects of microbial metabolites or other compounds which exist under natural conditions. Nematodes, such as *H. schachtii* (beet cyst-nematode), with a wide host range, are stimulated by root exudates of many hosts as well as by many synthetic and inorganic compounds. *G. rostochiensis* (potato cyst-nematode), having a limited host range, is affected only by exudates from a few solanaceous plant species and a few artificial hatching agents.

Because of the dependency of *G. rostochiensis* on root exudates for hatching, it might seem feasible to use root and rhizosphere-soil leachates to induce hatch in the absence of a host plant, thus creating a control measure. A question, however, is whether exudates or leachates can persist in soil long enough to be effective. Tsutsumi (1976) suggested that potato-root diffusate can persist for 100 days in soil after removal of plants. Perry et al. (1981) collected soil leachates at intervals before and after removal of host plants and tested these for hatching stimulus of the potato cyst nematode. Greatest total hatch occurred with leachate from plants after 12 weeks growth and the hatching stimulus, though declining slowly, persisted for 8 weeks after removal of plants from the soil.

That plant parasitic nematodes are attracted to roots has been recognized for many years. Since Linford (1939) first observed that the larvae of root-knot nematodes congregated in the region of cell elongation behind the root cap, speculation has continued regarding the true nature of the reactions involved between nematodes and root exudates. The attraction varies with species of both plant and nematode and probably is attributable to a combination of factors associated with root metabolism and with microbial metabolites in the regions of greatest root exudation. Whatever the mechanisms involved, the physiology, behavior, and distribution of nematodes are affected in ways contributive to subsequent root infection.

7.4.3 Rhizosphere Factors Influencing Disease Resistance

Having examined the criteria for saprophytic survival of pathogens, the conditions for disease to occur, and the attributes of a successful parasite, we can now begin to form a mental picture of the actual and potential roles of root exudates (Chap. 3) and microbial interactions (Chap. 5) in both the preinfection behavior of pathogens and the establishment of disease. Though we normally think of the influence of the rhizosphere on disease as being manifested primarily in the preinfection phase of the pathogen life cycle, microbial activity in the rhizosphere can affect plant growth (Chap. 6) and therefore affect disease proneness and the act of parasitism. Some of the features of the rhizosphere relating to plant susceptibility or resistance are not very different from those enunciated by R. W. Lewis (1953) in the nutrition hypothesis of parasitism and thoroughly discussed by Bateman (1978). The first hypothesis suggests that a parasite grows in a particular host or organ because nutrients essential for its life are there, and fails to grow in resistant hosts where these nutrients are not available either in required kind or quantity. The second hypothesis proposes that, depending upon the environment, nutrient compounds may be either stimulatory or inhibitory. To certain elements of the balance hypothesis of parasitism, Garber (1956) added the mechanisms of host defense and suggested the nutrition-inhibition hypothesis of pathogenicity, which shows that a parasite encounters within the host either a favorable or an unfavorable environment. This suggests that the unfavorable environment involves factors in addition to inadequate nutrition. Thus, "Pathogenicity, susceptibility, resistance, and pathogenesis are all processes or attributes governed by the genetic characteristics of living systems and the physical environment in which they operate" (Bateman 1978). A pathogen lying in the rhizosphere or at the rhizoplane also is subjected to elements of a multiple-component system first prior to and then at the time and site of infection. The nature of these environments determines pathogen virulence.

Given an adequate inoculum density or parasite population and exudate-induced germination of pathogen propagules at the root surface or in the laimosphere, infection is highly probable but not inevitable. Infection, or the host-pathogen cellular association, is established only when the other criteria for inoculum potential are satisfied. Even a plant which is lacking in the genetic characteristics for resistance may escape infection if environmental factors of the root sphere are not optimal for pathogenic activity. Resistance can be induced or lessened by the quantitative and qualitative nature of the root-surface flora and the metabolic activities of specific nonpathogenic microorganisms in response to root-exudate stimuli. Genetic research has shown that the host genotype may play a major role in determining the characteristics of the rhizosphere populations of bacteria in wheat, largely through control of the quality of root exudates (Atkinson et al. 1975). Polygenic resistance (that controlled by several genes as opposed to monogenic resistance controlled by a single gene) is more sensitive to environmental fluctuations and is primarily responsible for any resistance in host plants to less specialized parasites such as *Fusarium solani* and *Phytophthora* spp. (K. F. Baker and Cook 1982). It is likely that much of the resistance of plants to root-infecting organisms is associated with host-microbe interactions that have not been genetically defined.

The familiar resistance characteristics, such as mechanical barriers (thick cell walls or cuticle) and host-produced, fungitoxic chemicals (phenolics, chlorogenic acid, quinones, tannins, etc.) are supplemented by host-synthesized phytoalexins (Ingham 1972) in response to pathogen invasion or to nonpathogen activity at the root surface. Common broad-spectrum antibiotics of this kind are pisatin produced by pea and phaseolin produced by bean. Certain metabolites, such as actinomycin D, nitromycin C, gliotoxin, cycloheximide, chloramphenicol, and culture filtrates of nonpathogenic fungi are potent inducers of pisatin production in garden pea (Hadwiger and Schwochau 1969). Apparently, microbial metabolites at certain concentrations interfere with gene-control mechanisms in the host cells resulting in high metabolic activity and protein synthesis. Some saprophytes in the rhizoplane become weak parasites when conditions are favorable. Perhaps these organisms might be more damaging to plants except that their efforts to colonize living plant cells activate the phytoalexin defense mechanism of the potential host.

Symbiotic organisms, generally considered to be only beneficial, may in some instances induce more severe disease. Such a case was described in which *Phytophthora* root rot of soybeans (*Glycine max*), caused by *Phytophthora megasperma* f. sp. *glycinea,* was more severe on plants inoculated with *Rhizobium japonicum* than on plants not inoculated (Beagle-Ristaino and Rissler 1983). It was suggested that zoospores of the pathogen may be attracted to nodule exudates, resulting in stimulated growth in the rhizosphere, or dormant oospores may be induced to germinate. Other examples could be cited in which *Rhizobium* nodulation was accompanied by less severe disease. The interaction apparently is subject to variation according to the physiological nature of the host plant and the pathogen involved.

Mycorrhizal fungi are part of the rhizosphere and rhizoplane flora before the symbiotic relationship is established with plant roots. The initiation of this mutualistic association to form either ectomycorrhizae or endomycorrhizae (Chap. 4) does not end the interactions between mycorrhizal fungi and other root-surface microorganisms. Besides imparting nutritional benefits to both plant and fungal symbiont, mycorrhizae may influence disease resistance in specific plants. Considering that nearly all plant roots are mycorrhizal, and root-disease fungi have evolved with this condition for thousands of years, it might seem that pathogens should have adapted to a compatible coexistence with mycorrhizal fungi. However, there is increasing evidence that mycorrhizae serve as natural deterrents against root pathogens. The ectomycorrhizae, which form a fungal sheath or mantle over the root surface, seem especially suited for a protective role. Ectomycorrhizal fungi may influence disease resistance in plants by: (a) utilizing surplus nutrients at the root-soil interface, thereby reducing the stimulatory effect on pathogens; (b) secreting antibiotics; and (c) promoting, along with the root, a favorable environment for protective (antagonistic) rhizosphere microorganisms (Zak 1964).

The formation of ectomycorrhizae involves the short feeder roots of plants which are also the primary targets of so-called feeder-root pathogens such as species of *Phytophthora, Pythium, Rhizoctonia,* and *Fusarium.* Cortical root cells are enclosed in the Hartig net of the mycorrhizal fungus and the dense hyphal network or mantle over the root surface along with soil and microorganisms. Thus, for these roots the term mycorrhizosphere, rather than rhizosphere, more accurately describes the zone of soil affected. While part of the mycorrhizal benefit to plants relates to increased availability and absorption of nutrients that promote physiological resistance, other factors, as itemized by Zak (1964) and expanded upon by Marx (1972, 1975) and Schenk (1981), provide some protection from parasites.

Many species of ectomycorrhizal fungi, including *Suillus luteus, Clitocybe diatreta, Lactarius deliciosus, Scleroderma bovista, Thelephora terrestris,* and *Rhizopogon vinicolor,* produce antibiotics in vitro which inhibit species of either *Pythium* or *Phytophthora.* Other fungal symbionts, such as *Pisolithus tinctorius,* seem to be less inhibitory to these pathogens. One of the common antibiotics produced is a polyacetylene compound, diatretyne nitrile. Short roots of pine seedlings with ectomycorrhizae that produce diatretyne are less susceptible to infection by *Phytophthora cinnamomi* zoospores. Short roots without such ectomycorrhizae tend to be highly susceptible. Mycelial mantles formed by fungal symbionts that do not produce antibiotics also may protect roots from infection by *P. cinnamomi,* the mantle serving purely as a mechanical obstruction. Cortical cells of the root symbiont also respond to infection by ectomycorrhizal fungi, and produce volatile and nonvolatile organic compounds. Pine roots infected by the fungal symbiont, *Boletus variegatus,* produced volatile terpenes and sesquiterpenes in concentrations several times greater than amounts produced by nonmycorrhizal roots (Krupa and Fries 1971). Grown in pure culture, the fungal symbiont also produced the fungistatic volatile compounds isobutanol and isobutyric acid. Other ectomycorrhizae known to produce terpenes include those formed by *Pisolithus tinctorius, Thelephora terrestris,* and *Cenococcum graniforme* on pine. Whereas these volatiles inhibit growth of several pathogenic fungi, including *P. cinnamomi,* they also inhibit growth of the fungal symbionts. Herein probably lies an explanation for the apparent balance in nature between parasitism and mycorrhiza-induced resistance of host plants.

Ectomycorrhizae develop a microbial barrier to root-infecting pathogens along with the mechanical barrier of the fungal symbiont (Marx 1972). Microbial populations in the ectomycorrhizosphere vary in quantity and kind according to the specific fungal symbiont involved. Thus, it is believed that the influence of different ectomycorrhizal fungi on the mycorrhizosphere flora may influence infection by root pathogens. Microbial populations vary among ectomycorrhizae and are different from those of nonmycorrhizal roots and nonrhizosphere soil. The probability is suggested that mycorrhizal roots support a flora that may inhibit pathogen activity by competition or antibiosis, thus providing a degree of biological control. Little is yet known about the nature of mycorrhizal root exudates as compared with substances from nonmycorrhizal roots; however, some indirect evidence suggests that ectomycorrhizae may not induce the same chemical attraction to *Phytophthora* zoospores nor stimulate zoospore germination and germtube growth as is commonly observed around roots in vitro.

Since the parasitic activity of many nematodes is limited to the juvenile tissues of the feeder roots, interactions with mycorrhizal fungi can be expected to occur. Indeed, these interactions are largely quite different from those involving pathogenic fungi. While ectomycorrhizae serve as biological deterrents against fungal pathogens, nematodes are most likely to adversely affect mycorrhizal development (Marx 1972). Several mycophagous and plant-parasitic nematodes feed directly upon ectomycorrhizal fungi before or during initiation of the symbiotic association, and some plant-parasitic nematodes can feed upon the developed mycorrhizae. Such feeding may reduce the deterrent or protective benefit of the ectomycorrhizal mantle and increase susceptibility of pine roots to *Phytophthora cinnamomi*. The feeding behavior of some endoparasitic nematodes, such as lance (*Hoplolaimus coronatus*), suggests that ectomycorrhizae may provide more favorable feeding sites than nonmycorrhizal roots. Only occasional reports can be found where fungal symbionts of mycorrhizae produce toxins lethal or suppressive to nematodes.

The vesicular arbuscular (VA) mycorrhizae also may increase or decrease the general resistance of plants to attack by pathogenic fungi and nematodes (Schenck and Kellam 1978; Schonbeck 1979; Schenck 1981). Species of *Glomus, Gigaspora,* and other VA endophytes occur in the rhizosphere as resting spores; from these, hyphae grow along the host roots and form appressoria from which infection pegs penetrate the cell walls. Thereafter, growth of the fungus is entirely intra- and intercellular in the cortical tissue. Thus, no protective fungal mantle is formed on the root surface as in the ectomycorrhizae. The VA-mycorrhizal fungi are host-dependent and generally are beneficial, providing the plant with phosphorus and other nutrients. However, mycorrhizal infection induces certain chemical, physiological, and morphological alterations in the host plant that can change resistance to disease. In most instances, diseases caused by root-infecting fungi (*Olpidium brassicae, Phytophthora megasperma, Thielaviopsis basicola, Cylindrocladium scoparium,* etc.) are less severe in VA mycorrhizal plants. Similarly, numbers of plant parasitic nematodes (*Heterodera* spp. and *Meloidogyne* spp.) in roots and their harmful effects on plants are generally reduced. Reasonable speculation can be made regarding the mechanisms of resistance of VA mycorrhizal roots to invasion by pathogenic fungi, since the mycorrhiza strengthens the cell walls by increasing lignification and production of other polysaccharides; and certain pathogen-inhibiting compounds (phenols, ethylene, etc.) may be produced (Schonbeck 1979). On the other hand, detrimental effects of VA mycorrhizae on nematodes are not yet explained, and seem inconsistent with the ability of plant-parasitic nematodes to readily penetrate the mantle of ectomycorrhizae.

Despite the common occurrence of endomycorrhizae in field crops, the positive role of root-knot nematodes (*Meloidogyne* spp.) in the predisposition of some plants to attack by pathogenic bacteria and fungi is well known. Examples are increased Granville wilt disease (*Pseudomonas solanacearum*) of tobacco in the presence of *Meloidogyne incognita* (Lucas et al. 1955) and increased severity of various *Fusarium* wilt diseases in soil infested by species of this nematode genus (N. T. Powell 1971). The mechanisms involved in these interactions apparently go beyond the simple process of nematode penetration providing avenues of entry for other pathogens. Morphological and physiological changes occur in the host plant, sometimes converting a previously disease-resistant plant to a susceptible one.

7.5 Pathogenesis Without Parasitism

Some microorganisms at the root-soil interface are pathogenic but not parasitic. These nonparasitic organisms (exopathogens) cause disease but do not live within tissues of the affected plant; thus the term "host" is inappropriate and the susceptible plant is referred to as the "suscept". An exopathogen, such as *Rhizoctonia solani* also may be parasitic where exopathogenesis leads to the establishment of parasitism by inducing pathogenesis in advance of the parasitic association with suscept tissue.

Since pathogenesis infers the involvement of a pathogen, we must exclude the traditional nonparasitic diseases brought on by unfavorable environment and nutritional disorders. These, however, can have a profound influence on root exudates and microbial activities in the rhizosphere. Stress factors on plants not only induce more exudate release but also may provide the necessary favorable conditions for a pathogenic saprophyte to synthesize toxins injurious to the predisposed suscept. Poor plant performance due to nutrient deficiency alone is considered to be a disease, and such a condition can be created by competitive microbial activity at the root surface and a decline in available nitrogen, phosphorus, potassium, and minor elements. A very important phenomenon contributing to plant injury without involvement of a parasitic organism is allelopathy. Where plant root systems are in close proximity, organic acids, phenolic compounds and other substances released from one plant may injure another (Putnam and Duke 1978) and predispose it to attack by weak pathogens. In other instances, compounds released may protect roots from invasion by parasites.

A number of plant diseases have been attributed to nonparasitic plant pathogens (Woltz 1978). One of the oldest cases is frenching in tobaccos characterized by leaf rosetting and severe stunting of affected plants. Many probable causes have been suggested, primarily relating to certain nutrient deficiencies, poor soil aeration, and soil pH changes. In the opinion of Woltz, the disease may be due to an unstable organic toxin which is produced by a widely distributed soil organism; the factor can be destroyed by soil sterilization with heat or chemicals.

The role of toxins in the nonparasitic stunting of tobacco has been established in a series of investigations by Hendrix and associates in Kentucky (as reviewed by Woltz 1978). Both *Phytophthora cryptogea* and *P. parasitica* induce tobacco diseases, the former by nonparasitic means and the latter as a potent parasite, the black shank pathogen.

The relation of rhizosphere to exopathogenesis is evidenced by the fact that these pathogens, being largely saprophytic in nature, are highly subject to changes in metabolic activity with changes in the immediate environment. The extent to which exopathogen toxins, or volatiles such as ethylene (A. M. Smith 1976), can accumulate to effective levels and persist is dependent on other microbial activities in pockets or microsites of the rhizosphere and along the root surface where nutrients and oxygen levels vary. Pathogenesis in the absence of physical contact between pathogen and host can be demonstrated by artificial means. Wyllie (1962) placed soybean seedlings inside dialysis membrane (cellulose acetate) bags and covered the root system with sterile sand. These bags then were inserted into a sand-cornmeal medium infested with *R. solani*. Recovery and examination of the roots after 4 to 7 days revealed necrosis of primary and secondary roots resulting from a toxin produced by the fungus in the artificial rhizosphere and diffusable through the membrane.

7.6 Disease Control

Control of soil-borne plant pathogens and root diseases is achieved by (a) seed and soil treatment with pesticides, (b) use of biological control agents along with bioenvironmental management, and (c) breeding for resistance. The success of these measures depends upon having a thorough understanding of the cyclic life processes of specific pathogens. Most of the rhizosphere-microbe interactions discussed in previous chapters relate in some way to the efficacy of these disease control practices.

7.6.1 Use of Pesticides

The interactions between fungicides or nematicides and rhizosphere phenomena relating to root disease may take on many forms, depending upon the method of application. Chemicals toxic to soil microorganisms are usually classed as: (a) general toxicants which are active against a wide range of organisms, and (b) specific toxicants with a narrow spectrum of activity (Kreutzer 1963). Some general toxicants are fumigant biocides (methyl bromide, chloropicrin, methyl isothiocyanate, carbon disulfide, etc.) and would be expected to overwhelm the rhizosphere biota; however, such materials are usually applied in advance of planting dates, thus reducing the opportunity for interaction with the rhizosphere. More specific chemicals, particularly those applied as seed and in-furrow soil treatments, are in direct contact with early developing root systems where either the primary compound or its degradation products may inhibit or stimulate microorganisms in the root zone. Though there is little experimental evidence of what occurs between toxicants and microbes at microsites along root surfaces, logic suggests a potential for influencing spore germination (or soil fungistasis) and the ectotrophic growth of fungal pathogens. Chemicals applied as dips to ornamental cuttings and garden transplants may interact directly with the rhizosphere biota. Systemic compounds (benzimidazoles, oxathins, etc.), when used as seed treatments, can be taken up and translocated through developing root systems and into the foliage. The systemic nematicide oxyamyl can be used as a seed treatment for preventing nematode damage to crop plants (Truelove et al. 1977). The chemical in aqueous solution was taken up by seeds of bean, corn, soybean, watermelon, and cucumber. It was found that leachate from the treated seeds reduced the numbers of parasitic and free-living nematodes in the surrounding soil and lessened nematode damage to seedling roots.

Toxicants applied to foliage can alter the behavior of root-infecting organisms. Though downward movement of applied compounds does not occur as readily as upward movement from seed or soil, the use of various adjuvants may improve both uptake and translocation (Erwin 1973). Fungal populations, including the wheat root-rot organism *Helminthosporium sativum (Cochliobolus sativus)*, can be reduced in the rhizosphere by foliar applications of chloramphenicol (Jalali 1976). In tests with the systemic nematicide oxamyl, relatively high rates (10,000 ppm) applied as leaf sprays to tomato plants reduced the number of *Meloidogyne javanica* larvae penetrating roots and the percentage of larvae that entered the roots (Atilano and Van Gundy 1979).

The modern approach to pest control involves integrated programs to combat fungi, nematodes, insects, and weeds. It is to be expected then that a wide variety of soil- and foliar-applied compounds may act upon soil organisms for which the chemicals were not primarily intended. Nematicides affect fungi, fungicides affect nematodes and nontarget fungi or bacteria, and herbicides may affect all of these as well as host physiology (Rodríguez-Kábana and Curl 1980). The enormous complexity of the induced interactions around roots resulting from these nontarget effects defies a clear understanding of the real and potential effects on pathogens at the root surface. The implications are that specific toxicants or their degradation products may shift the balance within a microhabitat of microbial activity to favor one organism over another and either promote a favorable biological control effect upon a primary pathogen or interfere with the efficiency of intended chemical control. Chemicals applied to soil at fungicidal dosages may increase populations of bacteria and actinomycetes. Herbicides, and other toxicants as well, applied to soil or foliage, can either stimulate or inhibit propagule production and germination of pathogenic fungi in the rhizosphere. Thus, there is little doubt that a relationship exists between pesticide alteration of the root environment and pathogen behavior or disease, but the nature of these interactions is largely undefined.

7.6.2 Biological Control of Bacterial and Fungal Pathogens

Some serious root diseases cannot be controlled economically with chemical pesticides alone, and few resistant crop varieties have been developed. For such dis-

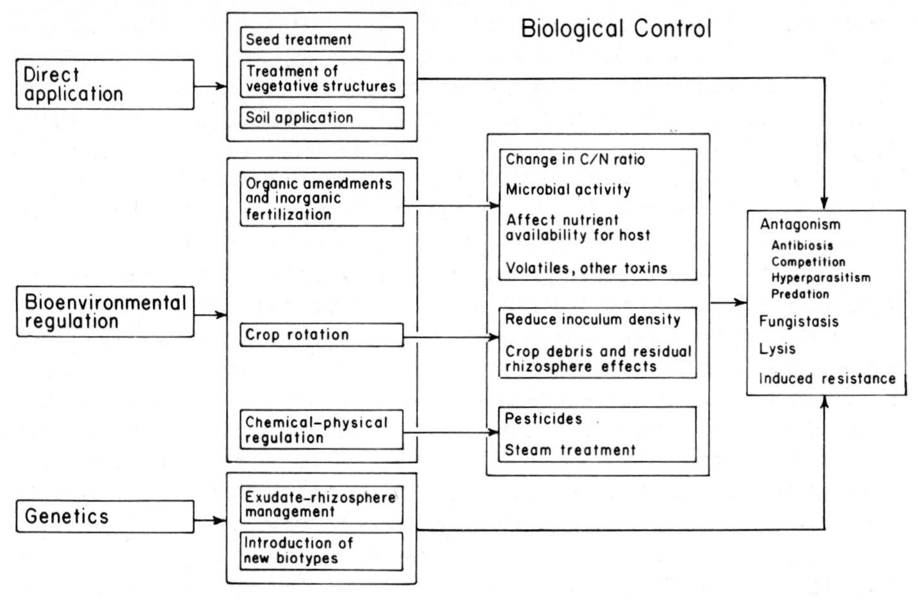

Fig. 7.4. The major approaches to biological control and some interactions involved in the control process

eases, biological control measures have become increasingly important. The role of the host plant in biological control of root-infecting organisms (Cook and K. F. Baker 1983) is based largely on the plant-mediated interactions that occur between pathogens and roots or between pathogens and antagonistic microorganisms in the rhizosphere and rhizoplane. The earlier discussions in this chapter on pathogen populations (7.2) and on growth and survival in the rhizosphere (7.3) also relate to this topic.

The major approaches to biological control and their ultimate relation to the root environment are outlined schematically in Fig. 7.4. Methods employed within the system are either direct or indirect. The direct procedure refers to inoculation or infestation of plant parts or soil with pure cultures of antagonistic microorganisms. Examples are: (a) seed inoculation with bacteria or fungal spores, (b) treatment of vegetative structures (roots and cuttings or transplants) by dipping in suspensions of biocontrol agents, and (c) direct application to soil. Indirect procedures involve: (a) bioenvironmental management with organic amendments, fertilization, crop rotation, and pesticides to augment biological control, and (b) plant breeding to promote a pathogen-suppressive rhizosphere. The underlying principle of all biocontrol tactics is antagonism (competition, antibiosis, lysis, hyperparasitism, predation), which is the ultimate desired product of microbial activity at the root–soil interface. This antagonistic influence varies according to cultural practices and soil amendments that affect the growth and vigor of plants.

The number of successes in biological control, leading to commercial formulations and application of agents, are few indeed. However, the potential for direct use of biological agents and the augmentation of indigenous antagonists by altering the soil ecosystem seem greater than ever before as environmental and economic demands have spurred a more intensified search for effective antagonists and application methods. Recently, new promising microbial agents have been discovered and tested for control of plant diseases caused by fungi (Papavizas and Lewis 1981; Kommedahl and Windels 1981; Ayers and Adams 1981), bacteria (Moore 1981), and nematodes (Kerry 1981 a). Effective antagonists are most likely to be found in the rhizosphere and rhizoplane of plants growing in disease-suppressive (decline) soils and in the root zones of resistant or escape plants in pathogen-infested areas. In the *Phymatotrichum* root-rot region of the southwestern United States, individual healthy cotton plants often persist within otherwise disease-devastated fields, suggesting that some natural, protective biological agents must be involved. Monocotyledonous plants are rarely affected by *P. onmivorum,* perhaps because the chemical nature of their root exudates promotes the activities of specific suppressive microorganisms. Candidate organisms isolated from sources such as these are tested for their inhibitory capabilities against specific pathogens and for their root-colonization ability and persistence in the rhizosphere of specific crop plants.

Seed inoculation with bacteria or fungal spores offers considerable promise for reproducible biological control against seed-borne and soil-borne pathogens. Kommedahl and Windels (1981) have provided a list of bacterial and fungal antagonists that have been used experimentally or commercially to control a number of pathogens and diseases. Among the bacteria showing high potential as seed protectants are *Agrobacterium radiobacter, Bacillus subtilis,* and *Pseudomonas fluorescens.* Common fungal antagonists suitable for coating seeds are species of *Penicillium, Trichoderma, Gliocladium,* and *Chaetomium. Laetisaria arvalis,* earlier identified as *Corticium* sp., also has shown potential for effective seed treatment, as demonstrated on sugar beet in greenhouse tests to control *Rhizoctonia solani* (Odvody et al. 1980). Seeds were coated by dipping them in a methyl cellulose solution and rolling them in a mycelia-sclerotia inoculum preparation of the antagonist.

Application to seeds offers an immediate competitive advantage for antagonists to colonize the spermosphere and the young seedling root, at least for a short time, permitting the seedling to escape the pre-emergence and early post-emergence damping-off phases of disease caused by unspecialized parasites such as *Rhizoctonia solani* and *Pythium* spp. Biocontrol agents seem to be most effective under environmental conditions that are not optimal for plant growth. For example, cotton and corn seeds planted too early in cold soils are more likely to survive seed rot and seedling disease if the seeds are coated with appropriate antagonists before planting. Some biological agents actually inhibit seed germination because of competitive deprivation of essential oxygen and nutrients at the seed surface. This can sometimes be overcome by adding another beneficial organism, such as the nonsymbiotic nitrogen-fixer *Azotobacter chroococcum* (Lynch 1978).

Microorganisms applied to seeds of various crop plants can be readily recovered from the rhizosphere for a short time, and in some cases for several weeks after planting. Thus, some reproduction in the rhizosphere is indicated, but whether permanent establishment occurs remains a question for most antagonists. Distribution of bacteria or fungal spores from treated seeds along the emerging roots may be largely passive as the organisms are moved in water films (Kommedahl and Windels 1981). Generally, fungal spores seem to survive longer than bacteria, since they are more tolerant of physical and chemical changes in the root environment. Bowen and Rovira (1976) have discussed the limiting factors involved in the colonization of root surfaces by bacteria and fungi.

Agrobacterium radiobacter has been developed commercially in Australia and subsequently used in the other countries for control of crown gall caused by *A. tumefaciens* in nursery plantings (Moore and Warren 1979). The nonpathogenic biocontrol agent, referred to as strain 84 (K84) occurs in stone-fruit nurseries in Australia along with the pathogenic form. The two forms are indistinguishable morphologically and physiologically except that strain K84 produces a highly specific antibiotic (a nucleotide bacteriocin) called agrocin 84. Commercial cultures are supplied in peat by techniques similar to those developed for *Rhizobium*. Seeds, cuttings, or roots of young plants, dipped in the bacterial suspension just before planting, are protected by the rapidly developing saprophyte. The nonpathogenic K84 is resistant to the antibiotic. When K84-inoculated peach seeds were planted in field soil, the biocontrol organism predominated around roots and plant crowns, suggesting a natural affinity for the rhizosphere and laimosphere (Kerr 1972). Kerr (1980) and Cooksey and Moore (1982) have discussed the mechanisms involved in the biocontrol function and acknowledged the probability that mutations may occur, altering agrocin 84 production by strain K84 or conferring resistance to agrocin 84 by *A. tumefaciens*.

The studies by Cooksey and Moore (1982) revealed that some biological control activity by strain K84 may be due to factors other than agrocin 84 production, since a mutant of *A. radiobacter* strain 84 that no longer produced the antibiotic also reduced infection when seedlings were inoculated 24 h before the pathogen was added. The mechanism of biological control by strain 84 may be related to a physical blockage of infection sites due to competition for infection-site binding (attachment of the bacterial lipopolysaccharide to the host cell wall). The findings of Du Plessis et al. (1985) agree with those of Cooksey and Moore (1982), and suggest that the host plant probably affects the mechanism of control. For example, an agrocin 84-insensitive pathogenic test strain of *A. tumefaciens* was completely suppressed by K84 on tomato but not on chrysanthemum plants.

Other bacteria have been successfully used as seed inoculants to improve plant health and yields (Cook and K. F. Baker 1983). Notable among these is *Bacillus subtilis,* which has provided significant protection of cereals and other crops from diseases caused by *Rhizoctonia solani, Pythium,* spp., and *Fusarium* spp. While the disease-control potential of *B. subtilis* seed treatment has been demonstrated in the field a number of times since 1960 (for example, Chang and Kommedahl 1968; Kommedahl and Mew 1975), only recently has a commercial formulation been used with success by farmers. Among participating farmers in Alabama, 17% experienced yield increases of 15% after peanut seed treatment with a bacterial preparation termed Quantum-4000 (Backman et al. 1984). The antibiotic and hormone-producing bacterial agent competitively colonized roots and persisted throughout the life of the peanut plant. *Pseudomonas fluorescens,* isolated from the rhizosphere of cotton, produces a potent antibiotic (pyrrolnitrin), and effectively protects cotton seedlings from infection by *Rhizoctonia solani* when the bacterium is used as a seed treatment (Howell and Stipanovic 1979). In all uses, the efficacy of control depends on the competitive ability of the control agent to establish in the rhizosphere and persist for a length of time sufficient to protect the developing seedling.

A number of abundantly sporulating fungi isolated from the rhizosphere of various plants have been used experimentally as seed protectants, and some have suitable characteristics for commercial formulation. A good example is *Penicillium oxalicum* which, when applied to *Pisum sativum* seed, functioned in a protective role against preemergence damping-off of seedlings in field soil (Windels 1981; Windels and Kommedahl 1982). Plant stand and vine and pod weights were equal to those of plants from fungicide (captan)-treated seed. The fungal spores, utilizing readily available nutrients from the seed exudate, germinate and reproduce sufficiently to protect the seed, but mycelium apparently fails to develop along the roots and in the rhizosphere to the extent that would be required to control post-emergence damping-off or root rot.

Treatment of vegetative structures with antagonistic microorganisms is another effective direct method of biological control. Usually, transplant seedlings and cuttings are dipped in a bacterial or fungal spore suspension before planting in field or nursery soils. For example, *Bacillus subtilis* has been effective against *Fusarium roseum* in transplanted carnation cuttings (Aldrich and Baker 1970) and also controls *Botryodiplodia solani-tuberosi* in potato seed pieces (Thirumalachar and O'Brien 1977). *Agrobacterium radiobacter* strain 84 provided exceptionally good control of crown gall when roots of peach seedlings were dipped in a suspension of the bacteriocinogenic strain before transplanting to pathogen-infested soil (Htay and Kerr 1974). As with seed inoculants, the effectiveness of microorganisms on vegetative structures depends on their compatibility with or resistance to other rhizosphere activities. *Pseudomonas fluorescens* and *P. putida* survived for at least 1 month on treated potato seedpieces planted in field soil, and were the predominant bacteria in potato root rhizosphere 2 months later. The bacteria enhanced growth and increased tuber yield (Burr et al. 1978).

The first successful biological control method leading to field application with commercially produced inoculum was devised by Rishbeth (1963) in England to prevent infection of pine trees by the root- and butt-rotting fungus, *Fomes annosus*, (*Heterobasidion annosum*). A formulation of the oidia of *Peniophora gigantea*, a minor pathogen and common inhabitat of forest organic matter, when applied to freshly cut stump surfaces, prevents stump invasion by the air-borne basidiospores of *Fomes*. This method is somewhat in contrast to other examples of biological control by treating vegetative structures, because the antagonist in this case has little opportunity to interact with the rhizosphere. When *Peniophora* is applied to the stumps, *F. annosus*, being a root-inhabiting parasite of low competitive ability, is either excluded from stump surface colonization or its advance through the decaying inner wood of host roots is limited by the more aggressive antagonist.

Application of biological agents to soil rather than to plant parts often leaves one to speculate on the role of the rhizosphere in the control benefit. The best chance for success with this method may be realized when antagonists are applied to plant-propagative media or seed beds before or immediately after planting, especially if the soil has been previously steamed to reduce competition at the seed and root surfaces. *Bacillus subtilis* has been successfully used in such procedures to protect ornamental plantings from *Rhizoctonia solani, Pythium ultimum,* and *Fusarium roseum* (K. F. Baker and Cook 1982). Myxobacteria of the *Cytophaga* group produce powerful proteolytic and chitinolytic enzymes. These organisms, when introduced into peat for container-grown tree seedlings, actively colonized the rhizosphere of seedlings and reduced chlorosis, stunting and mortality which usually accompany damping-off (Hocking and Cook 1972). Perforation and lysis of *R. solani* hyphae and *Cochliobolus miyabeanus* conidia in field soils have been attributed to one or more species of myxobacteria. However, the mechanisms of disease control in relation to specific rhizosphere factors have not been determined.

There is increasing evidence that mycoparasitism plays an important role in the mechanisms of biological control (Lumsden 1981; Ayers and Adams 1981). This action, the parasitism of one fungus upon another, must be at least partially involved in the many instances of disease control by soil inoculation with *Trichoderma* spp. and certain other known fungal antagonists. Effective protection of a host plant may depend on such action occurring in the rhizoplane at the infection site.

Two mycoparasites, both characterized by their pathogenicity on specific sclerotium-forming fungi, are of special interest even though neither has been extensively studied in the rhizosphere. *Coniothyrium minitans* and *Sporidesmium sclerotivorum* are natural destructive agents of sclerotia and of hyphae associated with sclerotium formation. Sclerotia become filled with mycelia of the invading mycoparasites and enzymatic lysis follows; the host cytoplasm disintegrates and the sclerotial cell walls collapse. The extensive work of Huang (1977, 1980), Ahmed and Tribe (1977), and others has provided convincing experimental evidence of the biological control potential of *C. minitans,* which parasitizes and kills sclerotia that develop on roots of *Sclerotinia*-infected plants and continues to destroy sclerotia inside the roots and stems. Inoculum of the mycoparasite for greenhouse and field application was prepared by culturing the fungus on an autoclaved mixture of barley, rye, and sunflower seeds, and this was applied to soil artificially infested with sclerotia of *Sclerotinia sclerotiorum.* The number of primary infection loci was reduced in mycoparasite-treated soil and yields were increased by virtue of a reduction in inoculum density (Huang 1980). Pycnidial dust of *C. minitans* also has been used effectively as a seed dressing or seed-furrow treatment for control of *Sclerotinia cepivorum,* cause of the white rot in onion (Ahmed and Tribe 1977).

Sporidesmium sclerotivorum, cultured on living sclerotia in moist sand and applied to field soil, can destroy 95% of the sclerotia of *Sclerotinia minor* within 10 weeks or less (Ayres and Adams 1979 a). Macroconidia of the mycoparasite germinate readily in soil adjacent to sclerotia and even at distances of up to 9 mm from sclerotia, but fail to germinate in soil without sclerotia (Ayers and Adams 1979 b). Presently, there is no evidence that *S. sclerotivorum* can exist as a saprophyte in nature, since it has only been isolated from sclerotia. Nevertheless, it is wide-spread and has been found in soil samples from at least ten states in the

USA. Since diffusible compounds from sclerotia seem to be essential for germination of the macroconidia, it would be surprising if similar compounds were not present also in root exudates or synthesized by some microorganisms in the rhizoplane.

These two mycoparasites, which together can destroy and significantly reduce the sclerotial inoculum of a number of extremely important root pathogens, deserve more intensified study of their ecology at infection sites within plant-root systems.

In many biological control trials, mass inoculation of field soil with large quantities of antagonist inoculum delivered on various organic materials has been used in efforts to overwhelm and suppress the activities of plant pathogens. *Trichoderma* spp. and *Gliocladium roseum* are prominent among the agents used to control *Rhizoctonia solani, Pythium* spp., *Sclerotium rolfsii* and other pathogens. Mass inoculation of soil is effective but rarely practical, and offers little probability of duplication of the results in different fields and in different years. Effective control, and opportunities to determine the mechanisms involved, are most likely to be realized where formulations and application methods allow delivery of antagonist inoculum directly to the vicinity of seeds, roots, or the plant crown (laimosphere).

Following the successful field control of sclerotial blight (*Sclerotium rolfsii*) of tomato by mass inoculation of soil with *Trichoderma harzianum* (Wells et al. 1972), Backman and Rodríguez-Kábana (1975) devised a practical procedure for application of the antagonist to field soil for control of *S. rolfsii* on peanuts. Diatomaceous earth (Celatom) granules were impregnated with blackstrap molasses to serve as a culture medium for production of *Trichoderma* inoculum in quantity. A dried preparation of the inoculum was then mixed with more molasses-impregnated granules and broadcast over field soil 70 days after planting peanuts, the critical period of disease occurrence. Disease control was equivalent to that obtained with the fungicide PCNB. *Trichoderma,* an aggressive, toxin-producing mycoparasite, grows rapidly from the granule food base and colonizes field organic matter which harbors the overwintering sclerotia of *S. rolfsii.* Some sclerotia are parasitized and destroyed by *Trichoderma.* The primary control, however, lies in the prevention of early summer colonization of the host laimosphere by the pathogen, protection of the young fruit-initiating pegs upon entering the soil, and reduction of the chance for pathogen invasion of the geocarposphere and infection of the peanut fruit. The formulation of *Trichoderma* inoculum into granules offers considerable promise for delivery in-furrow along with seed using conventional farm machinery.

Other systems for the application and augmentation of biological control agents have been devised primarily to promote antagonist proliferation in the rhizosphere. Lewis and Papavizas (1984) demonstrated that the successful establishment and subsequent proliferation of a variety of isolates of *Trichoderma* spp., *Gliocladium* spp., *Talaromyces flavus,* and *Aspergillus ochraceus* in natural soils depended upon the kind of propagule used in the preparation and the availability of a suitable food base. The test agents proliferated abundantly in soil when added as young mycelium on wheat bran, but not when conidia alone were used. Intimate contact of hyphae with the bran substrate was essential for proliferation to occur. A lignite-stillage carrier system also has been recommended for the application of *Gliocladium virens* and *Trichoderma harzianum* to soil to control *Rhizoctonia solani* damping-off in peanuts (Jones et al. 1984). Whereas carriers such as bran, vermiculite and peat may be too light or of unsuitable particle size for proper mechanical delivery, lignite can be ground to any particle size and offers greater density for application. The liquid stillage is a sorghum fermentation by-product of ethanol production. Lignite has been widely accepted as a carrier for applying inoculum of *Rhizobium* to legume seeds. Sclerotia-producing biocontrol agents, such as the Basidiomycete *Laetisaria arvalis,* may be especially well suited for distribution and competitive survival in soil. This agent is readily cultured on simple media and, when applied to sugar beet field soil as air-dried mycelium and sclerotia with or without a supplementary food base, increases to high populations and reduces the inoculum density of *R. solani* (Larsen et al. 1985). However, inconsistency of results between fields and between regions continues to plague the investigator, largely because of insufficient information about the concentration or distribution of agent propagules in relation to root surfaces to be protected.

Bioenvironmental management of indigenous antagonists in field soils involves organic amendments, crop rotation and other cultural practices, and chemical or physical soil treatments designed to increase antagonist populations and reduce disease severity (Papavizas and Lewis 1981; Palti 1981). Such procedures are usually performed in advance of planting dates; therefore, their effects on the rhizosphere-pathogen association are indirect, largely reflecting changes in host-plant physiology and vigor along with reduced inoculum density of pathogens available for infection.

Bioenvironmental control procedures are designed to induce certain selective mechanisms of antagonism (Lewis and Papavizas 1974; Papavizas and Lumsden 1980). While most biocontrol investigations have not determined the interactions between soil treatments and rhizosphere influence in establishing the desired control effect, these interactions must inevitably occur during the process of pathogen inhibition at the root surface and potential infection site. Among the selective mechanisms sought in bioenvironmental control are: reduced pathogen populations, prevention of propagule germination by increasing soil fungistasis, negation of soil fungistasis followed by germ-tube lysis, and toxicity to pathogen propagules by decomposition products of organic amendments. The rhizosphere effect is most evident when these mechanisms are associated with alteration of host-plant physiology.

Organic amendments have provided significant control of a number of root diseases, the mechanisms involved varying with the pathogen and the kind of amendment used. The addition to soil of cellulose and other organic materials that provide a high C/N ratio can reduce the severity of bean root rot caused by *Fusarium solani* f. sp. *phaseoli* (Papavizas et al. 1968). In nonamended soils, chlamydospores of the pathogen germinate in the rhizosphere of germinating bean seeds as a result of nutritional stimulation that nullifies soil fungistasis. In cellulose-amended soils, however, chlamydospores do not germinate in the rhizosphere, presumably because root exudates and microbial metabolites are not sufficient to overcome the increased fungistatic effect that accompanies cellulose decomposition and intensified microbial activity in the soil (J. A. Lewis and Papavizas 1974). This prolonging of the dormancy of propagules by enhanced fungistasis may lead to the exhaustion of propagule energy and finally death, thus reducing the inoculum density.

Some organic residues induce, rather than inhibit, propagule germination. This effect actually may be a desirable biocontrol feature if propagules are stimulated to germinate rapidly followed by immediate lysis of germ tubes before new propagules can be formed. Substances believed to be responsible for induced germination are found in plant tissues and in root exudates. It has been suggested that lecithin, vitamin E acetate, or unsaturated fatty acids from freshly added alfalfa or corn may explain why germination of *Thielaviopsis basicola* spores occurs in fungistatic soils amended with these materials (Lewis and Papavizas 1974).

Frequently, the same organic amendment may either enhance fungistasis and limit propagule germination or negate fungistasis and promote germination, depending on other soil factors. Thus, there is always a measure of uncertainty regarding the exact mechanisms involved. Nevertheless, since the early attempts to control potato scab in Canada and *Phymatotrichum* root rot of cotton in the southwestern U.S.A. with various organic residues, many instances of successful biological control have been recorded (Cook and K. F. Baker 1983). Legume residues in particular, being rich in available nitrogen as well as carbon compounds, promote intensive microbial activity which subsequently affects pathogen behavior in the spermosphere and rhizosphere of planted crops. During the decomposition of such residues volatiles may be released, triggering the germination of propagules which are then lysed; the release of ammonia also contributes to soil pH changes which influence the biological balance around seedling roots if seeds are planted in soil where residue decomposition is still active. Thus, the benefits of organic amendments are evident, but the real value will be fully recognized only when mechanisms are better understood in relation to root physiology and pathogen response at the root surface.

Crop rotation and biological control of plant diseases are closely related (Curl 1963). While maintaining desirable fertility for growth, crop sequence regimes also are aimed at denying plant pathogens a suitable host for a period sufficient to reduce the inoculum density, as pathogen propagules die of senility or are destroyed by saprophytic microorganisms. Certain crops leave residual rhizosphere effects that promote better growth of succeeding crops. These favorable effects can be attributed in part to the microbial populations supported by root systems of the preceding crops; crop refuse, of course, also provides substrates of varying C/N ratios that promote microbial activity. Among microbial populations are antagonistic species that impose some biological control on root-infecting organisms, and other species that produce plant-growth-stimulating factors enhancing plant vigor. In Chapter 4 we have seen numerous examples of the differential effects of plant species on the quantitative and qualitative nature of rhizosphere populations. Early studies in Ohio (Kommedahl and Brock 1954; Menon and Williams 1957) revealed that mold counts per gram of soil were consistently higher in soil continuously cropped to oats than in soil cropped to corn or wheat. Corn grew best on oat soil, resulting in fewer *Fusarium*-infected plants than on corn or wheat soil. Many other, similar, examples have been documented through the years.

While the practice of rotating crops continues to be essential in many farming operations, monoculture of certain crops along with minimal tillage can be beneficial. The take-all disease (root- and foot-rot) of wheat and barley caused by *Gaeumannomyces graminis* can be controlled by short-term rotations with nonsusceptible potatoes, corn, alfalfa, or oats; however, disease severity also declines after about 4 years of monoculture with wheat, suggesting a build-up of natural biological control agents (Cook 1981). Admittedly, one can only speculate as to how much of this benefit is derived from the accumulative rhizosphere effect of living roots from year to year and how much from the residual effects of crop debris incorporated back into the soil. Both must be important to the increase of antagonistic populations.

The microbiological degradation of organic amendments may release ammonium-nitrogen, subsequently resulting in the accumulation of nitrates through nitrification and thus affecting plant growth; or the accumulation of NH_3 and nitrites can be toxic to plants. In either case, the physiology and root development of plants are affected, altering the nature of rhizosphere components that influence root-infecting organisms. At this point, we are reminded of the early work of Snyder et al. (1959), who found that mature barley amendment providing a high C/N ratio controlled root rot of bean caused by *Fusarium solani* f. sp. *phaseoli,* whereas soybean and alfalfa residues with low C/N ratios increased root-rot severity. Apparently, the accelerated microbial activity accompanying a high level of available carbon may lead to nitrogen exhaustion, depriving the pathogen of an adequate source of required exogenous nitrogen. It seems logical to believe that C/N ratios along root surfaces also vary with the qualitative nature of exudates from specific plants under the influence of organic residues, thus affecting pathogen activity at potential infection sites.

Directly affecting the rhizosphere are inorganic fertilizers, particularly NH_4-N and NO_3-N sources, which alter the soil pH in the root zone and influence the pathogenic activities of various root-infecting fungi (Smiley 1975). Diseases caused by *Phymatotrichum omnivorum, Thielaviopsis basicola, Gaeumannomyces graminis, Verticillium albo-atrum* and *Streptomyces scabies* are favored by soil alkalinity; they are suppressed by NH_4-N and enhanced by NO_3-N. Certain other pathogens and diseases are favored by NH_4-N, or they are favored by environmental factors that are unfavorable for growth of the host plant. The suppressive influence of NH_4-N on take-all disease of wheat and the ineffectiveness of NO_3-N to control the disease were found to be correlated with the rhizosphere pH (Smiley and Cook 1973; Smiley 1974). Absorption and assimilation of NH_4-N by roots in poorly buffered soils causes a sharp reduction in the rhizosphere pH to a value less favorable to the pathogen and more conducive to antagonism by other microorganisms.

The enhancement or augmentation of indigenous antagonists by sublethal chemical or physical treatments also constitutes bioenvironmental control. Since soils are usually treated prior to planting, the rhizosphere is not directly altered, but rather the developing root system of the subsequently planted crop establishes a rhizosphere microbial population that is dominated by the most competitive species during recolonization of the treated soil. A number of examples have been cited in which the activities of microbial antagonist were promoted by soil treatments nonlethal to pathogens (McKeen 1975; Papavizas and Lewis 1981; K.F. Baker and Cook 1982). Treatments with fungicides, herbicides, or aerated steam have resulted in control of a number of diseases, even though the rates or duration of treatments used did not kill the pathogens; only weakening of a pathogen is necessary to shift the microbiological balance in favor of more aggressive saprophytes. Broad spectrum fumigants have been especially effective in this kind of disease control. Methyl bromide, chloropicrin, and carbon disulfide have been used to favor species of *Trichoderma, Penicillium, Aspergillus,* pseudomonads, and other microorganisms for indirect control of *Armillaria mellea* in citrus soils, *Gaeumannomyces graminis* in wheat, *Ganoderma* sp. in tea, and others. *Trichoderma* species often increase following sodium azide treatments. Common fungicides such as thiram and captan can increase *Trichoderma* or bacterial populations and suppress damping-off fungi. Nontarget effects of herbicides or sublethal heating of soil with steam frequently lead to the proliferation of antagonists and control of pathogen activity. Any of these practices is certain to influence the biological, chemical and physical nature of the rhizosphere, though the specific interactions that occur between host and pathogen at the infection site under these conditions are still largely unknown.

7.6.3 Role of the Microflora in Nematode Control

Biological control of plant parasitic nematodes by natural enemies (Mankau 1980) can be related to the rhizosphere by association and implication, though deliberate experimentation within this zone has rarely been performed. Viruses and rickettsias probably cause diseases of nematodes, but too little is presently known about either this role or their biology in the rhizosphere to permit an in-depth discussion. The most promising of the prokaryotes as biological control agents of nematodes is *Bacillus penetrans.* This organism, being an obligate nematode parasite, would not be expected to respond directly to root influence. However, perhaps the parasite becomes concentrated in the root zone indirectly. Following infection of the nematode host, endospores develop and ultimately fill the body cavity, then are released upon decomposition of the dead nematode and remain free in the soil until contacted by another nematode. Many nematodes, including plant parasitic forms known to be parasitized by *B. penetrans,* congregate around roots, thus providing a higher density of available hosts for the bacterial parasite. When potted soil containing *B. penetrans*-infected *Meloidogyne* populations is repeatedly recropped for several nematode generations, levels of the bacterium tend to build up, resulting in reduced nematode populations.

Biological control of root-knot nematodes can be demonstrated by infesting soil with spores of *B. penetrans* (Stirling 1984). Roots of plants in spore-treated soil have less nematode galling than roots in untreated soil, and growth of the plants is enhanced. However, successful exploitation of this organism for biological control of plant parasitic nematodes in the field may not be realized until more is known about the survival and dispersal of the bacterium in natural soil and cultural methods are devised for producing large quantities of inoculum.

Soil fungi belonging to many divergent orders and families are antagonists of nematodes and include the nematode-trapping or predacious fungi, endoparasitic fungi, parasites of cysts and eggs, and fungi that produce toxic metabolites. The major genera of predacious Hyphomycete fungi are *Arthrobotrys, Dactylella, Dactylaria,* and *Monocrosporium* which may represent over 100 species. These are well known for their specialized adaptations (constricting hyphal loops and adhesive knobs) for trapping nematodes and are considered to be potential biological control agents. However, much essential information is yet needed about their relationships to biotic and abiotic factors in soil, cropping effects, and microhabitats. Drawing upon a background of considerable experience with these fungi, Mankau (1980) observed that some predacious fungi appear to be rhizosphere organisms, while others probably should be classified as nonrhizosphere. He noted that their relationships to plant rhizospheres may be important with respect to their biological control capabilities. Because some plant parasitic nematodes tend to concentrate around roots, predacious fungi favored by a rhizosphere habitat would be most likely to influence plant-parasitic nematode populations. Mankau has observed that *A. dactyloides, Dactylaria brochopaga, M. ellipsosporium* and *M. gephyropagum,* which are restricted in saprophytic capability, are consistently found to be closely associated with plant roots. While the presence of nematodes is necessary to initiate the formation of trapping organs, the fungi also require an organic energy source other than nematodes in order to maintain a predaciously active state (Cooke 1962). The decomposition of sucrose, a component of root exudates, stimulates an increase in both the population of free-living nematodes and the activity of nematode-trapping fungi.

The Chytrid fungus *Catenaria auxiliaris* and the Oomycete *Nematophthora gynophila,* as well as the Hyphomycete *Verticillium chlamydosporium,* attack females and eggs of the cereal cyst-nematode *Heterodera avenae,* resulting in significantly lower population levels of the nematode in a wide range of soils (Kerry 1980). *Verticillium chlamydosporium* also is a principal egg parasite of the cyst nematode *H. schachtii* in Europe (Tribe 1979) and the root-knot nematode *Meloidogyne arenaria* in Alabama U.S.A. (Morgan-Jones et al. 1981). *Paecilomyces lilacinus* is an active parasite of *Meloidogyne incognita* (Jatala et al. 1980; Morgan-Jones et al. 1984). This highly competitive, saprophytic fungus has been field-tested, either as in-furrow applications with organic matter substrates or as seed treatments, by Dr. Jatala and colleagues in many developing countries worldwide, with some exciting results. Less studied are *Nematoctonus* spp. which, like some of the nematode-trapping fungi, produce polysaccharide toxins that immobilize nematodes before they are infected (Giuma et al. 1973). The relationships of rhizosphere factors to the build-up and behavior of these fungal parasites of nematodes have not been determined. Intensive cropping of susceptible cereals on land infested with *H. avenae* seems to contribute to increased populations of *N. gynophila;* this fungus appears to be an obligate parasite of cyst-nematodes and would not be likely to increase in the absence of its host.

A number of field observations have indicated that nematode-destroying fungi are implicated in the natural biological control of plant-parasitic nematodes. This is apparent when populations of specific nematodes fail to increase or the viability of eggs declines over time under intensive culture of susceptible crops, while at the same time populations of nematode or egg parasites increase. Little progress has been made, however, in the direct exploitation of the microflora for control of nematodes under field conditions, largely because of the relatively weak competitive nature of the prospective biocontrol agents when applied to natural soil.

A good opportunity for biological control of nematode diseases of plants lies in bioenvironmental management through selected cropping regimes or the use of organic amendments. It is common knowledge that the root systems of some crop plants promote higher populations of certain plant-parasitic nematodes, whereas more resistant crops do not favor nematode build-up. Whether the suppressive effect on populations is due entirely to the absence of a favorable host or in part to an increase in numbers of nematode-trapping or parasitic fungi remains to be determined. Organic amendments have been effective for control of certain plant-parasitic nematodes, though the mechanisms of control in relation to rhizosphere influence are far from clear. Soil amendments with chitinous materials or with peanut and cotton-seed oil cakes of narrow C/N ratios have been effective in controling root-knot nematodes (Mian et al. 1982; Mian and Rodríguez-Kábana 1982). Elements of a specific mycoflora capable of parasitizing eggs of *Meloidogyne arenaria* and *H. glycines* develop in response to chitin (Godoy et al. 1982) and various other amendments. Crustacean chitin, oil-seed cakes, and chicken litter, in the process of decomposition, release ammoniacal nitrogen in the soil through the action of proteolytic and deaminating enzymes produced by microorganisms. This can lead to soil pH changes and the accumulation of nitrates which may affect the life cycles of nematode-destroying fungi at the root surface.

7.6.4 Role of the Soil Fauna in Biological Control

Protozoa, nematodes, and microarthropods ingest bacteria, fungal spores, and mycelial fragments; some are also predators of plant-parasitic nematodes. Yet we know little about their biological control potential or their activities in the rhizosphere. Mycophagous, vampyrellid amoebae are widely distributed in agricultural soils and may play a role in the natural control of pathogenic fungi by destruction of spore populations (Old and Patrick 1979). Populations of protozoa generally are higher around plant roots (Chap. 4), presumably in response to a more abundant bacterial food supply. Sporozoans of the nonflagellate protozoa have been reported as parasites of nematodes, but largely of the nonstylet forms; the stylet canal of plant-parastic nematodes is believed to be too small for entry of protozoan cysts.

Tardigrades can be found in root and soil samples, and a number of observations have implicated them in significant reductions of plant-parasitic nematode populations, even though they are nonselective feeders (Norton 1978).

The biomass of predacious nematodes is substantial in soils generally and in the rhizosphere. Though their biocontrol potential is unknown, nematodes in the Mononchida, Dorylaimida, and Diplogasteroidea are largely predacious but omnivorus with individual species feeding on fungi, algae, nematodes, and other soil

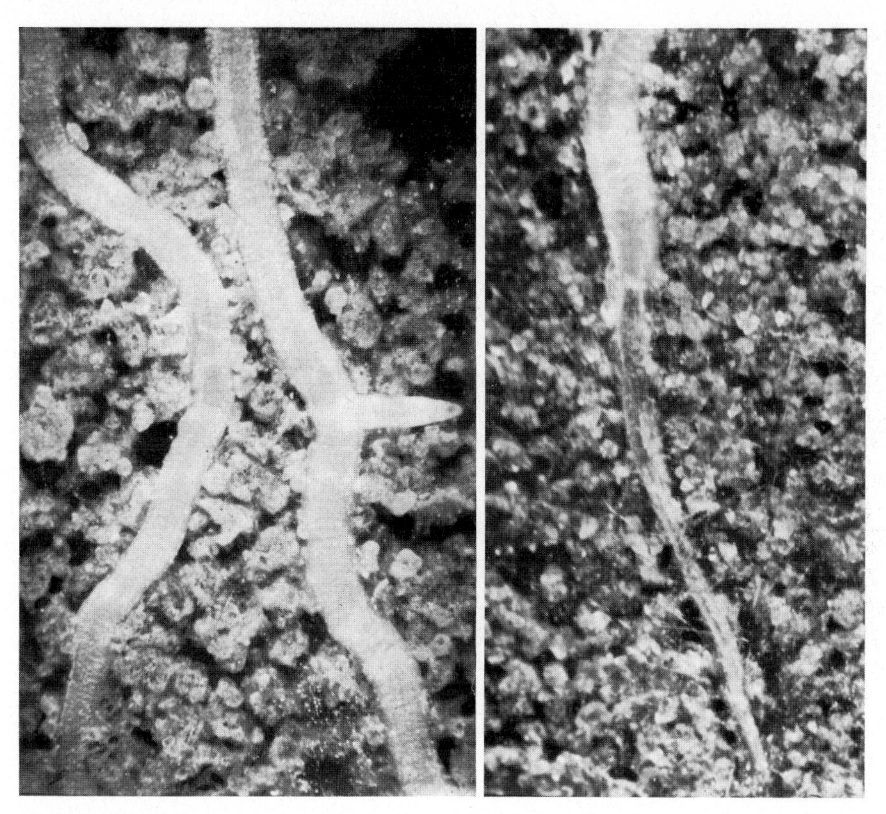

Fig. 7.5. Collembolan protection of roots from infection by *Rhizoctonia solani*. (*Left*) Roots from pathogen-infested soil with mycophagous *Collembola*; (*right*) diseased root from pathogen-infested soil without *Collembola*

microfauna (Mankau 1980). The Mononchida have attracted most research attention in terms of their natural populations and biology, but no reasonable judgement of their biocontrol value can be made based on the meager information now available.

Rhizosphere-inhabiting microarthropods (Acarina and Collembola) may prove to be formidable biocontrol agents. Many species are mycophagous and congregate where fungi are most available. Others are predacious on nematodes. Species of collembolan genera, *Proisotoma* and *Onychiurus*, extracted from the rhizosphere of field cotton in Alabama were observed to feed destructively on *R. solani* (see Fig. 5.8) and a number of other root-infecting fungi (Curl 1979). In laboratory experiments, Collembola protected cotton roots from infection by *R. solani*. Roots growing in previously sterile soil infested with a chopped oat-grain inoculum of the pathogen (0.05 g kg^{-1} of soil) suffered high mortality, whereas Collembola added in numbers of 1,600–4,800 kg^{-1} of soil to the same kind of plant culture provided significant protection (Fig. 7.5). Subsequently, effective control of pre- and post-emergence damping-off of cotton seedlings was obtained by adding Collembola to *Rhizoctonia*-infested field soil in pots. Insect

Fig. 7.6. Biological control of *Rhizoctonia* disease in seedling cotton by soil infestation with mycophagous Collembola. (*Left to right*) Unamended field soil, 75% seedling emergence; field soil with *Collembola* (2,000 kg^{-1} soil), 99% emergence; field soil with supplementary *Rhizoctonia* inoculum but without *Collembola*, 48% emergence; field soil with both added *Rhizoctonia* inoculum and *Collembola*, 88% emergence

numbers of 1,000–2,000 kg^{-1} soil reduced the disease severity index value by almost 50% (Fig. 7.6). These animals are unspecialized feeders and may consume other fungi; however, the abundantly sporulating Hyphomycetes (*Penicillium, Aspergillus, Trichoderma,* etc.) seem to be less suitable food sources than are the nonsporulating or sparsely sporulating pathogens.

Collembola can reduce the populations of nematodes substantially, at least under experimental conditions. An extensive study by Gilmore (1972) probably established the status of our present knowledge of the predatory action of Collembola on nematodes. It was estimated that a single collembolan might consume 150 nematodes daily. Since these predators are unspecialized in their feeding habits, they would not be expected to distinguish between plant-parasitic nematodes and saprophytic forms.

7.6.5 Host Resistance

Breeding for resistance is aimed primarily at altering certain morphological and physiological characteristics of the host leading to the exclusion or inhibition of pathogens. Some features altered are cell-wall and cuticle thickness, suberin or lignin content, fungitoxic compounds, host-enzyme systems, and metabolic pathways. However, in the case of root-infecting organisms, the nature of the environment created by the rhizosphere of different plant varieties can play a critical role in resistance or susceptibility.

A background in basic genetics relating to pathogenesis has been provided in three preceding volumes of the Springer-Verlag Advanced Series in Agricultural Sciences (Wheeler 1975; Robinson 1976; Vanderplank 1978). Generally, two types of resistance are recognized, though the terminology used to describe them varies among researchers. Resistance which is controlled by a single gene (monogenic) or only a few genes (oligogenic) is often referred to as vertical resistance. This type is usually specific for a certain race of pathogen and is associated qualitatively with variation in the host; it may confer potential for immunity. Among soil-borne pathogens, the highly specialized vascular wilt organisms are most subject to control by a single dominant gene.

Polygenic, or horizontal, resistance is nonspecific and most effective against variants of a pathogen; it is not lost through adaptation by the pathogen. Therefore, this type of resistance may offer some opportunity for controling the less specialized root-infecting organisms such as *Phytophthora* spp., *Pythium* spp., *Aphanomyces euteiches, Rhizoctonia solani,* and *Fusarium solani.* The foregoing generalizations should be taken with caution, however, since exceptions are common. For example, breeding programs in the search for resistance to root rots of various dicotyledonous plants have revealed both vertical and horizonal resistance to certain unspecialized pathogenic fungi (D. H. Wallace and Wilkinson 1975). Success in breeding for root-rot resistance apparently is derived largely from trial-and-error research for each plant species.

Since most soil organisms are not pathogenic, it follows that most plants (perhaps all plants) possess a degree of general resistance which is not well defined genetically but is associated with the physiological factors that govern host vigor. Environmental conditions that favor plant growth and maximum vigor contribute to general disease resistance, whereas conditions that impose stress upon plant growth may predispose the plant to infection. K. F. Baker and Cook (1982) discussed these points in view of the relation of polygenic and general resistance to biological control. Horizontal and general resistance are most subject to modification by the physical, chemical, and biotic features of the rhizosphere that were discussed in earlier chapters. The rhizosphere serves as the connecting link between the predisposing elements, or the growth-promoting elements, of the soil environment and genetic resistance. Wheeler (1975) reminds us that the events which occur during pathogenesis reflect the combined activities of the genetic systems of the plant and the pathogen. Aggressiveness and virulence of a pathogen may be associated with either vertical or horizontal resistance.

Genetic management of the rhizosphere seeks to manipulate the physicochemical and microbiological status of the rhizosphere to reduce pathogen activity and disease induction. The promise of success is based on the assumption that the quantity and quality of root exudates can be varied in the genotype to govern specific microbial populations and their behavior. A pathogen must first penetrate the complex floral sheath around roots before the true resistance of host cells is encountered. Herein lies the point of integration of biological and genetic control mechanisms.

Differential rhizosphere effects of resistant and susceptible plants have been observed many times. Krigsvold et al. (1982) clearly showed that microsclerotia of *Cylindrocladium crotalariae* germinated significantly better in the rhizosphere of a susceptible peanut plant cultivar than in the rhizosphere of a resistant cultivar; the difference apparently was related to a differential in carbon levels of the root exudates. Pea seed and seedling exudates of plant introduction accessions with the A-gene for anthocyanin production contain the anthocyanidin (anthocyanin-aglycone) pigment delphinidin which is fungistatic to conidial germination of *Fusarium solani* f. sp. *pisi* (Kraft 1977). However, the pathogen can germinate in the presence of delphinidin if glucose is available in sufficient amounts. Thus, these plants can be susceptible if they exude sufficient sugar to negate the inhibitor effect. Higher pathogen populations are frequently observed around roots of susceptible plants, suggesting a more favorable environment for the organism. On the other hand, if one or more microbial species antagonistic to the pathogen are favored to proliferate in the presence of root exudates, the resulting biocontrol potential might outweigh the stimulation of a pathogen population. Continuing speculation on what probably occurs or what might occur, rather than discourse on what does occur at the root surface, is due to the inability of researchers to distinguish and measure microbial activities under field conditions or to relate with confidence in vitro results to a natural environment. Nevertheless, new hope is generated by studies such as that of Atkinson et al. (1975), who found some interesting relationships between host (wheat) reaction to root rot (*Cochliobolus sativus*) and rhizosphere characteristics. The total bacterial population in the rhizosphere of S-615 (susceptible variety) was twice that in the rhizosphere of two other lines, Apex (resistant) and S-A5B (resistant due to substitution of a chromosome pair from Apex). None of the S-615 bacteria exhibited antibiosis to *C. sativus* in standard tests, whereas 20% of those tested from resistant lines inhibited the pathogen. Line S-615 differed from S-A5B by only one chromosome. Seed-bacterization tests with rhizosphere antagonists failed to establish a cause and effect relationship between the incidence of bacteria antibiotic in vitro to *C. sativus* and resistance to root rot. However, this does not exclude the probability of rhizosphere-mediated biocontrol in situ, since effective antagonists in field soil need not exhibit antibiosis in vitro. Since the percentages of different physiological groups of bacteria in rhizosphere populations vary with plant species and varieties, it is predictable that percentages of antibiotic producers also will vary. Laboratory tests cannot tell us with certainty what happens in a field soil, but the potential for a natural relationship between rhizosphere populations and resistance of a variety to root pathogens seems real. Admittedly, the assumption must be made that the necessary antagonistic (biocontrol) organisms are present in the soil in which the host is planted, or that such organisms occur naturally on the planted seed.

Another example of genetic manipulation of the rhizosphere comes from an organized regional project on rhizosphere ecology in the southwestern United States. A Texas A & M University research program referred to as the MAR (multi-adversity resistance) system is designed to develop cotton cultivars with resistance to root rot (*Phymatotrichum omnivorum*) and to other major diseases and insects (L. S. Bird 1982).

The MAR procedure identifies genotypes that have seed and root exudates differing in quantity and quality of carbohydrates, calcium, potassium, sodium, and magnesium. Certain combinations of these nutrients are believed to favor seed- and root-colonizing microorganisms that provide a natural protective biological system through competition and other forms of antagonism. Cultivars selected from annual field-test performance trials have shown levels of resistance sufficient to reduce root-rot losses when used in conjunction with crop management practices aimed at reducing the inoculum density of the pathogen.

Other sources of resistance can be found in either natural or applied processes that promote a host response leading to the suppression of pathogen activity. Induced resistance has been most evident in relation to foliar diseases in cases where disease reaction was altered by various physical, chemical, or biological agents (Wheeler 1975). For example, resistance can be induced by first inoculating with a potential pathogen to which a plant is resistant and later inoculating with a normally virulent form. This type of resistance is generally attributed to a hypersensitive reaction of host tissue and the production of phytoalexins (Paxton 1975). These antibiotics are normally present in plants in extremely small concentrations until stress or injury is imposed by a pathogen. This physiological stimulus then results in greater antibiotic production. The reaction also may occur in response to phytotoxic compounds produced by saprophytic microorganisms, and there is reason to believe that much resistance is induced naturally by microbial activity in the rhizosphere.

Burden et al. (1974) identified phytoalexin-like, fungitoxic compounds in root exudates of *Pisum sativum* and *Phaseolus vulgaris* and suggested that such compounds might contribute to natural resistance of plants to fungal and bacterial pathogens. Constituents of cotton-root exudate are known to affect growth and proteolytic enzyme production by *Verticillium albo-atrum* (Booth 1974). A *Verticillium*-tolerant cotton cultivar exuded 3.5 times more choline from roots than did a susceptible cultivar. Alanine was exuded in greater quantity by the susceptible cultivar. Alanine increased growth (dry weight) of the pathogen in vitro by 320% and stimulated polygalacturonase activity, whereas choline added to the medium significantly reduced growth and suppressed enzyme activity. Differences in pathogen response to these exudate constituents may be indicators of more complex physiological activities related to disease resistance.

Many volatile compounds, which are common components of the soil atmosphere, affect growth of both plants and pathogens, either adversely or by stimulation. Among the volatiles released by germinating seeds and seedlings are ethylene, ethanol, methanol, formaldehyde, acetaldehyde, formic acid, and propylene (Vančura and Stotzky 1976). When these inhibit plant pathogens and prevent or delay infection, or slow the disease process after penetration, the relation to resistance is clear. In other cases, a volatile such as ethylene may be produced by microorganisms in response to root exudates at anaerobic or microaerobic sites in the rhizosphere (A. M. Smith 1976). These anaerobic microislands occur where intense microbial activity and root growth create oxygen sinks, which then favor the proliferation of microaerophilic organisms such as certain species of *Clostridium*. Ethylene and other volatiles have been implicated in the natural phenomenon of soil fungistasis (Chap. 5), which prevents pathogen propagules from germinating, thereby contributing to host protection; this can be viewed as a factor in resistance since the protection is indirectly mediated by the potential host. Benzyl isothiocyanate (BITC), a highly bioactive volatile compound in the rhizosphere of growing papaya plants, is lethal at 0.01 M to the mycelium of *Phytophthora palmivora,* a serious root-rot pathogen. However, the studies of Tang and Takenaka (1983), using a continuous exudate trapping system for undisturbed rhizosphere, suggest that the rate of BITC released may not be sufficient alone to account for observed differences in resistance between cultivars. It should be noted that all pathogens do not respond alike to a specific volatile; some may be stimulated to germinate and grow, thus contributing to the pathogenic phase.

It is evident that a high degree of resistance to root disease is induced naturally. Without this resistance and the associated elements of natural biological control, pathogenic organisms could grow unimpeded throughout the root system of a susceptible plant, as freely as in pure culture on a favorable medium. Since no method has been devised to duplicate the complex biological and physical environment around roots in natural soil, the declared role of phytoalexins, ethylene, or other compounds in resistance is based largely on assumptions made from in vitro studies. Nevertheless, while more specific information is being sought to distinguish between resistance mechanisms and biological control at the root surface, efforts to manipulate the rhizosphere for disease control are at the forefront of research activity.

Chapter 8 Current Trends and Projected Emphasis

8.1 General Overview

When Lorenz Hiltner coined the term rhizosphere in 1904 he could not have envisioned the stability of the term nor the worldwide research effort that would center upon rhizosphere ecology in relation to agriculture. Apparently, Hiltner never used the term again in any subsequent publications. Nevertheless, that single spark was sufficient to ignite the imagination of other biologists, ultimately resulting in the accumulation of a wealth of information derived from both basic and applied research.

We have seen in Chapter 4 that a large, diverse, and unstable population of soil microorganisms exists in the rhizosphere. Living plant roots penetrate the soil profile, creating a totally different environment by virtue of root exudates, sloughed epidermal cells, and soil physical changes that promote or alter microbial activity. Microbial metabolites also contribute to this unique environment as they stimulate or inhibit associated microorganisms and plant root growth. Nematodes and other small soil animals enter the melee, some feeding on plant roots and others exploiting the microflora as a food source. Thus, in field soil the rhizosphere is indeed "where the action is". This action begins in the spermosphere of the germinating seed and intensifies in the rhizoplane and rhizosphere of the developing plant as the root-system density increases and the rhizospheres of many individual roots merge. Following death of the plant, a true rhizosphere effect ceases to exist as an opportunistic microbial community begins the decomposition process.

Rhizosphere microorganisms have been appropriately labeled by M.E. Brown (1975) as "opportunists, bandits, or benefactors". Certainly, many common soil microbes are opportunistic as they exploit the amino acids, carbohydrates, and other nutrients in root exudates for their metabolic needs. Some microorganisms competitively rob plants of nutrients, produce phytotoxic substances, or induce root and vascular diseases. But at the same time, many rhizosphere inhabitants return something of value to the plant (Chap. 6). Thus, the rhizosphere effect is largely a reflection of the influence of microbial activity on two broad processes that ultimately determine crop productivity (Fig. 8.1). These are: (1) nutrient availability and uptake in relation to plant growth, and (2) plant pathogen activities and root disease. These processes are, of course, interrelated and subject to regulation by the genetic constitution of both plant and microbe.

In this chapter, we emphasize certain major advances made in rhizosphere investigations relative to plant growth and health, and attempt to focus upon current trends and the opportunities that lie ahead. Certain points of relevance to previous chapters have been delayed for discussion at this time, since they repre-

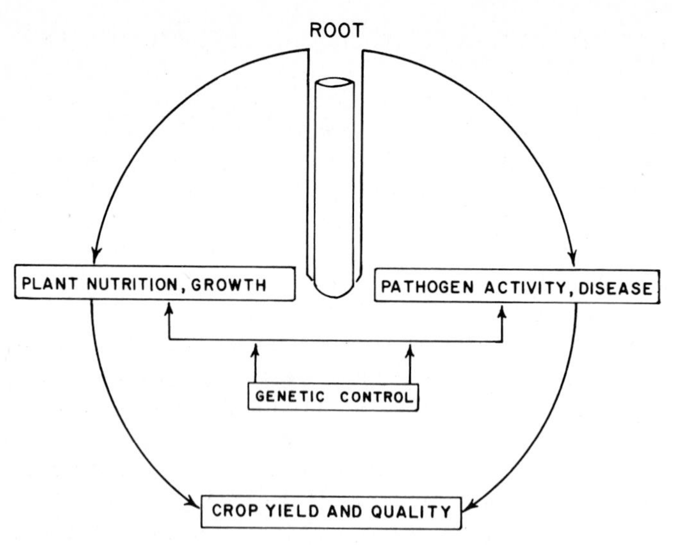

Fig. 8.1. A schematic representation of the influence of rhizosphere ecology on the two major processes relating to crop production and quality

sent new developments in rhizosphere research and tell us something of the future emphasis that may be expected. To some extent, this chapter also represents an updating of certain subject matter discussed earlier.

8.2 Status of the Technology

The current status of our understanding of the rhizosphere phenomena that affect crop production is largely a reflection of the advances made in research methodology. Though a great volume of valuable information has been accrued with simple model systems, most of the data provide only a basis for speculation as to what actually occurs in natural field environments. These basic studies must be continued and expanded, but at the same time systems need to be devised to utilize or simulate natural conditions for testing the validity of results obtained under more rigidly controlled conditions.

Much has been learned about the biology and microbiology of the rhizosphere, particularly the quantitative and qualitative nature of root exudates and corresponding microbial populations. Components of exudates collected from highly artificial systems have been identified and tested for their effects on rhizosphere microorganisms. However, root exudates are readily metabolized by microorganisms in natural environments; thus the soil solution bathing the roots of growing plants must contain many intermediate compounds as yet unidentified. In a natural root-soil environment these intermediates, along with other organic matter decomposition products and microbial metabolites, contribute to a soil solution of extreme complexity. Little is known about the synergistic action of these components on rhizosphere microorganisms, or on root morphology and physiology in relation to root growth and function.

Only a small fraction of the natural microbial population of the rhizosphere is represented in isolations obtained using selective media. Therefore, both the quantitative and qualitative assessments of bacterial and fungal populations are incomplete. As a result, most studies of microbial interactions with plant roots or with other microorganisms have been performed with a narrow spectrum of the common microflora. Ways must be found to extract, or study in situ, many other species that exist in the rhizosphere but do not grow on standard nutrient media; some of these are obligate parasites of other organisms.

New methods or improved techniques are constantly being sought for the direct observation of root growth and the microorganisms on root surfaces or soil particles. Experimental plants are commonly grown in glass-front (or Plexiglas-front) boxes of various design for direct observation of root development. The most modern development of this principle is the root observation laboratory or rhizotron for observing roots under field conditions (Bohm 1979; Huck and Taylor 1982). An old method, recently reactivated for new uses in field observations, employs a root-observation tube or borescope referred to as a mini-rhizotron (Bohm 1974). A hole of about 70 mm in diameter and 110 cm in depth is bored either vertically or at an angle into a field-soil profile before planting, and a slightly smaller diameter Plexiglas tube about 130 cm long is inserted, leaving part of the tube aboveground. The tube interior is equipped with magnifying lenses and a light source suitable for observing and photographing roots that grow over the exterior surface of the tube. A version of this device has been used effectively in the blackland soils of Texas to observe the mycelial strands and sclerotia of *Phymatotrichum omnivorum* on roots of cotton (Rush et al. 1984).

The scanning electron microscope (SEM) has been extremely useful in demonstrating the distribution and spatial relationships of microorganisms in soils and on plant-root surfaces. The value of this instrument is well illustrated in *Ultrastructure of the Root-Soil Interface* by Foster et al. (1983). Campbell and Porter (1982) describe a low-temperature SEM technique with which soils and roots can be prepared by rapid freezing for examination in a frozen, hydrated state that preserves water films and reveals natural root-soil interface microhabitats of organisms. Electron microbeam analysis and SEM have been used for making quantitative, point-count determinations of phosphorus, potassium, calcium, and magnesium at root-soil interfaces of peanut and soybean plants (K. H. Tan and Nopamornbodi 1981). This procedure showed soil to contain lower concentrations of P, K, and Ca than are found in root tissue, but the Mg content was higher in the soil. On the basis of a sharp break in element distribution with distance from the root, the authors estimated that the rhizosphere zone was 0.20 mm wide. This kind of study could by advantageous if coupled with microbial observations in relation to nutrient availability and uptake, particularly with regard to mycorrhizal associations.

Advances have been made in methods for estimating microbial population density and biomass in soils and rhizospheres. Newman and Watson (1977) presented a mathematical model for predicting the abundance of microorganisms in the rhizosphere (as microbial dry weight cm^3 soil) in relation to distance from the root surface and time of initial exudate release. Such a model could be used for predicting the effects of soil physical and biological changes on microbial and substrate (exudate) concentrations. J. P. E. Anderson and Domsch (1978) suggest that a quantitative measurement of microbial biomass can be obtained by estimating the amount of carbon in living, nonresting soil microbial populations. Following the addition of increasing amounts of glucose to soil, the maximum initial respiratory response (ml CO_2 unit^{-1} soil h^{-1}) is determined and converted to biomass (mg C unit^{-1} soil) through a regression equation. Martens (1982) devised an apparatus that could be used to study the quantitative relationships between rhizosphere microbial populations and root exudates. The key component of the apparatus is a polyethylene molecular-filter core which, when impregnated with synthetic substrate, slowly releases inorganic and organic materials into soil contained in an evacuated incubation chamber. Respired CO_2 can be collected and soil samples taken at regular distances from the "root" surface for the determination of microbial populations or biomass.

Enzyme methodology offers additional potential for the assessment of microbial activity, based on the assumption that the microflora provides the major source of many soil enzymes. Invertase, urease, β-glucosidase, and phosphatase activities are commonly higher in rhizosphere than in nonrhizosphere soil, and such activities vary with plant density and composition. That soil enzyme activity can be correlated with microbial proliferation has been alluded to in a number of investigative reports edited by R. G. Burns (1978). Soil-enzyme analysis has been used to practical advantage in field investigations. In crop-rotation experiments, xylanase (Rodríguez-Kábana 1982) and catalase (Rodríguez-Kábana and Truelove 1982) activities were high under wheat, soybean, and winter legumes and low under cotton and winter fallow. These results were interpreted as reflecting greater microbial activity, and thus increased enzyme activity, in soil with crops of high root density compared with crops of lower root density. The qualitative nature of carbohydrates released in root exudates was believed to be a major contributing factor.

Documentation of nematode numbers and distribution has been based largely on mechanical methods of extraction and, therefore, may represent the true population more nearly than is the case for bacteria and fungi. However, a common problem with both nematodes and the microflora is assessing whether they are active at sampling time or in a quiescent state as unhatched nematode eggs or as dormant propagules. Other members of the soil small fauna, particularly the protozoa and microarthropods, have been grossly neglected and not adequately considered as interacting rhizosphere components.

Major technological challenges lie ahead to determine, in natural environments, the interactions among chemical, physical, and biological components of the rhizosphere and their impact on plant performance. The methodology for determining such complex interactions and for elucidating the underlying mechanisms is not likely to be found within a single discipline, but will require a multidisciplinary approach, particularly featuring contributions in molecular biology that may better define the pathways and mechanisms of intergeneric genetic exchange among bacteria in the rhizosphere. Whereas a bacterial cell in isolation has a relatively small genetic information content and a limited adaptive potential, bacteria in their natural habitats are essentially in a genetically "open" system with the potential for cell-to-cell transfer of genes made possible by various vector systems, such as bacteriophage (Reanney et al. 1982). The extent to which gene flow occurs among members of natural microbial communities is not known. However, the probability seems great indeed in the vicinity of plant roots where diverse and large populations of the microflora, fed by root exudates, have many opportunities for physical association. Surely, herein must be some of the mechanisms of resistance to antibiotics and the competitiveness of some microorganisms subsequently affecting the growth of plants or the behavior of root pathogens.

8.3 Relating to Plant Growth

It seems appropriate at this time to reexamine the role of the rhizosphere in growth-related functions of plant roots with a look toward some current research trends. We have seen (Chap. 6) that microorganisms which confer benefits on plant growth are primarily in two categories: (1) those physically linked with plant roots in symbiotic relationships (the rhizobia and mycorrhizal fungi) and (2) the nonsymbiotic microorganisms that fix atmospheric nitrogen, produce growth-regulatory substances, or bring about changes in the chemical composition of soil, ultimately enhancing nutrient availability and adsorption. Included in the second category also are microorganisms that degrade phytotoxic substances produced by other organisms, thereby releasing plants from growth inhibition by such substances.

8.3.1 Challenge of Symbiosis

Symbiotic nitrogen fixation in leguminous plants has been well defined, and knowledge on this subject is being applied regularly to field-crop production. However, the precise mechanisms involved in the *Rhizobium*-host relationship need further clarification, particularly with reference to the establishment and survival of rhizobia in the rhizosphere prior to host infection. An adequate population of *Rhizobium* in the rhizosphere is essential for infection to occur, and root exudates are directly or indirectly responsible for proliferation of the organism.

Root exudates contain tryptophan which is oxidized to indole-3-acetic acid (IAA) by rhizobia and other bacteria; the IAA, together with unidentified cofactors, presumably from plant roots, induces the curling of root hairs which usually precedes infection. Also, IAA serves to alleviate some of the known detrimental effects of nitrate on root-hair infection. Polygalacturonase produced by roots acts upon root cells, creating conditions favorable for *Rhizobium* invasion. The selectivity frequently exhibited by *Rhizobium* strains for roots of specific legume hosts may be due in part to the concentrations of biotin, thiamine, and homoserine supplied in root exudates. Further, chemotaxis of *Rhizobium* toward specific organic compounds in root exudates may account for the selective colonization of host roots (Kush and Dadarwal 1981).

A number of rhizosphere-related factors can interfere with the reproduction and activities of rhizobia in a natural environment. Along with low soil pH and toxic decomposition products, antagonism by other microorganisms needs to be given more serious attention. The impact of predation by protozoa or nematodes and parasitism by *Bdellovibrio* or bacteriophages on *Rhizobium* may be greater than is generally thought. Ramirez and Alexander (1979) showed that *Rhizobium* colonization of the rhizosphere can be enhanced through chemical control of the protozoan population.

More detailed information on rhizobia and the process of nodule formation can be found in publications by Dart and Day (1975), Nutman (1975), Dazzo (1980), Atlas and Bartha (1981), and Bauer (1981). These writers also point out other deficiencies in our knowledge of rhizosphere-rhizoplane relationships to the establishment and function of rhizobia.

Nitrogen fixation in nonleguminous plants also is of considerable economic significance (Becking 1975). The endophytes involved in this type of symbiosis with dicotyledenous plants are either actinomycetes or rhizobia and "*Rhizobium*-like" organisms, predominantly the former. The actinomycete hyphae infect cortical cells of the host root and produce growth factors which stimulate the initiation and development of a root primordium. The endophyte then invades the meristematic cells of the primordium, eventually resulting in the formation of clusters of lobed nodules within which the actinomycete produces nitrogenase and fixes atmospheric nitrogen. The nitrogen economy of soils in many areas of the world is highly dependent on this type of nitrogen fixation. A greater research effort is needed to determine the specific rhizosphere conditions necessary for maximum efficiency of nitrogen fixation in nonleguminous plants before this natural process can be exploited for wide use in agriculture.

Symbiotic fungi associated with plant roots in the formation of mycorrhizae of many forest and field-crop plants show great potential for use in managed programs of plant culture where the nutrient status (especially phosphorus) of soil is low. It is well established that mycorrhizae can increase the rate of nutrient uptake by plants. Among the mechanisms suggested for ectomycorrhizal benefit to plants (Nye and Tinker 1977) are altered root morphology, resulting in an increase in absorptive capacity, and the prolonging of the absorptive life of short roots having a mycorrhizal sheath. Much remains to be learned about the interactions of the ectomycorrhizal fungi with other rhizosphere microorganisms, and the effects of environmental stress factors, in the establishment of fungus-root associations.

Even less well defined are the mechanisms involved in the establishment and function of the endomycorrhizae, which are associated with most crop plants other than those in the Cruciferae and Chenopodiaceae. Though the fungal symbiont in this case forms no sheath or mantle around roots, well-developed mycelia extend several millimeters from the root surface into the rhizosphere, and often beyond, serving in effect as a hyphal bypass for phosphate to move through the zone of nutrient depletion at the root-soil interface. This may serve as a supplement to root-hair function, as the mycorrhizae seem to be most beneficial to those host plants with few root hairs.

The full potential for production, application, and management of mycorrhizal-fungus inocula will be realized only when the interrelationships of symbiotic fungi, root exudates, nonsymbiotic microorganisms, and environmental stress factors are better understood. Despite gaps in our knowledge, however, extensive plans are being promoted for the commercial production and application of mycorrhizal fungi to field and nursery soils. Research toward this goal for both ecto- and endomycorrhizae generally follows four sequential steps: (1) isolation or recovery of mycorrhizal fungi from natural habitats, (2) production of mycorrhizal inocula, (3) inoculation of plants, and (4) evaluation of plant response. The procedures have been described in a compilation of methods for mycorrhizal research edited by Schenck (1982). The nature of the rhizosphere environment plays a prominent role in the success or failure of efforts to establish mycorrhizal infections in natural soil. To assure a competitive advantage for root colonization, the inoculum of a mycorrhizal fungus must be applied directly to seed or in close proximity to roots, or placed so that developing roots will come in contact with the fungus.

Ectomycorrhizal inoculum may consist simply of infested soil or humus collected from pine plantations. However, defined inocula in the form of basidiospores and cultured vegetative mycelia of fungi such as *Suillus granulatus, Rhizopogon luteolus, Thelephora terrestris,* and *Pisolithus tinctorius* are more suitable for developing commercial formulations that may be compatible with the rhizosphere. The U.S. Department of Agriculture Forest Service, in collaboration with commercial companies, has conducted evaluation tests for several years on the effectiveness of different formulations of *Pisolithus tinctorius* for use in nurseries of container-grown forest tree seedlings (Cordell and Marx 1980; Marx and Kenney 1982). Field applications of ectomycorrhizal fungi also appear promising in afforestation practices and in the revegetation of areas denuded by coal-mining operations.

Use of the vesicular-arbuscular (VA) type of mycorrhiza in agricultural practice has been curtailed by the inability of scientists to culture the fungal symbionts in nutrient media free of host roots. However, the potential benefits that might be derived from use of these Zygomycete endophytes could be even greater than for the ectomycorrhizae, since the VA mycorrhizal association occurs in a much larger number of economically important plant species.

The source of inoculum to be used for starter cultures of VA mycorrhizal fungi is the rhizosphere. Spores are obtained by wet-sieving of rhizosphere soil samples and removal from soil detritus with a microspatula or Pasteur pipette under a dissecting microscope. The spores are then used to establish mycorrhizal pot cultures of various grasses (Ferguson and Woodhead 1982). Chopped whole-pot cultures or the spores and sporocarps of the fungi are then used in appropriate forms for application to soil or for seed pelleting (Menge and Timmer 1982). Successful inoculation of seedlings with efficient mycorrhizal fungi has been reported for experiments in greenhouse pots using nonsterilized soil, and in field tests. Species of *Glomus* and *Gigaspora* have been used frequently in such tests because of their relatively strong competitive ability in the presence of indigenous mycorrhizal fungi and other rhizosphere microorgaisms. In New Zealand, the drilling of inoculum with or below barley seed in the field has resulted in significant growth responses even in soil not considered deficient in phosphorus or potassium (C. Ll. Powell et al. 1980). Whatever the method of applying inoculum, the final limiting factors in achieving infection are, as with the ectomycorrhizae, associated with phenomena common in the rhizosphere. The environmental deterrents in natural soil which interfere with the establishment and management of mycorrhizal inoculum are essentially the same as those encountered by plant pathologists attempting to establish biological control systems for root disease. It is a matter first of selecting the most efficient strain of fungal symbiont, then applying a sufficient inoculum density with a suitable substrate to induce a competitive advantage for colonization of the host rhizoplane.

In experimental work it is essential to assess the extent of root colonization by mycorrhizal fungi, since one cannot assume that inoculum added to soil or seed will lead to infection and establishment of the symbiotic association. A wide variety of techniques has been used for the quantitative evaluation of mycorrhizae, thus limiting the possibility of standardization. Efforts are being made to improve techniques that may be more widely acceptable for working with both the ectomycorrhizae (Grand and Harvey 1982) and endomycorrhizae (Giovannetti and Mosse 1980; Biermann and Linderman 1981; Kormanik and McGraw 1982).

Bowen (1978) referred to certain "shortfalls" in mycorrhizal effectiveness attributable to various forms of antagonism by the native soil microflora and the disruption of the ectomycorrhizal fungus sheath or the endomycorrhizal hyphae and vesicles by mycophagous nematodes. The role of hyperparasites (Ross and Daniels 1982) in the success or failure of soil and seed inoculation with fungal symbionts must be better understood. Also scarcely studied for their potentially disruptive capacity are the mycophagous microarthropods that are attracted to the rhizosphere microflora food source.

8.3.2 Interactions in Symbiosis

The competitive interactions between strains of *Rhizobium* or between indigenous and introduced mycorrhizal fungi are generally recognized as problems in attempting to establish symbiotic relationships with crop plants. Only recently, other interactions, particularly between rhizosphere bacteria and mycorrhizal fungi, have been noted as important factors influencing plant growth. In Chapter 6, we alluded to the synergistic interactions between VA mycorrhizae and *Rhizobium* in which nitrogen fixation by root nodules may be enhanced by mycorrhiza-aided phosphorus absorption. Redente and Reeves (1981) showed that sweetvetch (*Hedysarum boreale*) plants inoculated with a combination of a mycorrhizal fungus (*Glomus fasciculatum*) and a species of *Rhizobium* grew taller and produced more biomass than plants infected with *Rhizobium* alone. However, this relationship is not always so simple, since plant growth also can be inhibited by intersymbiont competition for phosphorus and photosynthate, resulting in impaired nitrogen fixation and nitrogen stress upon the host plant (Bethlenfalvay et al. 1982).

8.3.3 Nonsymbiotic Activities Reexamined

Rhizosphere microorganisms not physically linked in a symbiotic interchange with plant roots are usually referred to as nonsymbiotic. However, the relationship between roots and the free-living, nitrogen-fixing bacteria is sometimes referred to as associative symbiosis. Other rhizosphere bacteria and fungi, which are believed to impart some growth-promoting benefit to plants in response to nutrients in root exudates, have been called symbiotic organisms. Indeed, in some cases, a very thin line separates the free-living beneficial microorganisms in the rhizosphere from those with a strong affinity for root surfaces and cells. Nevertheless, until a better understanding is acquired of the relationships of rhizoplane inhabitants to root function it may be wise to stay with familiar terminology.

In recent years, a great awareness of the potential for use of free-living, nitrogen-fixing bacteria in promoting higher yields of grass and grain crops has stimulated a surge in research efforts, largely relating to the role of rhizosphere or rhizoplane interactions in establishing the associative process. The information obtained has been summarized in two volumes edited by Vose and Ruschel (1979). In at least some instances it has been established that a suitable carbon source, such as that supplied in root exudates, is required to achieve the maximum nitrogen-fixing benefit to plants. For example, glucose was found to be a limiting factor in nitrogenase activity associated with the rhizosphere of marram grass (*Ammophila arenaria*) seedlings inoculated with *Azotobacter chroococcum* (Abdel Wahab and Wareing 1980). Yet, questions continue to be raised concerning the precise role of root exudates in associative nitrogen fixation. While root exudates may comprise 7–13% of the total dry matter content of plants growing in a sterile soil system (D. A. Barber and Martin 1976), this readily available carbon source in natural soil is rapidly depleted by the rhizosphere population, leaving the probability that water-soluble exudate alone cannot provide sufficient substrate carbon for significant root nitrogenase activity (Beck and Gilmour 1983). Other, more complex, moieties of the soil, such as sloughed root cells and decaying organic matter at the root-soil interface, must contribute to the energy requirement of free-living N-fixing bacteria in wheat, barley, and other grain crops. As researchers probe into the root environmental factors relating to the establishment of species or strains in the rhizosphere-rhizoplane, increasing interest is evident in the nitrogen-fixing capacity of *Azospirillum* spp. Results have indicated a requirement by *Azospirillum* strains for resistance to antibiotics, especially to streptomycin, for infection to occur in cereal roots (Baldani and Dobereiner 1981). This is taken as indicating a selection mechanism of plants for certain microorganisms, and suggests new possibilities for the introduction and manipulation of selected nitrogen-fixing bacteria.

Most bacteria in the rhizosphere do not fix atmospheric nitrogen, but nevertheless promote plant growth by other mechanisms. For example, phosphate-solubilizing bacteria that also produce plant-growth regulators can be found among the species of several rhizosphere-inhabiting genera (Barea et al. 1976). Most attention recently has been focused upon the rhizosphere pseudomonads in the *Pseudomonas fluorescens* and *P. putida* groups which promote plant growth indirectly by antagonistic activities contributing to disease suppression (Suslow and Schroth 1982b; Schroth and Hancock 1982). Serious efforts are being made to separate these efficacious bacteria from other rhizosphere species that are toxigenically deleterious to root function, and to devise procedures for the genetic and ecological manipulation of processes applicable to plant culture. Considering the complexity of the natural environment in which these microorganisms operate, and the myriad chances for interaction with other key organisms such as nitrogen fixers, mycorrhizal fungi, plant pathogens, and microphagous fauna the challenges seem both endless and awesome.

8.3.4 The Root-Shoot Relationship

Like the shoot system, the form of the root system is ultimately determined by a plant's genome. Quite small genetic changes, such as those existing between the various cultivars of crop species, can produce dramatic differences in root morphology; breeding for improved and more efficient root systems has become the prime area of research for many plant breeders and agronomists. However, we must not lose sight of the fact that while the total genetic make-up of a species determines the basic morphology and size of the root system, realizing this full genetic potential is entirely dependent upon a constant import into the root of materials originating in the shoot system. The most obvious of these materials are, of course, the carbohydrates manufactured through the process of photosynthesis; but there are other essential materials, such as the hormone IAA, constantly being transported basipetally from shoot to root, which are responsible for developmental changes in roots such as the onset of secondary growth. The roots, in turn, manufacture compounds essential for developmental changes in shoot, and these are moved acropetally within the plant. Superimposed upon these genetic and biochemical determinants of plant form are the effects of the environmental conditions under which a plant is growing. The soil environment, with its potential for wide fluctuations in water and nutrient status, pH, and oxygen tension, can profoundly influence the growth and physiology of roots which will, in turn, be reflected in the vigor and performance of the shoot system. Similarly, as we pointed out in Chapter 2, the conditions to which the shoot system is exposed can affect the morphology of roots in ways which cannot be explained entirely in terms of their effects on photosynthesis. More directly, a number of the presently used agricultural chemicals when applied to shoot systems are translocated basipetally to the roots and from there they may be exuded into the soil (see Chap. 3). Once in the soil these chemicals clearly are in a position to influence the normal relationships pertaining between the root and its associated microorganisms.

Perhaps because of the immediately apparent role of the root system for anchorage and water and mineral-nutrient absorption, plant physiologists have always been cognisant of the intimate relationship existing between the activities of roots and shoots. In the past, however, most scientists concerned with the interactions occurring between plant roots and soil microbial populations have tended to view the root system as an isolated, discrete entity influenced primarily, if not entirely, by the physical and chemical conditions pertaining in the soil. A failure to appreciate that the behavior of roots, and hence the interactions of roots with their associated microorganisms, will also be affected by the activity of the shoots and the conditions under which they are growing, has undoubtedly stood in the way of progress in rhizosphere studies. In an area of research already fraught with enormous technical difficulties, the prospect of also having to consider the physiological status of the shoot system while studying root-microorganism interactions may appear truly daunting. However, the potential for the shoot system to affect the nature of the rhizosphere relationship, and the activities occurring within the rhizosphere, is now well established and can no longer be ignored. Scientists should accept this new challenge and devise techniques and experiments which will answer questions not solely about the root system in relation to microorganisms, but about the whole plant in relation to soil microorganisms. If the challenge is accepted, we may expect that new concepts and truths will emerge, adding immeasurably to our knowledge of plant growth and providing information invaluable to future advances in agriculture.

8.4 Opportunities in Plant Pathology

Despite the many years of rhizosphere research as summarized in this book, the most difficult and rewarding investigations may yet lie ahead. Perhaps no other investigative discipline stands to benefit more from future intensified research on root-soil relationships than plant pathology. The rhizosphere holds the key to a better understanding of host – parasite relationships in root diseases and in the mechanisms responsible for biological and, in some cases, chemical control.

8.4.1 Modeling the Rhizosphere

Although they have been accepted, the terms rhizosphere and rhizoplane continue to generate controversy and varied concepts with regard to the spatial extent of their influence and the precise effects of these zones on the inoculum density-disease incidence (ID-DI) relationship. A number of researchers have turned to mathematical modeling to distinguish theoretically between rhizosphere and rhizoplane effects in relation to pathogen propagule germination and root infection. Modeling is an attempt to reduce the complex root-soil system to simple expressions that may explain, for example, how close a propagule must be to the host surface for infection to occur. The simplicity of the process, however, may be more apparent to some than others, depending upon one's mathematical expertise and general attitude toward modeling.

In modeling terms, the distinction between rhizosphere and rhizoplane effects is based largely on the original mathematical interpretation of Baker et al. (1967) and on subsequent models suggested by Baker (1971) to show the theoretical relationships between ID and disease. In these studies, the rhizosphere has been viewed as consisting of a cylinder of constant diameter enveloping roots and, theoretically, given a fixed infection court and nonmotile inoculum, an increase in ID in the rhizosphere soil would be expected to increase infection in the host in direct proportion to the change in inoculum (see Baker and Drury 1981). Disease incidence then plotted against ID would have a slope of 1. If inoculum is added to soil, the proportional increase in ID in the rhizoplane is said to be less than in the rhizosphere. Therefore, Baker et al. (1967) proposed that the number of propagules per unit area rather than propagules per unit volume would be the critical factor for predicting the number of successful infections that might occur. Rhizoplane infection is indicated by the Baker models when the slope of the linear relationship between ID and DI is 0.67. As evidenced in the review by Gilligan (1983a), a number of critics (Gilligan 1979; Leonard 1980; Grogan et al. 1980; Ferriss 1981) have questioned the validity of the surface-density models and prefer to recognize a volume-density phenomenon, since earlier definitions and studies of the rhizoplane generally included adhering soil and debris particles as part of the zone, suggesting that the rhizoplane ist not entirely planar. Ultrastructure studies (Foster et al. 1983) have indicated that the rhizoplane is highly sculptured and intimately associated with the soil. Griffin (1969) observed that highest germination of *Fusarium oxysporum* chlamydospores occurred within a 0 to 5-mm distance of peanut root surfaces, and he regarded this as a rhizoplane effect.

As stated by Ferriss (1983), all models simplify reality and, consequently, all are inaccurate to some extent, their value depending on the intended purpose; or, as Baker and Drury (1981) aptly put it, the essence of every model is within the frame. Equations for calculating rhizosphere and spermosphere widths and for studying relationships among rhizoplane, inoculum density, and disease incidence require that certain assumptions be made; thus, disagreements among investigators are based largely on differences in fundamental assumptions (see Drury et al. 1983). Calculations, for example, are usually based on the assumption that inoculum and infection sites are distributed at random along a root when, in fact, inoculum distribution frequently is discontinuous (clumped). The problem of how to test for and correct for possible nonrandomness of infection by soilborne pathogens has been addressed in studies exemplified by those of Gilligan (1983b) and Griffin and Tomimatsu (1983).

Mathematical modeling of the rhizosphere-rhizoplane is still in its infancy. Although these exercises have spawned a continuing controversy over mathematical interpretations, the difference gaps are closing, and strategies for further modeling are emerging along with an increasing realization of the potential application of modeling to rhizosphere investigations relating to root infection. While most modeling activity has centered upon the inoculum density-disease relationship, with host and environmental parameters being held constant, factors relating to the saprophytic phase of soilborne pathogens are equally important and eventually must be considered along with inoculum density when predicting the potential for root infection. Bouhot (1979) has discussed the competitive saprophytic ability-disease relationship as a predictive system. Also relating to the performance of pathogen propagules, Cushman (1982) has attempted to explain through mathematical models the importance of nutrient availability and transport between the rhizosphere and the nonrhizosphere soil matrix.

Time will determine whether Kranz's (1974) assessment that modeling "disciplines research and organizes knowledge" is correct. Certainly, greater precision of this kind will be needed to reveal the exact nature of primary, and perhaps secondary, infections of roots and help explain the mechanisms involved in biological or conventional disease control practices.

8.4.2 New Challenges in Biological Control

Current research indicates that biological control of root-infecting pathogens is rapidly coming to the forefront of attention in many areas, stimulated by the discovery of new control agents and by the development of more effective application or augmentation procedures (Chap. 7). Seed treatment with preparations of either fungal spores (Harman et al. 1981; Windels and Kommendahl 1982) or rhizosphere bacteria (Kloepper and Schroth 1981) offers one of the most attractive and feasible methods of establishing biological agents directly in the spermosphere and rhizosphere to create a pathogen-suppressive environment. Successful biological control by seed bacterization or other means of application is dependent upon the rhizosphere-colonizing ability of candidate control agents. Then, for reproducible results, the mechanisms responsible for the control effect must be understood; the answers lie among the complex interactions associated with the root-soil interface. New rhizosphere research is now addressing some important questions relating to the colonization, distribution, and survival of pseudomonads on seeds and roots, implicating such factors as soil matric potential, HCN production, chemotaxis, nutrient levels, etc. For example, bacterial agents such as *Pseudomonas putida* (Dupler and Baker 1984) can be expected to require high soil moisture for optimal activity, competitive colonization, and survival. Knowledge of survival characteristics, coupled with application methods most suitable for specific control agents and specific crop plants, may then lead to meaningful biocontrol. The ultimate aim is for commercialization of the control agent and wide distribution for farmer use. Seed treatment with *Bacillus subtilis* has gained new emphasis in biological control efforts since a commercial preparation (Quantum 4000) has been successfully tested by peanut farmers (Backman et al. 1984). The antibiotic-producing bacterium is a good competitive colonizer of roots and apparently produces a hormone that stimulates plant growth.

Rhizosphere-inhabiting, fluorescent pseudomonads that promote plant growth can be isolated and applied to seeds for the suppression of other rhizosphere bacteria and fungi that inhibit plant growth. Kloepper et al. (1980) presented evidence that some fluorescent pseudomonads produce extracellular siderophores (low molecular weight chelating agents) that complex environmental iron, making it less available to certain other microorganisms. The importance of iron in microbial physiology is well known, that element playing a key role in the synthesis of microbial DNA.

The water-soluble fluorescent pigments of 156 *Pseudomonas* isolates were found to be fungistatic to several major plant pathogens including *Rhizoctonia solani, Sclerotinia sclerotiorum, Phymatotrichum omnivorum, Phytophthora megasperma,* and *Pythium aphanidermatum* (Misaghi et al. 1982). These pigments, or siderophores, are produced only in environments of low iron level. Since little is yet known about the levels of iron around roots, direct evidence for the production of siderophores by bacteria in the natural rhizosphere or rhizoplane is lacking. However, a great many aerobic and facultatively aerobic microbial species excrete siderophores (Neilands 1977), at least in vitro, suggesting an interesting potential for reducing soil-borne disease by increasing populations of these microorganisms in soils of limited iron availability.

Bacteriocins (Vidaver 1983) offer a new challenge for biological control of plant pathogenic bacteria. Since the development of agrocin 84 (see Chap. 7) for control of *Agrobacterium tumefaciens,* other bacteriocin-producing bacteria have been studied, but relatively little progress has been made toward the control of plant pathogens. The highly specific nature of these compounds tends to dampen the interest of commercial developers, who prefer compounds with wide-spectrum activities. However, perhaps economic justification will be realized for developing bacteriocin seed treatments for certain pathogenic bacteria with wide host ranges.

Hypovirulence has not been fully exploited as a biocontrol mechanism. This refers to the reduction of virulence or aggressiveness of a pathogen by mycoviruses. Hypovirulence is attributed to the dsRNA of the hypovirulent strain and this can be transferred from strain to strain through anastomosis or plasmogamy. Such a transmissible factor is common in the cytoplasm of the chestnut blight fungus (*Endothia parasitica*) and also has been recorded in soil-borne pathogens, *Rhizoctonia solani* (Castanho and Butler 1978) and *Gaeumannomyces graminis* var. *tritici* (Lapierre et al. 1970). Too little is yet known about the relation of rhizosphere phenomena to the potential for conversion of a resident pathogen population from the virulent to the hypovirulent phenotype.

Events that contribute to the positioning of microorganisms for host contact and host-microbe interaction (recognition) have been studied intensively for the bacteria (Pueppke 1984) and may pertain as well to the establishment of pathogenic fungi (Daly 1984) on host plants, and perhaps even the colonization of root surfaces and the recognition of pathogen hyphae or spores by biocontrol agents (Barak et al. 1985). Whereas the tropic (filamentous fungi) and the taxic (motile bacteria and zoospores) responses of organisms that migrate up an exudate concentration gradient to roots are not recognition events, they lead to host contact where specificity for a host then depends on surface components of both microbe and host. If these are complementary, stimulation of either a microbe or a host reaction, or both, occurs, leading to adherence. Free-living nematodes also utilize specific recognition mechanisms, leading to contact with host roots, and may even contribute to their own demise by responding to attraction events toward nematophagous fungi (Zuckerman and Jansson 1984). The role of exudates, and of other rhizoplane-rhizosphere factors, in the chain of events contributing to pathogen contact and post-contact interaction with root surfaces needs critical study in relation to the capacities of applied biocontrol agents to interrupt these events and displace the pathogen. For an excellent discussion of concepts and experimental approaches in host-microbe recognition and the environmental factors affecting recognition, the reader is referred to Lippincott and Lippincott (1984).

Until recently the use of natural enemies to control plant-parasitic nematodes has lagged behind the advances made toward biological control of other soil-borne pathogens. However, a new emphasis has been generated as additional parasites and predators have been revealed as natural deterrents to nematode activities (Kerry 1981 a). While it is not expected that any control method can eradicate plant-parasitic nematodes from field soil, it seems feasible that biologists could devise the necessary methodology for concentrating fungal nematode parasites such as *Catenaria auxiliaris, Nematophthora gynophila, Verticillium chlamydosporium, Acremonium strictum,* and *Paecilomyces lilacinus* in the rhizosphere at a time to coincide with cyst formation or egg hatch, before juveniles can enter root tissue of the host plant.

The common nematode-trapping fungi are naturally endowed with devices (hyphal loops and adhesive knobs) for capturing their prey, yet efforts made to utilize these agents for the protection of plant roots have been largely unsuccessful. It would be especially advantageous to find biological control organisms that can protect roots and enhance plant growth through more than one mechanism. For example, *Arthrobotrys oligospora, A. robusta,* and *A. superba* are not only effective nematode-trapping fungi but also parasitize the plant pathogen *Rhizoctonia solani* by specialized coil-form hyphae (Tzean and Estey 1978). Vesicular-arbuscular mycorrhizal fungi, while enhancing the nutrient status of plants, sometimes also contribute to a reduction in the incidence and severity of diseases caused by fungal and nematode pathogens (Dehne 1982; Hussey and Roncadori 1982). In some cases, however, disease is increased in mycorrhizal plants. Whether plant protection is enhanced or disease is promoted by the symbiosis is determined by the nutritional requirements of the specific plant and fungal symbiont along with the chemical, physical, and biological nature of the rhizosphere-rhizoplane.

The potential of mycophagous, small soil animals as biological control agents is viewed with skepticism by some but, without more specific information about their role in rhizosphere ecology, they cannot be disregarded. Evidence was presented in 1983 at both the International Congress of Plant Pathology in Australia and the Mycological Congress in Japan to implicate amoebae and collembolan microarthropods in the reduction of plant-pathogen inocula and the suppression of root disease. Whereas the melanin pigment in spores of many fungi confers resistance to microbial lysis, a number of amoeba genera can lyse pigmented propagules. Chakraborty et al. (1983) suggested that mycophagous amoebae of the genera *Gephyramoeba, Mayorella, Saccamoeba,* and *Thecamoeba,* isolated from field soil suppressive to take-all (*Gaeumannomyces graminis* var. *tritici*) of wheat, may contribute to the suppressive condition. Mycophagous Collembola not only suppress the pathogenic activities of *Rhizoctonia solani* (Chap. 7) but also ingest large numbers of chlamydosphores of *Fusarium oxysporum* f. sp. *vasinfectum* and graze significantly on germ tubes emerging from sclerotia of *Macrophomina phaseolina, Verticillium dahliae,* and *Sclerotium rolfsii* (Curl et al. 1985).

Nature offers an abundance of potential biological control agents if man can only design a plan and develop the necessary methodology for implementation. This may require in some cases the integration of biological and chemical procedures to create a rhizosphere environment most favorable for the function of either introduced or resident antagonists.

Genetic control of the rhizosphere to favor disease suppression is feasible. As stated by K.F. Baker and Cook (1982), the genetic resistance of certain crop plants to root-infecting fungi is mediated by the rhizosphere flora under the influence of root exudates. The virulence of a number of seed and root pathogens is affected, and can be altered, by the type and concentration of nutrients available to the inoculum. Such is the case for *Rhizoctonia solani* responding to different levels of carbon and nitrogen sources (Weinhold et al. 1969). Virulence of the pathogen on cotton hypocotyls increased with increasing concentration of glucose and asparagine, and the nutrient requirement for maximum virulence was greater than for vegetative growth of the fungus. The feasibility of genetic alteration of the quality and quantity of exudates from seeds and plant roots is suggested in the Texas investigations (L. S. Bird 1982) to control a variety of adversities in cotton. Some altered fluids released from roots are unfavorable nutritionally for certain pathogens, but favorable for microorganisms antagonistic toward the pathogens, thus providing a mechanism of resistance in the cotton plant.

Currently, an exciting area of biological research is genetic engineering with its enormous potential for improving the quality of life through sudden and dramatic advances in medicine and agriculture. The technology needed to directly transfer, as fragments of DNA, desirable traits from one organism to another is still in its infancy, but it grows daily at a remarkably rapid rate. With the discovery of more and better vectors capable of breaking into a cell's bank of genetic information and depositing new messages there, scientists of the future should be able to engineer almost at will organisms exhibiting a wide variety of new and desirable qualities which would be unattainable through the use of conventional breeding programs.

Crop-plant improvement is a prime target of much of the current genetic engineering research. Some projected goals, such as the transfer of the genes responsible for nitrogen fixation to a species such as corn, are probably far into the future, if attainable at all, but other more modest endeavors, such as producing crop species with a high level of resistance to specific herbicides, would seem to be just around the corner. Undoubtedly, with time, we will also see attempts made to transfer the resistance to pathogen invasion exhibited by the roots of certain species to crop plants which are at present highly susceptible to those same organisms. In many ways, therefore, the new technology of genetic engineering is going to change the nature of the plants we grow, and of necessity this will mean changes, albeit indirect, in the rhizosphere activities of those species.

We may also look to the time when, through the same techniques, changes in the activities of some species of rhizosphere microbes can be effected directly. If we could determine and characterize those components of certain microbial species which make them effective antagonists toward invading pathogens, it may be possible to augment this capability or even transfer the character to other species which are strong competitors for rhizosphere colonization. Already, molecular biology approaches are being used to track the movement of bacteria in the rhizosphere. Plasmids bearing *Escherichia coli* lac Z and lac Y genes (commonly used for the construction of fusions) have been maintained in fluorescent pseudomonad cells under rhizosphere conditions to provide a selectable marker for tracking the bacterium or plasmids in the environment (Drahos et al. 1984). This important experiment demonstrated that a bacterium containing a foreign gene could live and compete within the rhizosphere.

It is well recognized that certain species of microorganisms produce hormones or other materials which stimulate plant growth. It is conceivable that the genes for these products could be transferred to additional species present in the rhizosphere. The day may not be far away when a rhizosphere flora can be designed and produced which will contribute to plant growth and vigor and provide maximum protection in a hostile soil environment.

8.5 Focus: The Rhizosphere and Crop Production

The foregoing chapters have attempted to show the essential role of rhizosphere activities in plant development. We have seen that the root creates a dynamic environment of interacting chemical, physical, and microbial components which, along with photosynthesis and other physiological processes, determine the health, vigor and productive capacity of crop plants.

A large volume of literature has dealt with varied aspects of the rhizosphere, yet, because much of the accumulated information to date has been derived from artificial systems of plant culture, there still exists a great void between our basic understanding of rhizosphere phenomena and the mechanisms that lead to effects on crop yields in natural environments. Many agricultural experiments, even some on root growth and function, have not been designed to allow for the probability that microbial activity at the root-soil interface may contribute to the results obtained. The application of fertilizers and pesticides to field soil can be expected to alter the ecological nature of the root zone. Seed treatments with either chemical or biological pesticides will affect the spermosphere and, in some cases, the rhizosphere of developing seedlings. Crop rotations, irrigation, organic amendments, and soil compaction by farm machinery all ultimately affect root growth and the rhizosphere activities that further influence nutrient availability and absorption, water uptake, and the resistance of plants to disease. For the most part, adequate methodology has not been devised to monitor and measure these rhizosphere effects under field conditions.

Since the unique environment around roots holds the answers to so many questions relating to root development and function, a worldwide intensification of rhizosphere research could conceivably contribute enormous benefits toward the improvement of crop yields needed to feed an ever-increasing human population. The exciting challenges that lie ahead can be met effectively only through a cooperative, multidisciplinary, international effort. This was the principal message that emerged from the first international symposium on the root-soil interface held in Oxford, England (Harley and Russell 1979). Management of the rhizosphere for healthy, more efficient root systems is a promise of the future for crop production.

References

Abdel Wahab AM, Wareing PF (1980) Nitrogenase activity associated with the rhizosphere of *Ammophila arenaria* L. and effect of inoculation of seedlings with *Azotobacter*. New Phytol 84:711–721

Abraham TA, Herr LJ (1964) Activity of actinomycetes from rhizosphere and nonrhizosphere soils of corn and soybean in four physiological tests. Can J Microbiol 10:281–285

Adams PB, Ayers WA (1980) Factors affecting parasitic activity of *Sporidesmium sclerotivorum* on sclerotia of *Sclerotinia minor* in soil. Phytopathology 70:366–368

Adams PB, Ayers WA (1981) *Sporidesmium sclerotivorum:* Distribution and function in natural biological control of sclerotial fungi. Phytopathology 71:90–93

Agnihotri VP (1964) Studies on the aspergilli. XIV. Effect of foliar spray of urea on the aspergilli of the rhizosphere of *Triticum vulgare* L. Plant Soil 20:364–370

Agrios GN (1978) Plant pathology. Academic Press, London New York

Ahmed AHM, Tribe HT (1977) Biological control of white rot of onion (*Sclerotium cepivorum*) by *Coniothyrium minitans*. Plant Pathol 26:75–78

Alabouvette C, Pussard M, Pons R (1979) Isolation and characterization of a mycophagous amoeba: *Thecamoeba granifera* subsp. *minor*. In: Schippers B, Gams W (eds) Soil-borne plant pathogens. Academic Press, London New York, pp 629–633

Alderman SC, Hine RB (1982) Vertical distribution in soil of and induction of disease by strands of *Phymatotrichum omnivorum*. Phytopathology 72:409–412

Aldrich J, Baker R (1970) Biological control of *Fusarium roseum* f. sp. *dianthi* by *Bacillus subtilis*. Plant Dis Rep 54:446–448

Alexander M (1965) Nitrification. In: Bartholomew WV, Clark FE (eds) Soil nitrogen. Am Soc Agron, Madison, Wisc, pp 307–343

Alexander M (1982) Most-probable-number method for microbial populations. In: Page AL, Miller RH, Kenney DR (eds) Methods of soil analysis, 2nd edn, part 2. Chemical and microbiological properties. Am Soc Agron, Madison, Wisc, pp 815–820

Alexander M (1971) Microbial ecology. Wiley, New York

Alexander M (1977) Introduction to soil microbiology, 2nd edn. Wiley, New York

Allen EK, Allen ON (1958) Biological aspects of symbiotic nitrogen fixation. In: Ruhland W (ed) Encyclopedia of plant physiology Vol VIII. Springer, Berlin Heidelberg New York, pp 48–118

Allen RN, Newhook FJ (1973) Chemotaxis of zoospores of *Phytophthora cinnamomi* to ethanol in capillaries of soil pore dimensions. Trans Br Mycol Soc 61:287–302

Allison FE (1947) *Azobacter* inoculation of crops. I. Historical. Soil Sci 64:413–429

Altman J, Campbell CL (1977) Effect of herbicides on plant diseases. Annu Rev Phytopathol 15:361–385

Anderson JM, Healey IN (1972) Seasonal and interspecific variation in major components of the gut contents of some woodland Collembola. J Anim Ecol 41:359–368

Anderson JPE, Domsch KH (1978) A physiological method for the quantitative measurement of microbial biomass in soils. Soil Biol Biochem 10:215–221

Anderson TR, Patrick ZA (1978) Mycophagous amoeboid organisms from soil that perforate spores of *Thielaviopsis basicola* and *Cochliobolus sativus*. Phytopathology 68:1618–1626

Anderson WP (1976) Transport through roots. In: Luttge U, Pitman MG (eds) Transport in plants, part B. Encyclopedia of plant physiology, new series, vol II. Springer, Berlin Heidelberg New York, pp 129–156

Aronoff S, Dainty J, Gorham PR, Srivastava LM, Swanson CA (eds, (1975) Phloem transport. Plenum Press, New York

Asanuma S, Tanaka H, Yatazawa M (1978) Effects of soil microorganisms on the growth and on the nitrogen and phosphorus absorption of rice seedlings. Soil Sci Plant Nutr 24:207–219

Asanuma S, Tanaka H, Yatazawa M (1980) Effect of soil microorganisms on the growth of roots in rice seedlings. II. Nitrite formation by rhizoplane microorganisms and the effect of nitrite on the root development under aseptic conditions. Soil Sci Plant Nutr 26:63–70

Atilano RA, Van Gundy SD (1979) Systemic activity of oxamyl to *Meloidogyne javanica* at the root surface of tomato plants. In: Harley JL, Russell RS (eds) The soil-root interface. Academic Press, London New York, pp 339–349

Atkinson TG, Neal JL Jr, Larson RI (1975) Genetic control of the rhizosphere microflora of wheat. In: Bruehl GW (ed) Biology and control of soil-borne plant pathogens. Am Phytopathol Soc, St Paul, Minn, pp 116–122

Atlas RM, Bartha R (1981) Microbial ecology – fundamentals and applications. Addison-Wesley, Reading, Mass

Audus LJ (1964) Herbicide behavior in the soil. II. Interactions with soil microorganisms. In: Audus LJ (ed) The physiology and biochemistry of herbicides. Academic Press, London New York, pp 163–206

Aung LH (1974) Root-shoot relationships. In: Carson EW (ed) The plant root and its environment. Univ Press Virginia, Charlottesville, pp 29–61

Ayers, WA, Adams PB (1979a) Mycoparasitism of sclerotia of *Sclerotinia* and *Sclerotium* species by *Sporidesmium sclerotivorum*. Can J Microbiol 25:17–23

Ayers WA, Adams PB (1979b) Factors affecting germination, mycoparasitism, and survival of *Sporidesmium sclerotivorum*. Can J Microbiol 25:1021–1026

Ayers WA, Adams PB (1981) Mycoparasitism and its application to biological control of plant diseases. In: Papavizas GC (ed) Biological control in crop production. Granada, London, pp 91–103

Ayers WA, Thornton RH (1968) Exudation of amino acids by intact and damaged roots of wheat and peas. Plant Soil 28:193–207

Azćon-G de Aguilar C, Barea JM (1978) Effects of interactions between different culture fractions of "phosphobacteria" and *Rhizobium* on mycorrhizal infection, growth, and nodulation of *Medicago sativa*. Can J Microbiol 24:520–524

Backman PA, Rodríguez-Kábana R (1975) A system for the growth and delivery of biological control agents to the soil. Phytopathology 65:819–821

Backman PA, Turner JT, Crawford MA, Clay RP (1984) A new biological seed treatment fungicide for peanuts that increases yield. Highlights Agric Res. 31: no 1, p 4

Bagyaraj DJ, Menge JA (1978) Interaction between a VA mycorrhiza and *Azotobacter* and their effects on rhizosphere microflora and plant growth. New Phytol 80:567–573

Bagyaraj DJ, Manjunath A, Patil RB (1979) Interaction between a vesicular-arbuscular mycorrhiza and *Rhizobium* and their effects on soybean in the field. New Phytol 82:141–145

Baker DA (1984) Water relations. In: Wilkins MB (ed) Advanced plant physiology. Pitman, London, pp 297–318

Baker KF, Cook RJ (1982) Biological control of plant pathogens. Am Phytopathol Soc, St. Paul, Minn

Baker KF, Snyder WC (eds) (1965) Ecology of soil-borne plant pathogens. Univ Calif Press, Berkeley

Baker R (1965) The dynamics of inoculum. In: Baker KF, Synder WC (eds) Ecology of soil-borne plant pathogens. Univ Calif Press, Berkeley, pp 395–403

Baker R (1971) Analyses involving inoculum density of soilborne plant pathogens in epidemiology. Phytopathology 61:1280–1292

Baker R (1978) Inoculum potential. In: Horsfall JG, Cowling EB (eds) Plant disease, an advanced treatise, vol II. Academic Press, London New York, pp 137–157

Baker R, Drury R (1981) Inoculum potential and soilborne pathogens: The essence of every model is within the frame. Phytopathology 71:363–372

Baker R, Maurer CL, Maurer RA (1967) Ecology of plant pathogens in soil. VII. Mathematical models and inoculum density. Phytopathology 57:622–666

Balandreau J, Rinaudo G, Fares-Hamad I, Dommergues Y (1975) Nitrogen fixation in the rhizosphere of rice plants. In: Steward WDP (ed) Nitrogen fixation by free-living micro-organisms. Cambridge Univ Press, Cambridge, pp 57–71

Balasubramanian A, Rangaswami G (1969) Studies on the influence of foliar nutrient sprays on the root exudation pattern in four crop plants. Plant Soil 30:210–220

Baldani VLD, Döbereiner J (1981) Increase of antibiotic resistant bacteria in grass roots. In: Vose PB, Ruschel AP (eds) Associative N_2-fixation, vol II. CRC Press, Boca Raton, Fla, pp 33–37

Balis C, Kouyeas V (1979) Contribution of chemical inhibitors to soil mycostasis. In: Schippers B, Gams W (eds) Soil-borne plant pathogens. Academic Press, London New York, pp 97–106

Bamforth SS (1975) Rhizosphere protozoa from cacti in the Arizona desert. J Protozool 22:33A

Banihashemi Z, deZeeuw DJ (1975) The behavior of *Fusarium oxysporum* f. sp. *melonis* in the presence and absence of host plants. Phytopathology 65:1212–1217

Barak R, Elad Y, Mirelman D, Chet I (1985) Lectins: A possible basis for specific recognition in the interaction of *Trichoderma* and *Sclerotium rolfsii*. Phytopathology 75:458–462

Barber DA (1966) Effect of microorganisms on nutrient absorption by plants. Nature (London) 212:638–640

Barber DA (1967) The effect of micro-organisms on the absorption of inorganic nutrients by intact plants. I. Apparatus and culture technique. J Exp Bot 18:163–169

Barber DA (1968) Microorganisms and the inorganic nutrition of higher plants. Annu Rev Plant Physiol 19:71–88

Barber DA (1974) The absorption of ions by microorganisms and excised roots. New Phytol 73:91–96

Barber DA, Frankenburg UC (1971) The contribution of micro-organisms to the apparent absorption of ions by roots grown under sterile and nonsterile conditions. New Phytol 70:1027–1034

Barber DA, Gunn KB (1974) The effect of mechanical forces on the exudation of organic substances by the roots of cereal plants grown under sterile conditions. New Phytol 73:39–45

Barber DA, Martin JK (1976) The release of organic substances by cereal roots into soil. New Phytol 76:69–80

Barber DA, Bowen GD, Rovira AD (1976) Effects of microorganisms on absorption and distribution of phosphate in barley. Aust J Plant Physiol 3:801–808

Barber SA, Ozanne PG (1970) Autoradiographic evidence for the differential effect of four plant species in altering the calcium content of the rhizosphere soil. Soil Sci Soc Am Proc 34:635–637

Barclay GA, Crosse JE (1974) Populations of aerobic bacteria associated with the roots of apple and cherry plants. J Appl Bacteriol 37:475–486

Barea JM, Navarro E, Montoya E (1976) Production of plant growth regulators by rhizosphere phosphate-solubilizing bacteria. J Appl Bacteriol 40:129–134

Barea JM, Ocampo JA, Azcon R, Olivares J, Montoya E (1978) Effects of ecological factors on the establishment of *Azotobacter* in the rhizosphere. Environmental role of nitrogen-fixing blue-green algae and asymbiotic bacteria. Bull Ecol Res Commun (Stockh) 26:325–330

Barron GL (1977) The nematode-destroying fungi. Can Biol Publ, Guelph, Ontario

Barton R (1957) Germination of oospores of *Pythium mamillatum* in response to exudates from living seedlings. Nature (London) 180:613–614

Bateman DF (1978) The dynamic nature of disease. In: Horsfall JG, Cowling EB (eds) Plant disease. An advanced treatise, vol III. Academic Press, London New York, pp 53–83

Bauer WD (1981) Infection of legumes by rhizobia. Annu Rev Plant Physiol 32:407–449

Baya AM, Boethling RS, Ramos-Cormenzana A (1981) Vitamin production in relation to phosphate solubilization by soil bacteria. Soil Biol Biochem 13:527–531

Baylis GTS (1972) Fungi, phosphorus, and the evolution of root systems. Search 3:257–259

Beagle-Ristaino JE, Rissler JF (1983) Effect of *Rhizobium japonicum* nodulation on severity of Phytophthora root rot of soybean. Plant Dis 67:651–654

Beck SM, Gilmour CM (1983) Role of wheat root exudates in associative nitrogen fixation. Soil Biol Biochem 15:33–38

Becking JH (1975) Root nodules in non-legumes. In: Torrey JG, Clarkson DT (eds) The development and function of roots. Academic Press, London New York, pp 507–566

Bednarova M, Stanek M, Vančura V, Vesely D (1979) Microorganisms in the rhizosphere of wheat colonized by the fungus *Gaeumannomyces graminis* var. *tritici*. Folia Microbiol 24:253–261

Beijerinck MW (1888) Die Bakterien der Papilionaceenknöllchen. Bot Ztg 46:725–804. Also in Brock TD (1961) Milestones in microbiology. Prentice Hall, Englewood Cliffs, N J, pp 220–224

Beijerinck MW (1901) Über oligonitrophile Mikroben. Zentbl Bakteriol Parasitkd Abt 2, 7:561–582

Beijerinck MW, Delden A van (1902) Über die Assimilation des freien Stickstoffs durch Bakterien. Zentrbl Bakteriol Parasitkd Abt 2, 9:3–43

Bell DT, Koeppe DE (1972) Noncompetitive effects of giant foxtail on the growth of corn. Agron J 64:321–325

Benians GJ, Barber DA (1974) The uptake of phosphate by barley plants from soil under aseptic and non-sterile conditions. Soil Biol Biochem 6:195–200

Benson GL (1976) Effect of fluometuron, trifluralin and the rhizosphere of cotton seedlings on the inoculum potential of *Fusarium oxysporum* f. sp. *vasinfectum* (Atk.). Snyder and Hans. PhD dissertation, Auburn Univ, Alabama

Bethlenfalvay GJ, Pacovsky RS, Bayne HG, Stafford AE (1982) Interactions between nitrogen fixation, mycorrhizal colonization, and host-plant growth in the *Phaseolus-Rhizobium-Glomus* symbiosis. Plant Physiol 70:446–450

Bhaskaran R (1973) Effect of brassicol on auxin production by rhizosphere microflora of cotton. Science Cult 39:352–354

Biermann B, Linderman RG (1981) Quantifying vesicular-arbuscular mycorrhizae: a proposed method towards standardization. New Phytol 87:63–67

Bilai VI (1956) *Trichoderma* Pers.-Volatile antibiotics in fungi of the genus *Trichoderma*. Microbiology (Moscow) 25:458–465 (Engl transl)

Bird AF (1959) The attractiveness of roots to the plant parasitic nematodes *Meloidogyne javanica* and *M. hapla*. Nematologica 4:322–335

Bird AF (1962) Orientation of the larvae of *Meloidogyne javanica* relative to roots. Nematologica 8:275–287

Bird LS (1982) The MAR (multi-adversity resistance) system for genetic improvement of cotton. Plant Dis 66:172–176

Blair WC (1978) Interactions of soil fertility and the rhizosphere microflora of seedling cotton in relation to disease caused by *Rhizoctonia solani*. M S thesis, Auburn Univ, Alabama

Blake CD (1962) Some observations on the orientation of *Ditylenchus dipsaci* and invasion of oat seedlings. Nematologica 8:177–192

Blanchard FA, Diller VM (1950) Technique for growing plants with roots in a sterile medium. Plant Physiol 25:767–769

Boero G, Thien S (1979) Phosphatase activity and phosphorus availability in the rhizosphere of corn roots. In: Harley JL, Russell RS (eds) The soil-root interface. Academic Press, London New York, pp 231–242

Böhm W (1974) Mini-rhizotrons for root observations under field conditions. Z Acker-Pflanzenbau 140:282–287

Böhm W (1979) Methods of studying root systems. Springer, Berlin Heidelberg New York

Bollen WB (1961) Interactions between pesticides and soil microorganisms. Annu Rev Microbial 15:69–92

Bonner J, Galston AW (1944) Toxic substances from the culture media of guayule which may inhibit growth. Bot Gaz 106:185–198

Boosalis MG (1956) Effect of soil temperature and green-manure amendment of unsterilized soil on parasitism of *Rhizoctonia solani* by *Penicillium vermiculatum* and *Trichoderma* sp. Phytopathology 46:473–478

Boosalis MG (1964) Hyperparasitism. Annu Rev Phytopathol 2:363–376

Boosalis MG, Mankau R (1965) Parasitism and predation of soil microorganisms. In: Baker KF, Snyder WC (eds) Ecology of soil-borne plant pathogens. Univ Calif Press, Berkeley, pp 374–391

Booth JA (1969) *Gossypium hirsutum* tolerance to *Verticillium albo-atrum* infection. I. Amino acid exudation from aseptic roots of tolerant and susceptible cotton. Phytopathology 59:43–46

Booth JA (1974) Effect of cotton root exudate constituents on growth and pectolytic enzyme production by *Verticillium albo-atrum*. Can J Bot 52:2219–2224

Börner H (1960) Liberation of organic substances from higher plants and their role in the soil sickness problem. Bot Rev 26:393–424

Bouhot D (1979) Estimation of inoculum density and inoculum potential: Techniques and their value for disease prediction. In: Schippers B, Gams W (eds) Soilborne plant pathogens. Academic Press, London New York, pp 21–34

Boulter D, Jeremy JJ, Wilding M (1966) Amino acids liberated into the culture medium by pea seedling roots. Plant Soil 24:121–127

Bowen GD (1961) The toxicity of legume seed diffusates toward rhizobia and other bacteria. Plant Soil 15:155–165

Bowen GD (1969) Nutrient status effects on loss of amides and amino acids from pine roots. Plant Soil 30:139–142

Bowen GD (1978) Dysfunction and shortfalls in symbiotic responses. In: Horsfall JG, Cowling EB (eds) Plant disease. An advanced treatise, vol III. Academic Press, London New York, pp 231–256

Bowen GD (1980) Misconceptions, concepts and approaches in rhizosphere biology. In: Ellwood DC, Latham MJ, Hedger JN, Lynch JM, Slater JH (eds) Contemporary microbial ecology. Academic Press, London New York, pp 283–304

Bowen GD (1981) The root microorganism ecosystem. In: Biological and chemical interactions in the rhizosphere. Symp Proc Ecol Res Comm, Swed Nat Sci Res Counc, Stockholm, pp 3–42
Bowen GD, Rovira AD (1961) The effects of microorganisms on plant growth. I. Development of roots and root hairs in sand and agar. Plant Soil 15:166–188
Bowen GD, Rovira AD (1969) The influence of microorganisms on growth and metabolism of plant roots. In: Whittington WJ (ed) Root growth. Butterworths, London, pp 170–201
Bowen GD, Rovira AD (1976) Microbial colonization of plant roots. Annu Rev Phytopathol 14:121–144
Bowen GD, Theodorou C (1973) Growth of ectomycorrhizal fungi around seeds and roots. In: Marks GC, Kozlowski TT (eds) Ectomycorrhizae, their ecology and physiology. Academic Press, London New York, pp 107–150
Bowen GD, Theodorou C (1979) Interactions between bacteria and ectomycorrhizal fungi. Soil Biol Biochem 11:119–126
Braun HJ (1958) Untersuchungen über den Wurzelschwamm *Fomes annosus* (Fr.) Cooke. Forstwiss Zentralbl 77:65–88
Briggs GE, Robertson RN (1957) Apparent free space. Annu Rev Plant Physiol 8:11–30
Bristow JM (1974) Nitrogen fixation in the rhizosphere of freshwater angiosperms. Can J Bot 52:217–221
Brock TD (1966) Principles of microbial ecology. Prentice Hall, Englewood Cliffs, N J
Brookhouser LW, Weinhold AR (1979) Induction of polygalacturonase from *Rhizoctonia solani* by cotton seed and hypocotyl exudates. Phytopathology 69:599–602
Brouwer R (1953) Water absorption by the roots of *Vicia faba* at various transpiration strengths. II. Causal relation between suction tension, resistance and uptake. Proc K Ned Akad Wet Ser C 56:129–136
Brown ME (1972) Plant growth substances produced by microorganisms of soil and rhizosphere. J Appl Bacteriol 35:443–451
Brown ME (1973) Soil bacteriostasis limitation in growth of soil and rhizosphere bacteria. Can J Microbiol 19:195–199
Brown ME (1974) Seed and root bacterization. Annu Rev Phytopathol 12:181–197
Brown ME (1975) Rhizosphere microorganisms – opportunists, bandits or benefactors. In: Walker N (ed) Soil microbiology. Wiley, New York Toronto, pp 21–38
Brown ME, Burlingham SK, Jackson RM (1962) Studies on *Azotobacter* species in soil. II. Populations of *Azotobacter* in the rhizosphere and effects of artificial inoculation. Plant Soil 17:320–332
Brown ME, Jackson RM, Burlingham SK (1968) Growth and effects of bacteria introduced into soil. In: Gray TRG, Parkinson D (eds) The ecology of soil bacteria. Univ Liverpool Press, Liverpool, pp 531–551
Brown R, Edwards M (1944) The germination of the seed of *Striga lutea*. I. Host influence and the progress of germination. Ann Bot (London) N S 8:131–148
Brown RL, Tang CS, Nishimoto RK (1983) Growth inhibition from guava root exudates. Hortscience 18:316–318
Brown SL, Curl EA (1980) Rhizosphere effect of a herbicide-stressed plant, *Cassia obtusifolia*, on *Fusarium oxysporum* f. sp. *vasinfectum*. Phytopathology 70:565
Buddenhagen IW (1965) The relation of plant pathogenic bacteria to the soil. In: Baker KF, Snyder WC (eds) Ecology of soil-borne plant pathogens. Univ Calif Press, Berkeley, pp 269–284
Burden RS, Rogers PM, Wain RL (1974) Investigations on fungicides. XVI. Natural resistance of plant roots to fungal pathogens. Ann Appl Biol 78:59–63
Burges A (1939) Soil fungi and root infection. Broteria 8:64–81
Burns RE (1972) Environmental factors affecting root development and reserve carbohydrates of bermudagrass cuttings. Agron J 64:44–45
Burns RG (ed) (1978) Soil enzymes. Academic Press, London New York
Burr TJ, Schroth MN, Suslow T (1978) Increased potato yields by treatment of seedlings with specific strains of *Pseudomonas fluorescens* and *P. putida*. Phytopathology 68:1377–1383
Burstrom HG (1965) The physiology of plant roots. In: Baker KF, Snyder WC (eds) Ecology of soil-borne plant pathogens. Univ Calif Press, Berkeley pp 154–169
Butcher JW, Snider R, Snider RJ (1971) Bioecology of edaphic Collembola and Acarina. Annu Rev Entomol 16:249–288
Butler EE (1957) *Rhizoctonia solani* as a parasite of fungi. Mycologia 49:354–373

Buxton EW (1960) Effects of pea root exudate on the antagonism of some rhizosphere microorganisms towards *Fusarium oxysporum* f. sp. *pisi*. J Gen Microbiol 22:678–689
Caldwell BE (1969) Initial competition of root-nodule bacteria on soybeans in a field environment. Agron J 61:813–815
Cailloux M (1972) Metabolism and the absorption of water by root hairs. Can J Bot 50:557–573
Campbell R, Porter R (1982) Low-temperature scanning electron microscopy of micro-organisms in soil. Soil Biol Biochem 14:241–245
Campbell R, Rovira AD (1973) The study of the rhizosphere by scanning electron microscopy. Soil Biol Biochem 5:747–752
Cartwright PM, Snow D (1962) The influence of foliar applications of urea on the nodulation pattern of certain leguminous species. Ann Bot (London) N S 26:251–259
Castanho B, Butler EE (1978) *Rhizoctonia* decline: A degenerative disease of *Rhizoctonia solani*. Phytopathology 68:1505–1510
Chaboud, Annie (1983) Isolation, purification and chemical composition of maize root cap slime. Plant Soil 73:395–402
Chadwick AV, Burg SP (1967) An explanation of the inhibition of root growth caused by indole-3-acetic acid. Plant Physiol 42:415–420
Chakraborty S, Old KM, Warcup JH (1983) Amoebae from a take-all suppressive soil which feed on *Gaeumannomyces graminis tritici* and other soil fungi. Soil Biol Biochem 15:17–24
Chandra S, Chabot JF, Morrison GH, Leopold AC (1982) Localization of calcium in amyloplasts of root-cap cells using ion microscopy. Science 216:1221–1223
Chang I-P, Kommedahl T (1968) Biological control of seedling blight of corn by coating kernels with antagonistic microorganisms. Phytopathology 58:1395–1401
Chang-Ho, Yung (1970) The effect of pea root exudate on the germination of *Pythium aphanidermatum* zoospore cysts. Can J Bot 48:1501–1514
Chapman HD (1965) Chemical factors of the soil as they affect microorganisms. In: Baker KF, Snyder WC (eds) Ecology of soil-borne plant pathogens. Univ Calif Press, Berkeley, pp 120–141
Chatel DL, Parker CA (1972) Inhibition of rhizobia by toxic soil-water extracts. Soil Biol Biochem 4:289–294
Cholodny N (1930) Über eine neue Methode zur Untersuchung der Bodenmikroflora. Arch Mikrobiol 1:620–652
Christen AA (1975) Some fungi associated with Collembola. Rev Ecol Biol Sol 12:723–728
Christiansen K (1964) Bionomics of Collembola. Annu Rev Entomol 9:147–178
Christie P, Newman EI, Campbell R (1978) The influence of neighbouring grassland plants on each others' endomycorrhizae and root-surface microorganisms. Soil Biol Biochem 10:521–527
Chuang TY, Ko WH (1979) Propagule size: its value in the prediction of inoculum density and infection potential in soil. In: Schippers B, Gams W (eds) Soil-borne plant pathogens, Academic Press, London New York, pp 35–38
Clark FE (1947) Rhizosphere microflora as affected by soil moisture changes. Soil Sci Soc Am Proc 12:239–242
Clark FE (1949) Soil microorganisms and plant roots. Adv Agron 1:241–288
Clark FE (1965) The concept of competition in microbial ecology. In: Baker KF, Snyder WC (eds) Ecology of soil-borne plant pathogens. Univ Calif Press, Berkeley, pp 339–347
Clark FE (1968) The growth of bacteria in soil. In: Gray TRG, Parkinson D (eds) The ecology of soil bacteria. Univ Toronto Press, Toronto, pp 441–457
Clark FE, Beard WE (1965) Protozoa. In: Black CA, Evans DD, White JL, Ensminger LE, Clark FE (eds) Methods of soil analysis, part 2. Chemical and microbiological properties. Am Soc Agron, Madison, Wisc, pp 1513–1516
Clark FE, Durrell LW (1965) Algae. In: Black CA, Evans DD, White JL, Ensminger LE, Clark FE (eds) Methods of soil analysis, part 2. Chemical and microbiological properties, Am Soc Agron, Madison, Wisc, pp 1506–1512
Clark J, Gibbs RD (1957) Studies in tree physiology. IV. Further investigations of seasonal changes in moisture content of certain Canadian forest trees. Can J Bot 35:219–253
Clarke AE, Knox RB (1978) Cell recognition in flowering plants. Q Rev Biol 53:3–28
Clayton MF, Lamberton JA (1964) A study of root exudates by the fog-box technique. Aust J Biol Sci 17:855–866
Clements FE, Shelford VE (1939) Bio-ecology. Wiley, New York

Clowes FAL (1956a) Nucleic acids in root apical meristems of *Zea*. New Phytol 55:29–34
Clowes FAL (1956b) Localization of nucleic acid synthesis in root meristems. J Exp Bot 7:307–312
Clowes FAL (1958) Development of quiescent centres in root meristems. New Phytol 57:85–88
Clowes FAL (1959) Apical meristems of roots. Biol Rev 34:501–529
Clowes FAL (1961) Apical meristems. Blackwell's, Oxford
Cole CV, Elliott ET, Hunt HW, Coleman DC (1978) Trophic interactions in soils as they affect energy and nutrient dynamics. V. Phosphorus transformations. Microb Ecol 4:381–387
Coleman DC, Cole CV, Anderson RV, Blaha M, Campion MK, Clarholm M, Elliott ET, Hunt HW, Shaefer B, Sinclair J (1977) An analysis of rhizosphere-saprophage interactions in terrestrial ecosystems. In: Lohm U, Persson T (eds) Organisms as components of ecosystems. Swedish Natural Science Research Council, Stockholm, Ecol Bull No 25, pp 299–309
Coleman DC, Ingham RE, McClellan JF, Trofymow JA (1984) Soil nutrient transformations in the rhizosphere via animal-microbial interactions. In: Anderson JM, Rayner ADM, Walton DWH (eds) Invertebrate-microbial interactions. Cambridge Univ Press, Cambridge, pp 35–38
Coley-Smith JR, Cooke RC (1971) Survival and germination of fungal sclerotia. Annu Rev Phytopathol 9:65–92
Coley-Smith JR, King JE (1970) Response of resting structures of root-infecting fungi to host exudates: An example of specificity. In: Tousson TA, Bega RV, Nelson PE (eds) Root diseases and soil-borne pathogens. Univ Calif Press, Berkeley, pp 130–133
Cook RJ (1981) Biological control of plant pathogens: overview. In: Papavizas GC (ed) Biological control in crop production. Granada, London, pp 23–44
Cook RJ, Baker KF (1983) The nature and practice of biological control of plant pathogens. Am Phytopathol Soc, St Paul, Minn
Cook RJ, Papendick RI (1972) Influence of water potential of soils and plants on root diseases. Annu Rev Phytopathol 10:349–374
Cook RJ, Snyder WC (1965) Influence of host exudates on growth and survival of germlings of *Fusarium solani* f. *phaseoli* in soil. Phytopathology 55:1021–1025
Cooke RC (1962) The ecology of nematode-trapping fungi in the soil. Ann Appl Biol 50:507–513
Cooksey DA, Moore LW (1982) Biological control of crown gall with an agrocin mutant of *Agrobacterium radiobacter*. Phytopathology 72:919–921
Cordell CE, Marx DH (1980) Ectomycorrhizae: Benefits and practical application in forest tree nurseries and field plantings. Proc N Am For Tree Nursery Work Shop, Syracuse, N Y, pp 217–224
Crossett RN, Campbell DJ (1975) The effects of ethylene in the root environment upon the development of barley. Plant Soil 42:453–464
Crossett RN, Campbell DJ, Stewart HE (1975) Compensatory growth in cereal root systems. Plant Soil 42:673–683
Curl EA (1963) Control of plant diseases by crop rotation. Bot Rev 29:413–479
Curl EA (1979) Effects of mycophagous Collembola on *Rhizoctonia solani* and cotton seedling disease. In: Schippers B, Gams W (eds) Soil-borne plant pathogens. Academic Press, London New York, pp 253–269
Curl EA (1982) The rhizosphere: Relation to pathogen behavior and root disease. Plant Dis 66:624–630
Curl EA, Rodríguez-Kábana R (1977) Herbicide-plant disease relationships. In: Truelove B (ed) Research methods in weed science, 2nd edn. South Weed Sci Soc, Auburn, Alabama, pp 174–191
Curl EA, Snell JM (1981) Grazing patterns of mycophagous Collembola and assessment of biocontrol potential. Phytopathology 71:869 (Abstr)
Curl EA, Gudauskas RT, Harper JD, Peterson CM (1985) Effects of soil insects on populations and germination of fungal propagules. In: Parker CA, Rovira AD, Moore KJ, Wong PTW, Kollmorgen JF (eds) Ecology and management of soilborne plant pathogens. Am Phytopathol Soc, St. Paul, Minn, pp 20–23
Currier WW, Strobel GA (1976) Chemotaxis of *Rhizobium* spp. to plant root exudates. Plant Physiol 57:820–823
Cushman JH (1982) Nutrient transport inside and outside the root rhizosphere: Theory. Soil Sci Soc Am J 46:704–709
Cutler HG, Lefiles JH (1978) Trichodermin: Effects on plants. Plant Cell Physiol 19:177–182
Cutter EG (1971) Plant anatomy, part 2, Organs. Arnold, London

Cutter EG (1978) Plant anatomy, part 1. Cells and tissues, 2nd edn. Arnold, London
Daly JM (1984) The role of recognition in plant disease. Annu Rev Phytopathol 22:273–307
Dangerfield JA, Westlake DWS, Cook FD (1975) Quantitative assessment of the bacterial rhizosphere flora of *Pinus contorta* var. *latifolia*. Can J Microbiol 21:2034–2038
Dangerfield JA, Westlake DWS, Cook FD (1978) Characterization of the bacterial flora associated with root systems of *Pinus contorta* var. *latifolia*. Can J Microbiol 24:1520–1525
Darbyshire JF (1966) Protozoa in the rhizosphere of *Lolium perrenne* L. Can J Microbiol 12:1287–1289
Darbyshire JF, Greaves MP (1970) An improved method for the study of the interrelationships of soil microorganisms and plant roots. Soil Biol Biochem 2:63–71
Dart PJ (1971) Scanning electron microscopy of plant roots. J Exp Bot 22:163–168
Dart PJ, Day JM (1975) Non-symbiotic nitrogen fixation in soil. In: Walker N (ed) Soil microbiology. Wiley, New York, pp 225–252
Dart PJ, Mercer FV (1964) The legume rhizosphere. Arch Mikrobiol 47:344–378
Davey CB (1971) Nonpathogenic organisms associated with mycorrhizae. In: Hacskaylo E (ed) Mycorrhizae. Proc 1st N Am Conf Mycorrhizae. US Gov Print Off, Washington DC, pp 114–121
Davey CB, Papavizas GC (1960) Effect of decomposing organic soil amendments and nitrogen on fungi in soil and bean rhizosphere. Trans 7th Int Congr Soil Sci, pp 77–83
Davey CB, Papavizas GC (1961) Translocation of streptomycin from Coleus leaves and its effect on rhizosphere bacteria. Science 134:1368–1369
Davis JR, McDole RE (1979) Influence of cropping sequences on soil-borne populations of *Verticillium dahliae* and *Rhizoctonia solani*. In: Schippers B, Gams W (eds) Soil-borne plant pathogens. Academic Press, London New York, pp 399–405
Dazzo FB (1980) Microbial adhesion to plant surfaces. In: Berkley RCW, Lynch JM, Melling J, Rutter PR, Vincent B (eds) Microbial adhesion to surfaces. Wiley, New York, pp 312–328
Dazzo FB, Smith PH, Hubbell DH (1974) Changes in the rhizosphere effect of millet associated with sprinkler irrigation with animal wastes. J Environ Qual 3:270–273
De Bertoldi M, Rambelli A, Giovannetti M, Griselli M (1978) Effects of Benomyl and Captan on rhizosphere fungi and the growth of *Allium cepa*. Soil Biol Biochem 10:265–268
De Candolle AP (1832) Physiologie vegetale, vol III. Bechet Jeune, Lib Fac Med, Paris
Dehne HW (1982) Interaction between vesicular-arbuscular mycorrhizal fungi and plant pathogens. Phytopathology 72:1115–1119
de Mendonca M, Stanghellini ME (1979) Endemic and soilborne nature of *Erwinia carotovora* var. *atroseptica*, a pathogen of mature sugar beets. Phytopathology 69:1096–1099
Dennis C, Webster J (1971) Antagonistic properties of species-groups of *Trichoderma*. II. Production of volatile antibiotics. Trans Br Mycol Soc 57:41–48
Dittmer HJ (1937) A quantitative study of the roots and root hairs of a winter rye plant (*Secale cereale*). Am J Bot 24:417–420
Dobbs CG, Hinson WH (1953) A widespread fungistasis in soils. Nature (London) 172:197–199
Dobereiner J, Day JM (1975) Nitrogen fixation in the rhizosphere of tropical grasses. In: Stewart WDP (ed) Nitrogen fixation by free-living microorganisms. Cambridge Univ Press, Cambridge, pp 39–55
Dobereiner J, Day JM, Dart PJ (1972) Nitrogenase activity and oxygen sensitivity of the *Paspalum notatum-Azotobacter paspali* association. J Gen Microbiol 71:103–116
Dommergues YR, Krupa SV (eds) (1978) Interactions between soil microorganisms and plants. Elsevier, Amsterdam Oxford
Drahos D, Hemming B, McPherson S, Brackin J (1984) β-Galactosidase, a selectable non-antibiotic chromogenic marker for fluorescent pseudomonads. Phytopathology 74:800 (Abstr)
Drake M, Steckel JE (1955) Solubilization of soil and rock phosphate as related to root cation exchange capacity. Soil Sci Soc Am Proc 19:449–450
Drechsler C (1941) Predacious fungi. Biol Rev 16:265–290
Drury RE, Baker R, Griffin GJ (1983) Calculating the dimensions of the rhizosphere. Phytopathology 73:1351–1354
Duddington CL (1956) The predacious fungi: Zoopagales and Moniliales. Biol Rev 31:152–193
Duddington CL, Wyborn CHE (1972) Recent research on the nematophagous Hyphomycetes, Bot Rev 38:545–565

Dupler M, Baker R (1984) Survival of *Pseudomonas putida*, a biological control agent, in soil. Phytopathology 74:195–200

Du Plessis HJ, Hattingh MJ, Van Vuuren HJJ (1985) Biological control of crown gall in South Africa by *Agrobacterium radiobacter* strain K84. Plant Dis 69:302–305

Edwards CA, Thompson AR (1973) Pesticides and the soil fauna. Residue Rev 45:1–79

Egeraat AWSM van (1975a) Exudation of ninhydrin-positive compounds by pea-seedling roots: A study of the sites of exudation and of the composition of the exudate. Plant Soil 42:37–47

Egeraat AWSM van (1975b) The possible role of homoserine in the development of *Rhizobium leguminosarum* in the rhizosphere of pea seedlings. Plant Soil 42:381–386

Egeraat AWSM van (1979) Root exudates and microbial growth in the interface. In: Harley JL, Russell RS (eds) The soil-root interface. Academic Press, London New York, p 438

Eklund E, Sinda E (1971) Establishment and disappearance of introduced pseudomonads in the rhizoplane of peat grown cucumber plants. Plant Soil 35:495–504

Elliott ET, Coleman DC, Cole CV (1979) The influence of amoebae on the uptake of nitrogen by plants in gnotobiotic soil. In: Harley JL, Russell RS (eds) The soil-root interface. Academic Press, London New York, pp 221–229

Erwin DC (1973) Systemic fungicides: Disease control, translocation, and mode of action. Annu Rev Phytopathol 11:389–422

Erwin DC, Katznelson H (1961) Suppression and stimulation of mycelial growth of *Phytophthora cryptogea* by certain thiamine-requiring and thiamine-synthesizing bacteria. Can J Microbiol 7:945–950

Esau K (1969) The phloem. In: Zimmermann W, Ozenda P, Wulff HD (eds) Encyclopedia of plant anatomy, vol V, part 2. Borntraeger, Stuttgart

Esau K (1977) Anatomy of seed plants, 2nd edn. Wiley, New York

Estermann EF, McLaren AD (1961) Contribution of rhizoplane organisms to the total capacity of plants to utilize organic nutrients. Plant Soil 15:243–260

Evans J, Barnet YM, Vincent JM (1979) Effect of a bacteriophage on the colonisation and nodulation of clover roots by a strain of *Rhizobium trifolii*. Can J Microbiol 25:968–973

Evert RF (1977) Phloem structure and histochemistry. Annu Rev Plant Physiol 28:199–222

Evert RF (1982) Sieve tube structure in relation to function. BioScience 32:789–795

Fåhraeus G, Ljunggren H (1968) Pre-infection phases of the legume symbiosis. In: Gray TRG, Parkinson D (eds) The ecology of soil bacteria. Univ Toronto Press, Toronto, pp 396–421

Feldman LJ (1981) Root cap inhibitor formation in isolated root caps of *Zea mays*. J Exp Bot 32:779–788

Fenwick L (1973) Studies on the rhizosphere microflora of onion plants in relation to temperature changes. Soil Biol Biochem 5:315–320

Ferguson JJ, Woodhead SH (1982) Production of endomycorrhizal inoculum: Increase and maintenance of vesicular-arbuscular mycorrhizal fungi. In: Schenck NC (ed) Methods and principles of mycorrhizal research. Am Phytopathol Soc, St Paul, Minn, pp 47–54

Ferriss RS (1981) Calculating rhizosphere size. Phytopathology 71:1229–1231

Ferriss RS (1983) Calculating the dimensions of the rhizosphere – a response. Phytopathology 73:1355–1357

Filonow AB, Lockwood JL (1979) Conidial exudation by *Cochliobolus victoriae* on soils in relation to soil mycostasis. In: Schippers B, Gams W (eds) Soil-borne plant pathogens. Academic Press, London New York, pp 107–119

Fletcher WM (1960) The effects of herbicides on soil microorganisms. In: Woodford EK, Sagar GR (eds) Herbicides and the soil. Blackwell, Oxford, pp 20–62

Floyd RA, Ohlrogge AJ (1971) Gel formation on nodal root surfaces of *Zea mays*. Some observations relevant to understanding its action at the root-soil interface. Plant Soil 34:595–606

Foster RC (1981) The ultrastructure and histochemistry of the rhizosphere. New Phytol 89:263–273

Foster RC, Marks GC (1967) Observations on the mycorrhizae of forest trees. II. The rhizosphere of *Pinus radiata* D. Don. Aust J Biol Sci 20:915–926

Forster RC, Rovira AD (1976) Ultrastructure of wheat rhizosphere. New Phytol 76:343–352

Foster RC, Rovira AD, Cock TW (1983) Ultrastructure of the root-soil interface. Am Phytopathol Soc, St. Paul, Minn

Foy CL, Hurtt W (1967) Further studies on root exudation of exogenous growth regulators in *Phaseolus vulgaris* L. Abstr Proc Weed Soc Am, p 40

Franz H (1950) Bodenzoologie als Grundlage der Bodenpflege. Akademie, Berlin

Frederick LR (1965) Microbial populations by direct microscopy. In: Black CA, Evans DD, White JL, Emsminger LE, Clark FE (eds) Methods of soil analysis, part 2. Chemical and microbiological properties. Am Soc Agron, Madison, Wisc, pp 1452–1459

Frenzel B (1960) Zur Ätiologie der Anreicherung von Aminosäuren und Amiden im Wurzelraum von *Helianthus annuus* L. Ein Beitrag zur Klärung der Probleme der Rhizosphäre. Planta 55:169–207

Fries N (1973) Effects of volatile organic compounds on the growth and development of fungi. Trans Br Mycol Soc 60:1–21

Führer von E (1961) Der Einfluß von Pflanzenwurzeln auf die Verteilung der Kleinarthropoden im Boden, untersucht an *Pseudotritia ardua* (Oribatei). Pedobiologia 1:99–112

Garber ED (1956) A nutrition-inhibition hypothesis of pathogenicity. Am Nat 90:183–194

Gardner WK, Boundy KA (1983) The aquisition of phosphorus by *Lupinus albus* L. IV. The effect of interplanting wheat and white lupin on the growth and mineral composition of the two species. Plant Soil 70:391–402

Gardner WK, Barber DA, Parbery DG (1983) The aquisition of phosphorus by *Lupinus albus* L. III. The probable mechanism by which phosphorus movement in the soil/root interface is enhanced. Plant Soil 70:107–124

Garren KH (1964) Land plaster and soil rot of peanut pods in Virginia. Plant Dis Rep 48:349–352

Garrett SD (1956) Biology of root-infecting fungi. Cambridge Univ Press, Cambridge

Garrett SD (1970) Pathogenic root-infecting fungi. Cambridge Univ Press, Cambridge

Gaudet DA, Sands DC, Mathre DE, Ditterline RL (1980) The role of bacteria in the root and crown rot complex of irrigated sainfoin in Montana. Phytopathology 70:161–167

Gauze GF (1934) The struggle for existence. Williams and Wilkins, Baltimore

Gerdemann JW (1968) Vesicular-arbuscular mycorrhizae and plant growth. Annu Rev Phytopathol 6:397–418

Gerretsen FC (1948) The influence of microorganisms on the phosphate intake by the plant. Plant Soil 1:51–81

Gibbons GSB, Wilkins MB (1970) Growth inhibitor production by root caps in relation to geotropic responses. Nature (London) 226:558–559

Gilbert RG, Linderman RG (1971) Increased activity of soil microorganisms near sclerotia of *Sclerotium rolfsii* in soil. Can J Microbiol 17:557–562

Gilligan CA (1979) Modeling rhizosphere infection. Phytopathology 69:782–784

Gilligan CA (1983a) Modeling of soilborne pathogens. Annu Rev Phytopathol 21:45–64

Gilligan CA (1983b) A test for randomness of infection by soilborne pathogens. Phytopathology 73:1351–1354

Gilmore SK (1972) Collembola predation on nematodes. Search Agric 1:1–12

Giovannetti M, Mosse B (1980) An evaluation of techniques for measuring vesicular-arbuscular mycorrhizal infection in roots. New Phytol 84:489–500

Giuma AY, Hackett AM, Cooke RC (1973) Thermostable nematotoxins produced by germinating conidia of some endozoic fungi. Trans Br Mycol Soc 60:49–56

Glass ADM (1976) The allelopathic potential of phenolic acids associated with the rhizosphere of *Pteridium aquilinum*. Can J Bot 54:2440–2444

Godoy G, Rodríguez-Kábana R, Morgan-Jones G (1982) Parasitism of eggs of *Heterodera glycines* and *Meloidogyne arenaria* by fungi isolated from cysts of *H. glycines*. Nematropica 12:111–119

Goring CAI, Clark FE (1948) Influence of crop growth on mineralization of nitrogen in the soil. Soil Sci Soc Am Proc 13:216–266

Goto M, Makino T (1976) Induction of outgrowth formation on storage roots of sweet potato due to plant pathogenic bacteria. Phytopathology 66:28–33

Gottlieb D, Shaw PD (1970) Mechanism of action of antifungal antibiotics. Annu Rev Phytopathol 8:371–402

Graham PH (1963) Antigenic affinities of the root-nodule bacteria of legumes. Antonie van Leeuwenhoek J Microbiol Serol 29:281–291

Grand LF, Harvey AE (1982) Quantitative measurement of ectomycorrhizae on plant roots. In: Schenck NC (ed) Methods and principles of mycorrhizal research. Am Phytopathol Soc, St Paul, Minn, pp 156–164

Gray TRG, Williams ST (1971) Soil microorganisms. Hafner, New York

Gray TRG, Baxby P, Hill IR, Goodfellow M (1968) Direct observation of bacteria in soil. In: Gray TRG, Parkinson D (eds) The ecology of soil bacteria. Univ Toronto Press, Toronto, pp 171–197

Greaves MP, Darbyshire JF (1972) The ultrastructure of the mucilaginous layer on plant roots. Soil Biol Biochem 4:443–449

Green CD (1971) Mating and host finding behavior of plant nematodes. In: Zuckerman BM, Mai WF, Rohde RA (eds) Plant parasitic nematodes, vol II. Academic Press, London New York, pp 247–266

Green RJ Jr (1969) Survival and inoculum potential of conidia and microsclerotia of *Verticillium alboatrum* in soil. Phytopathology 59:874–876

Greig-Smith P (1964) Quantitative plant ecology. Butterworths, London

Griffin DM (1972) Ecology of soil fungi. Chapman and Hall, London

Griffin GJ (1969) *Fusarium oxysporum* and *Aspergillus flavus* spore germination in the rhizosphere of peanut. Phytopathology 59:1214–1218

Griffin GJ, Roth DA (1979) Nutritional aspects of soil mycostasis. In: Schippers B, Gams W (eds) Soilborne plant pathogens. Academic Press, London New York, pp 79–96

Griffin GJ, Tomimatsu GS (1983) Observed root infection pattern, infection efficiency and infection-disease incidence relationships of *Cylindrocladium crotalariae* in peanut field soil. Can J Plant Pathol 5:81–88

Griffin GJ, Hale MG, Shay FJ (1976) Nature and quantity of sloughed organic matter produced by roots of axenic peanut plants. Soil Biol Biochem 8:29–32

Grineva GM (1962) Excretion of plant roots during brief periods of anaerobiosis. Sov Plant Physiol 8:549–552 (Transl from: Fiziol Rast 8:686–691, 1961)

Grogan RG, Sall MA, Punja ZK (1980) Concepts for modeling root infection by soilborne fungi. Phytopathology 70:361–363

Guckert A, Breisch H, Reisinger O (1975) Interface Sol-Racine-I. Etude au microscope electronique des relations mucigelargile-microorganismes. Soil Biol Biochem 7:241–250

Hackett C (1968) A study of the root system of barley. I. Effects of nutrition on two varieties. New Phytol 67:287–299

Hadwiger LA, Schwochau ME (1969) Host resistance responses – an induction hypothesis. Phytopathology 59:223–227

Hale MG, Moore LD (1979) Factors affecting root exudation II: 1970–1978. Adv Agron 31:93–126

Hale MG, Foy CL, Shay FJ (1971) Factors affecting root exudation. Adv Agron 23:89–109

Hale MG, Lindsey DL, Hameed KM (1973) Gnotobiotic culture of plants and related research. Bot Rev 39:261–273

Hale MG, Moore LD, Griffin GJ (1978) Root exudates and exudation. In: Dommergues YR, Krupa SV (eds) Interactions between non-pathogenic soil microorganisms and plants, Elsevier, Amsterdam, pp 163–203

Hale MG, Moore LD, Griffin GJ (1981) Factors affecting root exudation and significance for the rhizosphere ecosystems. In: Biological and chemical interactions in the rhizosphere, Symp Proc Ecol Res Comm, Swed Nat Sci Res Counc, Stockholm, pp 43–71

Hall AD (1917) The book of the Rothamsted experiments. Dutton, New York

Halleck FE, Cochrane VW (1950) The effect of fungistatic agents on the bacterial flora of the rhizosphere. Phytopathology 40:715–718

Hallock DL, Garren KH (1968) Pod breakdown, yield, and grade of Virginia type peanuts as affected by Ca, Mg, and K sulfates. Agron J 60:253–257

Halsted BD (1888) Nitrogen appropriation in clovers: A hint. Bull Iowa State Agric Coll Bot Dep 28–33

Hameed KM, Saghir AR, Foy CL (1973) Influence of root exudates on *Orobanche* seed germination. Weed Res 13:114–117

Hamlen RA, Lukezic FL, Bloom JR (1972) Influence of age and stage of development on the neutral carbohydrate components in root exudates from alfalfa plants grown in a gnotobiotic environment. Can J Plant Sci 52:633–642

Hancock HG (1981) Effects of foliar fungicides on the soilborne microflora of peanuts. MS thesis, Auburn Univ, Alabama

Hansen JD, Curl EA (1964) Interactions of *Sclerotium rolfsii* and other microorganisms isolated from stolon tissue of *Trifolium repens*. Phytopathology 54:1127–1132

Harkes PAA (1973) Structure and dynamics of the root cap of *Avena sativa* L. Acta Bot Neerl 22:321–328

Harley JL (1965) Mycorrhiza. In: Baker KF, Snyder WC (eds) Ecology of soil-borne plant pathogens. Univ Calif Press, Berkeley, pp 218–230
Harley JL, Russell RS (eds) (1979) The soil-root interface. Academic Press, London New York
Harley JL, Waid JS (1955a) A method of studying active mycelia on living roots and other surfaces in the soil. Trans Br Mycol Soc 38:104-118
Harley JL, Waid JS (1955b) The effect of light upon the roots of beech and its surface populations. Plant Soil 7:96–112
Harman GE, Chet I, Baker R (1981) Factors affecting *Trichoderma hamatum* applied to seeds as a biocontrol agent. Phytopathology 71:569–572
Harmsen GW, Jager G (1962) Determination of the quantity of carbon and nitrogen in the rhizosphere of young plants. Nature (London) 195:1119–1120
Harmsen GW, Kolenbrander GJ (1965) Soil inorganic nitrogen. In: Bartholomew WV, Clark FE (eds) Soil nitrogen. Am Soc Agron, Madison, Wisc, pp 43–92
Harper SHT, Lynch JM (1980) Microbial effects on the germination and seedling growth of barley. New Phytol 84:473–481
Hattingh MJ, Louw HA (1969) Clover rhizoplane bacteria antagonistic to *Rhizobium trifolii*. Can J Microbiol 15:361–364
Hayes WA, Randle PE, Last FT (1969) The nature of the microbial stimulus affecting sporophore formation in *Agaricus bisporus* (Lange) Sing. Ann Appl Biol 64:177–187
Hayman DS (1970) The influence of cottonseed exudate on seedling infection by *Rhizoctonia solani*. In: Toussoun TA, Bega RV, Nelson PE (eds) Root diseases and soilborne pathogens. Univ Calif Press, Berkeley, pp 99-102
Healy WB, Lugwig TG, Losee, FL (1961) Soils and dental caries in Hawkes' Bay, New Zealand. Soil Sci 92:359–366
Hellriegel H, Wilfarth H (1888) Beilageheft zu der Z Ver Rübenzucker-Ind Dtsch Reichs 234
Hely FW, Bergersen FJ, Brockwell J (1957) Microbial antagonism in the rhizosphere as a factor in the failure of inoculation of subterranean clover. Aust J Agric Res 8:24–44
Henderson VE, Katznelson H (1961) The effect of plant roots on the nematode population of the soil. Can J Microbiol 7:163–167
Hendrix JW (1970) Sterols in growth and reproduction of fungi. Annu Rev Phytopathol 8:111–130
Henis Y, Ben-Yephet Y (1970) Effect of propagule size of *Rhizoctonia solani* on saprophytic growth, infectivity, and virulence on bean seedlings. Phytopathology 60:1351–1356
Heungens A (1968) The influence of DBCP on the soil fauna in azalea culture. Pedobiologia 8:281–288
Hiltner L (1904) Über neuere Erfahrungen und Probleme auf dem Gebiet der Bodenbakteriologie und unter besonderer Berücksichtigung der Gründüngung und Brache. Arb Dtsch Landwirt. Ges 98:59–78
Hiltner L, Störmer K (1903) Studien über die Bakterienflora des Ackerbodens. Arb Biol Abt K Gesundh 3:443–445
Hocking D, Cook FD (1972) Myxobacteria exert partial control of damping-off and root disease in container-grown tree seedlings. Can J Microbiol 18:1557–1560
Holland AA, Parker CA (1966) Studies on microbial antagonism in the establishment of clover pasture. II. The effect of saprophytic soil fungi upon *Rhizobium trifolii* and the growth of subterranean clover. Plant Soil 25:329–340
Homma Y (1984) Perforation and lysis of hyphae of *Rhizoctonia solani* and conidia of *Cochliobolus miyabeanus* by soil myxobacteria. Phytopathology 74:1234–1239
Hora TS, Baker R (1972) Soil fungistasis: Microflora producing a volatile inhibitor. Trans Br Mycol Soc 59:491–500
Horst RK, Herr LJ (1962) Effects of foliar urea treatment on numbers of actinomycetes antagonistic to *Fusarium roseum* f. *cerealis* in the rhizosphere of corn seedlings. Phytopathology 52:423–427
Howell CR, Stipanovic RD (1979) Control of *Rhizoctonia solani* on cotton seedlings with *Pseudomonas fluorescens* and with an antibiotic produced by the bacterium. Phytopathology 69:480–482
Htay K, Kerr A (1974) Biological control of crown gall: seed and root inoculation. J Appl Bacteriol 37:525–530
Huang HC (1977) Importance of *Coniothyrium minitans* in survival of sclerotia of *Sclerotinia sclerotiorum* in wilted sunflower. Can J Bot 55:289–295
Huang HC (1980) Control of sclerotinia wilt of sunflower by hyperparasites. Can J Plant Pathol 2:26–32

Huck MG, Taylor HM (1982) The rhizoton as a tool for root research. Adv Agron 35:1–35

Huck MG, Klepper B, Taylor HM (1970) Diurnal variations in root diameter. Plant Physiol 45:529–530

Hurtt W, Foy CL (1965) Some factors influencing the excretion of foliarly-applied dicamba and picloram from roots of Black Valentine beans. Plant Physiol (Suppl) 40:48

Husain SS, McKeen WE (1963) Interactions between strawberry roots and *Rhizoctonia fragariae*. Phytopathology 53:541–545

Hussey RS, Roncadori RW (1982) Vesicular-arbuscular mycorrhizae may limit nematode activity and improve plant growth. Plant Dis 66:9–13

Ingham JL (1972) Phytoalexins and other natural products as factors in plant disease resistance. Bot Rev 38:343–424

Ivarson KC, Katznelson H (1960) Studies on the rhizosphere microflora of yellow birch seedlings. Plant Soil 12:30–40

Ivarson KC, Mack AR (1972) Root-surface mycoflora of soybean in relation to soil temperature and moisture in a field environment. Can J Soil Sci 52:199–208

Iyer JG, Wilde SA (1965) Effect of Vapam biocide on the growth of red pine seedlings. J For 63:703–704

Jackson RM (1957) Fungistasis as a factor in the rhizosphere phenomenon. Nature (London) 180:96–97

Jackson RM (1965) Antibiosis and fungistasis of soil microorganisms. In: Baker KF, Snyder WC (eds) Ecology of soil-borne plant pathogens. Univ Calif Press, Berkeley, pp 363–369

Jalali BL (1976) Biochemical nature of root exudates in relation to root rot of wheat. III. Carbohydrate shifts in response to foliar treatments. Soil Biol Biochem 8:127–129

Jalali BL, Domsch KH (1975) Effect of systemic fungitoxicants on the development of endotrophic mycorrhiza. In: Sanders FE, Mosse B, Tinker PB (eds) Endomycorrhizae. Academic Press, London New York, pp 619–626

Jatala P, Kaltenbach R, Bocangel M, Devaux AJ, Campos R (1980) Field application of *Paecilomyces lilacinus* for controlling *Meloidogyne incognita* on potatoes. J Nematol 12:226–227

Jayman TCZ, Sivasubramaniam S (1975) Release of bound iron and aluminum from soils by the root exudates of tea (*Camellia sinensis*) plants. J Sci Food Agric 26:1895–1898

Jenny H, Grossenbacher K (1963) Root-soil boundary zones as seen in the electron microscope. Soil Sci Soc Am Proc 27:273–277

Jensen V, Holm E (1975) Associative growth of nitrogen-fixing bacteria with other micro-organisms. In: Stewart WDP (ed) Nitrogen fixation by free-living micro-organisms. Cambridge Univ Press, Cambridge, pp 101–119

Jensen WA, Kavaljian LG (1958) An analysis of cell morphology and the periodicity of division in the root tip of *Allium cepa*. Am J Bot 45:365–372

Joffe AZ (1969) The mycoflora of groundnut rhizosphere soil and geocarposphere on light, medium, and heavy soils and its relation to *Aspergillus flavus*. Mycopathol Mycol Appl 37:150–160

Johnen BG (1978) Rhizosphere microorganisms and roots stained with europium chelate and fluorescent brightner. Soil Biol Biochem 10:495–502

Johnson LF, Arroyo T (1983) Germination of oospores of *Pythium ultimum* in the cotton rhizosphere. Phytopathology 73:1620–1624

Johnson LF, Curl EA (1972) Methods for research on the ecology of soil-borne plant pathogens. Burgess, Minneapolis, Minn

Johnson TW, Sparrow FK Jr (1961) Fungi in oceans and estuaries. Cramer, New York

Johnston HW, Miller RB (1959) The solubilization of "insoluble" phosphate. The reaction between organic acids and tricalcium phosphate. N Z J Sci 2:109–120

Johnston WB, Olsen RA (1972) Dissolution of fluorapatite by plant roots. Soil Sci 114:29–36

Jones FGW (1959) Ecological relationships of nematodes. In: Holton CS Fischer GW, Fulton RW, Hart H, McCallan SEA (eds) Plant pathology, problems and progress, Univ Wisc Press, Madison, Wisc, pp 395–411

Jones RW, Pettit RE, Taber RA (1984) Lignite and stillage: Carrier and substrate for application of fungal biocontrol agents to soil. Phytopathology 74:1167–1170

Jordan VWL, Sneh B, Eddy BP (1972) Influence of organic soil amendments on *Verticillium dahliae* and on the microbial composition of the strawberry rhizosphere. Ann Appl Biol 70:139–148

Joshi MM, Wilt GR, Cody RM, Chopra BK (1974) Detection of a *Vibrio* sp. by the bacteriophagous nematode *Pelodera chitwoodi*. J Appl Bact 37:419–426

Juniper BE, Roberts RM (1966) Polysaccharide synthesis and the fine structure of root cap cells. J R Microsc Soc 85:63–72

Juniper BE, Groves S, Landau-Schachar B, Audus LJ (1966) Root cap and the perception of gravity. Nature (London) 209:93–94

Juo Pei-Show, Stotzky G (1970) Electrophoretic separation of proteins from roots and root exudates. Can J Bot 48:713–718

Kalininskaya TA (1967) Methods of isolating and culturing nitrogen-fixing bacterial associations. Mikrobiologiya 36:436–438

Kampert M, Strzelczyk E, Pokojska A (1975) Production of gibberellin-like substances by bacteria and fungi isolated from the roots of pine seedlings (*Pinus silvestris* L). Acta Microbiol Pol Ser B 7:157–166

Kandasamy D, Prasad NN (1979) Colonization by rhizobia of the seed and roots of legumes in relation to exudation of phenolics. Soil Biol Biochem 11:73–75

Katan J, Eshel Y (1973) Interactions between herbicides and plant pathogens. Residue Rev 45:145–177

Katznelson H (1946) The rhizosphere effect of mangels on certain groups of micro-organisms. Soil Sci 62:343–354

Katznelson H (1965) Nature and importance of the rhizosphere. In: Baker KF, Snyder WC (eds) Ecology of soil-borne plant pathogens. Univ Calif Press, Berkeley, pp 187–209

Katznelson H, Henderson VE (1962) Studies on the relationships between nematodes and other soil microorganisms. I. The influence of actinomycetes and fungi on *Rhabditis (Cephaloboides) oxycerca* de Man. Can J Microbiol 8:875–882

Katznelson H, Rouatt JW (1957) Studies on the incidence of certain physiological groups of bacteria in the rhizosphere. Can J Microbiol 3:265–269

Katznelson H, Lochhead, AG, Timonin MI (1948) Soil microorganisms and the rhizosphere. Bot Rev 14:543–587

Katznelson H, Rouatt JW, Payne TMB (1955) The liberation of amino acids and reducing compounds by plant roots. Plant Soil 7:35–48

Katznelson H, Rouatt JW, Peterson EA (1962) The rhizosphere effect of mycorrhizal and non-mycorrhizal roots of yellow birch seedlings. Can J Bot 40:377–382

Kaufman DD, Kearney PC (1970) Microbial degradation of *s*-triazine herbicides. Residue Rev 32:235–265

Kaufman DD, Williams LE (1964) Effect of mineral fertilization and soil reaction on soil fungi. Phytopathology 54:134–139

Kemble AR, Macpherson HT (1954) Liberation of amino acids in perennial rye grass during wilting. Biochem J 58:46–49

Kerr A (1972) Biological control of crown gall: Seed inoculation. J Appl Bacteriol 35:493–497

Kerr A (1980) Biological control of crown gall through production of agrocin 84. Plant Dis 64:25–30

Kerry B (1980) Biocontrol: Fungal parasites of female cyst nematodes. J Nematol 12:253–259

Kerry B (1981 a) Progress in the use of biological agents for control of nematodes. In: Papavizas GC (ed) Biological control in crop production. Granada, London, pp 79–90

Kerry B (1981 b) Fungal parasites: A weapon against cyst nematodes. Plant Dis 65:390–393

Kevan DK McE (1965) The soil fauna – its nature and biology. In: Baker KF, Snyder WC (eds) Ecology of soil-borne plant pathogens. Univ Calif Press, Berkeley, pp 33–51

Khew KL, Zentmyer GA (1974) Electrotactic response of zoospores of seven species of *Phytophthora*. Phytopathology 64:500–507

King JE, Coley-Smith JR (1969) Production of volatile alkyl sulphides by microbial degradation of synthetic alliin and alliin-like compounds, in relation to germination of sclerotia of *Sclerotium cepivorum* Berk. Ann Appl Biol 64:303–314

Kleinschmidt GD, Gerdemann JW (1972) Stunting of citrus seedlings in fumigated nursery soils related to the absence of endomycorrhizae. Phytopathology 62:1447–1453

Klingler J (1961) Anziehungsversuch mit *Ditylenchus dipsaci* unter Berücksichtigung der Wirkung des Kohlendioxyds, des Redoxpotentiale und anderer Faktoren. Nematologica 6:69–84

Klingler J (1963) Die Orientierung von *Ditylenchus dipsaci* in gemessenen künstlichen und biologischen CO_2-Gradienten. Nematologica 9:185–199

Klingler J (1965) On the orientation of plant nematodes and of some other soil animals. Nematologica 11:4–18

Kloepper JW, Schroth MN, Miller TD (1980) Effects of rhizosphere colonization by plant growth-promoting rhizobacteria on potato plant development and yield. Phytopathology 70:1078–1082

Kloepper JW, Schroth MN (1981) Development of a powder formulation of rhizobacteria for inoculation of potato see d pieces. Phytopathology 71:590–592

Knudson L (1920) The secretion of invertase by plant roots. Am J Bot 7:371–379

Knudson L, Smith RS (1919) Secretion of amylase by plant roots. Bot Gaz 68:460–466

Ko WH, Hora FK (1971) A selective medium for the quantitative determination of *Rhizoctonia solani* in soil. Phytopathology 61:707–710

Koch L (1887) Die Entwicklungsgeschichte der Orobanchen, mit besonderer Berücksichtigung ihrer Beziehungen zu den Kulturpflanzen. Winters, Heidelberg

Kommedahl T, Brock TD (1954) Studies on the relationship of soil microflora to disease incidence. Phytopathology 44:57–61

Kommedahl T, Mew IC (1975) Biocontrol of corn root infection in the field by seed treatment with antagonists. Phytopathology 65:296–300

Kommedahl T, Windels CE (1981) Introduction of microbial antagonists to specific courts of infection: seeds, seedlings, and wounds. In: Papavizas GC (ed) Biological control in crop production. Granada, London, pp 227–248

Kormanik PP, McGraw AC (1982) Quantification of vesicular-arbuscular mycorrhizae in plant roots. In: Schenck NC (ed) Methods and principles of mycorrhizal research. Am Phytopathol Soc, St. Paul, Minn, pp 37–45

Koske RE (1982) Evidence for a volatile attractant from plant roots affecting germ tubes of a VA mycorrhizal fungus. Trans Br Mycol Soc 79:305–310

Kovacs MF Jr (1971) Identification of aliphatic and aromatic acids in root and seed exudates of peas, cotton, and barley. Plant Soil 34:441–451

Kraft JM (1977) The role of delphinidin and sugars in the resistance of pea seedlings to Fusarium root rot. Phytopathology 67:1057–1061

Kraft JM, Burke DW (1974) Behavior of *Fusarium solani* f. sp. *pisi* and *Fusarium solani* f. sp. *phaseoli* individually and in combinations on peas and beans. Plant Dis Rep 58:500–504

Kramer PJ (1983) Water relation of plants. Academic Press, London New York

Kramer PJ, Bullock HC (1966) Seasonal variations in the populations of suberized and unsuberized roots of trees in relation to the absorption of water. Am J Bot 53:200–204

Kranz J (1974) Epidemics of plant disease: Mathematical analysis and modeling. Springer, Berlin Heidelberg New York

Krasilnikov NA (1958) Mikroorganizmy Pochvy i Vysshie rasteniya (Soil microorganisms and higher plants). Acad Sci USSR, Moscow

Kreutzer WA (1963) Selective toxicity of chemicals to soil microorganisms. Annu Rev Phytopathol 1:101–126

Kreutzer WA (1972) *Fusarium* spp. as colonists and potential pathogens in root zones of grassland plants. Phytopathology 62:1066–1070

Kreutzer WA, Baker R (1975) Gnotobiotic assessment of plant health. In: Bruehl GW (ed) Biology and control of soil-borne plant pathogens. Am Phytopathol Soc, St. Paul, Minn, pp 11–21

Krigsvold DT, Griffin GJ, Hale MG (1982) Microsclerotial germination of *Cylindrocladium crotalariae* in the rhizosphere of susceptible and resistant peanut plants. Phytopathology 72:859–864

Kruckelmann HW (1975) Effects of fertilizers, soils, soil tillage, and plant species on the frequency of *Endogone* chlamydospores and mycorrhizal infection in arable soils. In: Sanders FE, Mosse B, Tinker PB (eds) Endomycorrhizas. Academic Press, London New York, pp 511–525

Krupa S, Fries N (1971) Studies on ectomycorrhizae of pine. I. Production of volatile organic compounds. Can J Bot 49:1425–1431

Krupa S, Nylund J (1972) Studies on ectomycorrhizae of pine. III. Growth inhibition of two root pathogenic fungi by volatile organic constituents of ectomycorrhizal root systems of *Pinus silvestris* L. Eur J For Pathol 2:88–94

Kühnelt W (1950) Bodenbiologie mit besonderer Berücksichtigung der Tierwelt. Herold, Vienna

Kush AK, Dadarwal KR (1981) Root exudates as pre-invasive factors in the nodulation of chick pea varieties. Soil Biol Biochem 13:51–55

Lamont BB, McComb AJ (1974) Soil microorganisms and the formation of proteoid roots. Aust J Bot 22:681–688

Lapierre H, Lemaire JM, Jouan B, Molin G (1970) Mise en évidence de particles virales associées à une perte de pathogénicité chez le Piétin-échaudage des céreales, *Ophiobolus graminis* Sacc. C R Acad Sci Paris Ser D 271:1833–1836

Larsen HJ, Boosalis MG, Kerr ED (1985) Temporary depression of *Rhizoctonia solani* field populations by soil amendment with *Laetisaria arvalis*. Plant Dis 69:347–350

Läuchli A (1976) Apoplastic transport in tissues. In: Lüttge U, Pitman MG (eds) Transport in plants, part B. Encyclopedia of plant physiology, new series, vol II. Springer, Berlin Heidelberg New York, pp 3–34

Lee JS, Mulkey TJ, Evans ML (1983) Reversible loss of gravitropic sensitivity in maize roots after tip application of calcium chelators. Science 220:1375–1376

Lees H, Quastel JH (1946) Biochemistry of nitrification in soil. I. Kinetics of, and the effects of poisons on, soil nitrification, as studied by a soil perfusion technique. Biochem J 40:803–815

Leiser AT (1968) A mucilaginous root sheath in Ericaceae. Am J Bot 55:391–398

Leonard KJ (1980) A reinterpretation of the mathematical analysis of rhizoplane and rhizosphere effects. Phytopathology 70:695–696

Leszczyñska D (1970) The effect of some herbicides on yeast from the rhizosphere of cultivated plants. Meded Rijksfac Landbouwwet Gent 35:637–645

Lewis JA, Papavizas GC (1970) Evolution of sulfur-containing volatiles from decomposition of crucifers in soil. Soil Biol Biochem 2:239–246

Lewis JA, Papavizas GC (1971) Effect of sulfur-containing volatile compounds and vapors from cabbage decomposition on *Aphanomyces euteiches*. Phytopathology 61:208–214

Lewis JA, Papavizas GC (1974) Selective mechanisms of bioenvironmental control of root-infecting fungi. In: Papavizas GC (ed) The relation of soil microorganisms to soilborne plant pathogens. South Coop Ser Bull 183. Virginia Polytech Inst and State Univ, Blacksburg, Va, pp 35–38

Lewis RW (1953) An outline of the balance hypothesis of parasitism. Am Nat 87:273–281

Linder PJ, Mitchell JW, Freeman GD (1964) Resistance and translocation of exogenous regulating compounds that exude from roots. J Agr Food Chem 12:437–438

Linderman RG, Gilbert RG (1975) Influence of volatiles of plant origin on soil-borne plant pathogens. In: Bruehl GW (ed) Biology and control of soil-borne plant pathogens. Am Phytopathol Soc, St Paul, Minn, pp 90–99

Linford MB (1939) Attractiveness of roots and excised shoot tissues to certain nematodes. Proc Helminthol Soc Washington 6:11–18

Lippincott JA, Lippincott BB (1984) Concepts and experimental approaches to host-microbe recognition. In: Kosuge T, Nester EW (eds) Plant-microbe interactions, molecular and genetic perspectives, vol I. Macmillan, New York, pp 195–214

Lochhead AG (1959) Rhizosphere microorganisms in relation to root-disease fungi. In: Holton CS, Fischer GW, Fulton RW, Hart H, McCallan SEA (eds) Plant pathology problems and progress. Univ Wisc Press, Madison, Wisc, pp 327–338

Lochhead AG, Chase FE (1943) Qualitative studies of soil microorganisms. V. Nutritional requirements of the predominant bacterial flora. Soil Sci 55:185–195

Lochhead AG, Rouatt JW (1955) The rhizosphere effect on the nutritional groups of soil bacteria. Soil Sci Soc Am Proc 19:48–49

Lockwood JL (1964) Soil fungistasis. Annu Rev Phytopathol 2:341–362

Lockwood JL (1968) The fungal environment of soil bacteria. In: Gray TRG, Parkinson D (eds) The ecology of soil bacteria. Univ Toronto Press, Toronto, pp 44–65

Lockwood JL (1977) Fungistasis in soils. Biol Rev 52:1–43

Lockwood JL (1981) Exploitation competition. In: Wicklow DT, Carroll GC (eds) The fungal community, its organization and role in the ecosystem. Mycol Ser, vol II. Marcel Dekker, New York, pp 319–349

Loutit MW, Hillas J, Spears GFS (1972) Studies on rhizosphere organisms and molybdenum concentration in plants. II. Comparison of isolates from the rhizospheres of plants grown in two soils under the same conditions. Soil Biol Biochem 4:267–270

Loutit MW, Loutit JS, Brooks RR (1967) Differences in molybdenum uptake by micro-organisms from the rhizosphere of *Raphanus sativus* L. grown in two soils of similar origin. Plant Soil 27:335–346

Louw HA (1970) A study of the phosphate-dissolving bacteria in the root region of wheat and lupin. Phytophylactica 2:21–26

Louw HA; Webley DM (1959 a) The bacteriology of the root region of the oat plant grown under controlled pot culture conditions. J Appl Bacteriol 22:216–226
Louw HA, Webley DM (1959 b) A study of soil bacteria dissolving certain mineral phosphate fertilizers and related compounds. J Appl Bacteriol 22:227–233
Lowry OH, Rosebrough NJ, Farr AL, Randall RJ (1951) Protein measurement with the Folin phenol reagent. J Biol Chem 193:265–275
Lucas GB, Sasser JN, Kelman A (1955) The relationship of root-knot nematodes to Granville wilt resistance in tobacco. Phytopathology 45:537–540
Lukezic FL, Bloom JR, Levine RG (1969) Influence of top removal on the carbohydrate levels of alfalfa crowns and roots grown in a gnotobiotic environment. Can J Plant Sci 49:189–195
Lumsden RD (1981) Ecology of mycoparasitism. In: Wicklow DT, Carroll GC (eds) The fungal community, its organization and role in the ecosystem. Mycol Ser, vol II. Marcel Dekker, New York, pp 295–318
Luzzati V, Husson F (1962) The structure of the liquid-crystalline phases of lipid-water systems. J Cell Biol 12:207–219
Lyda SD (1974) Studies on *Phymatotrichum omnivorum* and *Phymatotrichum* root rot. In: Papavizas GC (ed) The relation of soil microorganisms to soilborne plant diseases. South Coop Ser Bull 183. Virginia Polytech Inst and State Univ, Blacksburg, Va, pp 69–73
Lyda SD (1978) Ecology of *Phymatotrichum omnivorum*. Annu Rev Phytopathol 16:193–209
Lynch JM (1978) Microbial interactions around imbibed seeds. Ann Appl Biol 89:165–167
Lyon TL, Wilson JK (1921) Liberation of organic matter by roots of growing plants. Cornell Univ Agric Exp Stu Mem 40:7–44
Macfadyen A (1957) Animal ecology: Aims and methods. Pitman, London Toronto
Macrae IC (1975) Effect of applied nitrogen upon acetylene reduction in the rice rhizosphere. Soil Biol Biochem 7:337–338
Macura J (1968) Physiological studies of rhizosphere bacteria. In: Gray TRG, Parkinson D (eds) The ecology of soil bacteria. Univ Toronto Press, Toronto, pp 379–395
Macura J (1974) Trends and advances in soil microbiology from 1924 to 1974. Pochvovedeniye 9:123–134 (Engl transl)
Malajczuk N, Bowen GD (1974) Proteoid roots are microbially induced. Nature (London) 251:316–317
Mankau R (1980) Biological control of nematode pests by natural enemies. Annu Rev Phytopathol 18:415–440
Manorik V, Belima NI (Engl transl 1969) Method of studying root exudates and of calculating their amounts. Fiziol Rast 2:358–364
Marks GC, Foster RC (1973) Structure, morphogenesis, and ultrastructure of ectomycorrhizae. In: Marks GC, Kozlowski TT (eds) Ectomycorrhizae, their ecology and physiology. Academic Press, London New York, pp 1–41
Martens R (1982) Apparatus to study the quantitative relationships between root exudates and microbial populations in the rhizosphere. Soil Biol Biochem 14:315–317
Martin JK (1971 a) ^{14}C-labelled material leached from the rhizosphere of plants supplied with $^{14}CO_2$. Aust J Biol Sci 24:1131–1142
Martin JK (1971 b) Influence of plant species and plant age on the rhizosphere microflora. Aust J Biol Sci 24:1143–1150
Martin JK (1973) The influence of rhizosphere microflora on the availability of ^{32}P myoinositol hexaphosphate phosphorus to wheat. Soil Biol Biochem 5:473–483
Martin JK (1977) Factors influencing the loss of organic carbon from wheat roots. Soil Biol Biochem 9:1–7
Martin P (1958) Einfluß der Kulturfiltrate von Microorganismen auf die Abgabe von Skopoletin aus den Keimwurzeln des Hafers (*Avena sativa* L.). Arch Mikrobiol 29:154–186
Marx DH (1972) Ectomycorrhizae as biological deterrents to pathogenic root infections. Annu Rev Phytopathol 10:429–454
Marx DH (1973) Mycorrhizae and feeder root diseases. In: Marks GC, Kozlowski TT (eds) Ectomycorrhizae, their ecology and physiology. Academic Press, London New York, pp 351–382
Marx DH (1975) The role of ectomycorrhizae in the protection of pine from root infection by *Phytophthora cinnamomi*. In: Bruehl GW (ed) Biology and control of soil-borne plant pathogens. Am Phytopathol Soc, St Paul, Minn, pp 112–115

Marx DH, Kenney DS (1982) Production of ectomycorrhizal fungus inoculum. In: Schenck NC (ed) Methods and principles of mycorrhizal research. Am Phytopathol Soc, St Paul, Minn, pp 131–146

Mason WH, Marshall NL (1983) The human side of biology. Harper and Row, New York, p 13

McDougall BM (1970) Movement of ^{14}C-photosynthate into the roots of wheat seedlings and exudation of ^{14}C from intact roots. New Phytol 69:37–46

McDougall BM, Rovira AD (1970) Sites of exudation of ^{14}C-labelled compounds from wheat roots. New Phytol 69:999–1003

McKeen CD (1975) Colloquium on integration of pesticide-induced and biological destruction of soil-borne pathogens: summary and synthesis. In: Bruehl GW (ed) Biology and control of soil-borne plant pathogens. Am Phytopathol Soc, St Paul, Minn, pp 203–206

McLaren AD, Skujins J (1968) The physical environment of micro-organisms in soil. In: Gray TRG, Parkinson D (eds) The ecology of soil bacteria. Univ Toronto Press, Toronto, pp 3–24

Melin E, Krupa S (1971) Studies on ectomycorrhizae of pine. II. Growth inhibition of mycorrhizal fungi by volatile organic constituents of *Pinus silvestris* (Scots pine) roots. Physiol Plant 25:337–340

Menge JA, Timmer LW (1982) Procedures for inoculation of plants with vesicular-arbuscular mycorrhizae in the laboratory, greenhouse, and field. In: Schenck NC (ed) Methods and principles of mycorrhizal research. Am Phytopathol Soc, St Paul, Minn, pp 59–68

Menge JA, Johnson ELV, Minassian V (1979) Effect of heat treatment and three pesticides upon the growth and reproduction of the mycorrhizal fungus *Glomus fasciculatus*. New Phytol 82:473–480

Menon SK, Williams LE (1957) Effect of crop, crop residues, temperature and moisture on soil fungi. Phytopathology 47:559–564

Merriman PR, Price RD, Kollmorgen JF, Piggott T, Ridge EH (1974) Effect of seed inoculation with *Bacillus subtilis* and *Streptomyces griseus* on the growth of cereals and carrots. Aust J Agric Res 25:219–226

Metz H (1955) Untersuchungen über die Rhizosphäre. Arch Mikrobiol 23:297–326

Meyer FH (1973) Distribution of ectomycorrhizae in native and man-made forests. In: Marks GC, Kozlowski TT (eds) Ectomycorrhizae, their ecology and physiology. Academic Press, London New York, pp 79–105

Mian IH, Rodríguez-Kábana R (1982) Soil amendments with oil cakes and chicken litter for control of *Meloidogyne arenaria*. Nematropica 12:205–220

Mian IH, Godoy G, Shelby RA, Rodríguez-Kábana R, Morgan-Jones G (1982) Chitin amendments for control of *Meloidogyne arenaria* in infested soil. Nematropica 12:71–84

Micheli P (1723) Relazione dell' erba detta da' Botanici Orobanche. Firenze

Miki NK, Clarke KJ, McCully ME (1980) A histological and histochemical comparison of the mucilages on the root tips of several grasses. Can J Bot 58:2581–2593

Milburn JA (1979) Water flow in plants. Longman, London New York

Millard WA, Taylor CB (1927) Antagonism of microorganisms as the controlling factor in the inhibition of scab by green manuring. Ann Appl Biol 14:202–215

Miller DE, Burke DW (1985) Effects of low soil oxygen on *Fusarium* root rot of beans with respect to seedling age and soil temperature. Plant Dis 69:328–330

Minchin PEH, McNaughton GS (1984) Exudation of recently fixed carbon by non-sterile roots. J Exp Bot 35:74–82

Misaghi IJ, Stowell LJ, Grogan RG, Spearman LC (1982) Fungistatic activity of water-soluble fluorescent pigments of fluorescent pseudomonads. Phytopathology 72:33–36

Mishra RP, Das SN (1975) Attendant changes in the bacterial flora of rice rhizosphere as influenced by fertilizer treatment. Plant Soil 42:359–365

Mishra RR, Kanaujia RS (1972) Investigations into rhizosphere mycoflora. XIII. Effect of foliar application of certain plant extracts on *Pennisetum typhoides* F. Burm Stapf & Hubb. Isr J Agric Res 22:3–9

Mishustin EN, Naumova AN (1962) Bacterial fertilizers, their effectiveness and mode of action. Mikrobiologiya 31:543–555

Miskovic K, Rasovic B, Starcevic L, Milosevic N (1977) Effect of nitrogen on dehydrogenase activity and the number of actinomycetes and fungi in soil and rhizosphere of corn. Mikrobiologiya 14:105–116

Mitchell JE (1979) The dynamics of the inoculum potential of populations of soil-borne plant pathogens in the soil ecosystem. In: Schippers B, Gams W (eds) Soil-borne plant pathogens. Academic Press, London New York, pp 3–20

Moawad M (1979) Ecophysiology of vesicular-arbuscular mycorrhiza in the tropics. In: Harley JL, Russell RS (eds) The soil-root interface. Academic Press, London New York, pp 197–209

Moghimi A, Lewis DG, Oades JM (1978a) Release of phosphate from calcium phosphates by rhizosphere products. Soil Biol Biochem 10:277–281

Moghimi A, Tate ME, Oades JM (1978b) Characterization of rhizosphere products especially 2-ketogluconic acid. Soil Biol Biochem 10:283–287

Molina JAE, Rovira AD (1964) The influence of plant roots on autotrophic nitrifying bacteria. Can J Microbiol 10:249–257

Molisch H (1937) Der Einfluß einer Pflanze auf die andere – Allelopathie. Fischer, Jena

Molz FJ (1976) Water transport through plant tissue: the apoplasm and symplasm pathways. J Theoret Biol 59:277–292

Molz FJ, Ferrier JM (1982) Mathematical treatment of water movement in plant cells and tissues: a review. Plant Cell Environ 5:191–206

Molz FJ, Ikenberry E (1974) Water transport through plant cells and cell walls: theoretical development. Soil Sci Soc Am Proc 38:699–704

Moore LW (1981) Recent advances in the biological control of bacterial plant pathogens. In: Papavizas GC (ed) Biological control in crop production. Granada, London, pp 375–390

Moore LW, Warren G (1979) *Agrobacterium radiobacter* strain 84 and biological control of crown gall. Annu Rev Phytopathol 17:163–179

Morgan-Jones G, Godoy G, Rodríguez-Kábana R (1981) *Verticillium chlamydosporium*, fungal parasite of *Meloidogyne arenaria* females. Nematropica 11:115–119

Morgan-Jones G, White JF, Rodríguez-Kábana R (1984) Fungal parasites of *Meloidogyne incognita* in an Alabama soybean field soil. Nematropica 14:93–96

Morré DJ, Jones DD, Mollenhauer HH (1967) Golgi apparatus mediated polysaccharide secretion by outer root cap cells of *Zea mays*. 1. Kinetics and secretory pathway. Planta 74:286–301

Mosse B (1973) Advances in the study of vesicular-arbuscular mycorrhiza. Annu Rev Phytopathol 11:171–196

Moursi AA (1962) The attractiveness of CO_2 and N_2 to soil arthropods. Pedobiologia 1:299–302

Mulder EG (1975) Physiology and ecology of free-living nitrogen-fixing bacteria. In: Stewart WDP (ed) Nitrogen fixation by free-living microorganisms. Cambridge Univ Press, Cambridge, pp 3–28

Munns DN (1978) Legume-rhizobium relations: Soil acidity and nodulation. In: Andrew CS, Kamprath EJ (eds) Mineral nutrition of legumes in tropical and subtropical soils. Commonwealth Sci Ind Res Org, Melbourne, pp 247–263

Nakas JP, Klein DA (1980) Mineralization capacity of bacteria and fungi from the rhizosphere-rhizoplane of a semiarid grassland. Appl Environ Microbiol 39:113–117

Nance JF, Cunningham LW (1951) Evolution of acetaldehyde by excised wheat roots in solutions of nitrate and nitrite salts. Am J Bot 38:604–609

Nash SM, Wilhelm S (1960) Stimulation of broomrape seed germination. Phytopathology 50:772–774

Neal JL Jr, Larson RI, Atkinson TG (1973) Changes in rhizosphere populations of selected physiological groups of bacteria related to substitution of specific pairs of chromosomes in spring wheat. Plant Soil 39:209–212

Neilands JB (1977) Siderophores: Biochemical ecology and mechanism of iron transport in enterobacteria. In: Raymond KN (ed) Bioinorganic chemistry-II, Adv Chem Ser, 162. Am Chem Soc, Washington DC, pp 3–32

Nesheim ON, Linn MB (1969) Deleterious effect of certain fungitoxicants on the formation of mycorrhizae on corn by *Endogone fasciculata* and on corn root development. Phytopathology 59:297–300

Neubauer R, Avizohar-Hershenson Z (1973) Effect of the herbicide, trifluralin, on *Rhizoctonia* disease in cotton. Phytopathology 63:651–652

Newman EI (1973) Permeability to water of the roots of five herbaceous species. New Phytol 72:547–555

Newman EI (1974) Root and soil water relations. In: Carson EW (ed) The plant root and its environment. Univ Press Virginia, Charlottesville, pp 363–440

Newmann EI (1976) Water movement through root systems. Philos Trans R Soc London Ser B 273:463–478
Newman EI, Bowen HJ (1974) Patterns of distribution of bacteria on root surfaces. Soil Biol Biochem 6:205–209
Newman EI, Rovira AD (1975) Allelopathy among some British grassland species. J Ecol 63:722–737
Newman EI, Watson A (1977) Microbial abundance in the rhizosphere: A computer model. Plant Soil 48:17–56
Nicholas DJD (1965) Influence of the rhizosphere on the mineral nutrition of the plant. In: Baker KF, Snyder WC (eds) Ecology of soil-borne plant pathogens. Univ Calif Press, Berkeley, pp 210–217
Nicoljuk VF (1964) Počnennye protisty i ich biologičeskoe znacenie. Pedobiologia 3:259–273
Nobel PS (1983) Biophysical plant physiology and ecology. Freeman, San Francisco
Norman AG (1955) The effect of polymyxin on plant roots. Arch Biochem Biophys 58:461–477
Norman AG (1961) Microbial products affecting root development. Trans 7th Int Congr Soil Sci 2:531–536
Northcote DH, Pickett-Heaps JD (1966) A function of the golgi apparatus in polysaccharide synthesis and transport in the root cap cells of wheat. Biochem J 98:159–167
Norton DC (1978) Ecology of plant-parasitic nematodes. Wiley, New York
Nusbaum CJ, Ferris H (1973) The role of cropping systems in nematode population management. Annu Rev Phytopathol 11:423–440
Nutman PS (1965) The relation between nodule bacteria and the legume host in the rhizosphere and in the process of infection. In: Baker KF, Snyder WC (eds) Ecology of soil-borne plant pathogens. Univ Calif Press, Berkeley, pp 231–247
Nutman PS (1975) Rhizobium in the soil. In: Walker N (ed) Soil microbiology. Wiley, New York, pp 111–131
Nye PH (1981) Changes of pH across the rhizosphere induced by roots. Plant Soil 61:7–26
Nye PH, Tinker PB (1977) Solute movement in the soil-root system. Univ Calif Press, Berkeley
O'Brien DG, Prentice EG (1930) An eelworm disease of potatoes caused by *Heterodera schachtii*. Scott J Agric 13:415–432
Ocampo JA, Barea JM, Montoya E (1978) Bacteriostasis and the inoculation of phosphate-solubilizing bacteria in the rhizosphere. Soil Biol Biochem 10:439–440
Odvody GN, Boosalis MG, Kerr ED (1980) Biological control of *Rhizoctonia solani* with a soil-inhabiting basidiomycete. Phytopathology 70:655–658
Ogan MT (1979) Potential for nitrogen fixation in the rhizosphere and habitat of natural stands of the wild rice *Zizania aquatica*. Can J Bot 57:1285–1291
Old KM, Nicolson TH (1975) Electron microscopical studies of the micro flora of roots of sand dune grasses. New Phytol 74:51–58
Old KM, Patrick ZA (1979) Giant soil amoebae, potential biocontrol agents. In: Schippers B, Gams W (eds) Soil-borne plant pathogens. Academic Press, London New York, pp 617–628
Overgaard Nielsen C (1949) Studies on the soil microfauna. II. Soil inhabiting nematodes. Natura Jutlandica 2:1–131
Palti J (1981) Cultural practices and infectious crop diseases. Springer, Berlin Heidelberg New York
Panosyan AK (1962) Interrelations of *Azotobacter* and other soil microorganisms. Izv Akad Nauk Arm SSR Biol Nauki 15:12–24 (1962) (Transl Engl summary in: Soils Fertil 25:378)
Papavizas GC (1970) Colonization and growth of *Rhizoctonia solani* in soil. In: Parmeter JR Jr (ed) Biology and pathology of *Rhizoctonia solani*. Univ Calif Press, Berkeley, pp 108–122
Papavizas GC, Davey CB (1961) Extent and nature of the rhizosphere of *Lupinus*. Plant Soil 14:215–236
Papavizas GC, Kovacs MF (1972) Stimulation of spore germination of *Thielaviopsis basicola* by fatty acids from rhizosphere soil. Phytopathology 62:688–694
Papavizas GC, Lewis JA (1981) Introduction and augmentation of microbial antagonists for the control of soilborne plant pathogens. In: Papavizas GC (ed) Biological control in crop production. Granada, London pp 305–322
Papavizas GC, Lumsden RD (1980) Biological control of soilborne fungal pathogens. Annu Rev Phytopathol 18:389–413
Papavizas GC, Lewis JA, Adams PB (1968) Survival of root-infecting fungi in soil. II. Influence of amendment and soil carbon-to-nitrogen balance on *Fusarium* root rot of beans. Phytopathology 58:365–372

Park D (1963) The ecology of soil-borne fungal disease. Annu Rev Phytopathol 1:241–258
Parker CA, Grove PL (1970) *Bdellovibrio bacteriovorus* parasitising *Rhizobium* in Western Australia. J Appl Bacteriol 33:253–255
Parkinson D (1955) Liberation of amino acids by oat seedlings. Nature (London) 176:35–36
Parkinson D, Taylor GS, Pearson R (1963) Studies on fungi in the root region. I. The development of fungi on young roots. Plant Soil 19:332-349
Parkinson D, Gray TRG, Williams ST (1971) Methods for Studying the Ecology of Soil Micro-organisms. IBP Handbook No 19. Blackwell, Oxford Edinburgh
Patel JJ (1974) Antagonism of actinomycetes against rhizobia. Plant Soil 41:395–402
Patterson DT (1981) Effects of allelopathic chemicals on growth and physiological responses of soybean (*Glycine max*). Weed Sci 29:53–59
Paull RE, Jones RL (1975) Studies on the secretion of maize root cap slime. II. Localization of slime production. Plant Physiol 56:307–312
Pavlica DA, Hora TS, Bradshaw JJ, Skogerboe RK, Baker R (1978) Volatiles from soil influencing activities of soil fungi. Phytopathology 68:758–765
Paxton JD (1975) Phytoalexins, phenolics, and other antibiotics in roots resistant to soil-borne fungi. In: Bruehl GW (ed) Biology and control of soilborne plant pathogens. Am Phytopathol Soc, St Paul, Minn, pp 185–192
Peacock FC (1961) A note on the attractiveness of roots to plant parasitic nematodes. Nematologica 6:85–86
Pearson R, Parkinson D (1961) The sites of excretion of ninhydrin-positive substances by broad bean seedlings. Plant Soil 13:391–396
Perry RN, Hodges JA, Beane J (1981) Hatching of *Globodera rostochiensis* in response to potato root diffusate persisting in soil. Nematologica 27:348–352
Persidsky DJ, Wilde SA (1960) The effect of biocides on the survival of mycorrhizal fungi. J For 58:522–524
Peterson EA (1958) Observations on fungi associated with plant roots. Can J Microbiol 4:257–265
Peterson EA (1959) Seed-borne fungi in relation to colonization of roots. Can J Microbiol 5:579–582
Peterson EA (1961) Observations on the influence of plant illumination on the fungal flora of roots. Can J Microbiol 7:1–6
Peterson EA, Katznelson H (1965) Studies on the relationships between nematodes and other soil microorganisms. IV. Incidence of nematode-trapping fungi in the vicinity of plant roots. Can J Microbiol 11:491–495
Pilet PE (1971) Root cap and georeaction. Nature (London) 233:115–116
Pilet PE (1976) Effects of gravity on the growth inhibitors of geostimulated roots of *Zea mays* L. Planta 131:91–93
Pilet PE (ed) (1977) Plant growth regulation. Springer, Berlin Heidelberg New York
Pilet PE, Elliott MC (1981) Some aspects of the control of root growth and georeaction: The involvement of indoleacetic acid and abscisic acid. Plant Physiol 67:1047–1050
Pitcher RS (1967) The host-parasite relations and ecology of *Trichodorus viruliferus* on apple roots as observed from an underground laboratory. Nematologica 13:547–557
Pitts G, Allam AI, Hollis JP (1972) *Beggiatoa:* occurrence in the rice rhizosphere. Science 178:990–991
Plhak F, Urbankova V (1969) Study of the effect of volatile substances from cereal roots. Biol Plant 11:226–235
Polonenko DR, Mayfield CI (1979) A direct observation technique for studies on rhizoplane and rhizosphere colonization. Plant Soil 51:405–420
Polonenko DR, Dumbroff EB, Mayfield CI (1983) Microbial responses to salt-induced osmotic stress. III. Effects of stress on metabolites in the roots, shoots and rhizosphere of barley. Plant Soil 73:211–225
Powell CLl, Groters M, Metcalfe D (1980) Mycorrhizal inoculation of a barley crop in the field. N Z J Agric Res 23:107–109
Powell NT (1971) Interactions between nematodes and fungi in disease complexes. Annu Rev Phytopathol 9:253–274
Preece FF, Dickinson CH (eds) (1971) Ecology of leaf surface microorganisms. Academic Press, London New York
Přikryl Z, Vančura V (1980) Root exudates of plants. VI. Wheat root exudation as dependent on growth, concentration gradient of exudates and the presence of bacteria. Plant Soil 57:69–83

Pueppke SG (1984) Adsorption of bacteria to plant surfaces. In: Kosuge T, Nester EW (eds) Plant-microbe interactions: Molecular and genetic perspectives, vol I. Macmillian, New York, pp 215–261

Putnam AR, Duke WB (1978) Allelopathy in agroecosystems. Annu Rev Phytopathol 16:431–451

Ramachandra-Reddy TK (1959) Foliar spray of urea and rhizosphere microflora of rice (*Oryza sativa* L.) Phytopathol Z 36:286–289

Rambelli A (1973) The rhizosphere of mycorrhizae. In: Marks GC, Kozlowski TT (eds) Ectomycorrhizae, their ecology and physiology. Academic Press, London New York, pp 299–349

Ramirez C, Alexander M (1981) Increased bacterial colonization of the rhizosphere by controlling the soil protozoan in the *Rhizobium*-legume system. In: Vose PB, Ruschel AP (eds) Associative N_2 fixation, vol I. CRC Press, Boca Raton, Fla, pp 69–86

Raney WA (1965) Physical factors of the soil as they affect soil microorganisms. In: Baker KF, Snyder WC (eds) Ecology of soil-borne plant pathogens. Univ Calif Press, Berkeley, pp 115–119

RAO, AS (1962) Fungal populations in the rhizosphere of peanuts (*Arachis hypogaea* L.). Plant Soil 17:260–270

Reanney DC, Roberts WP, Kelly WJ (1982) Genetic interactions among microbial communities. In: Bull AT, Slater JH (eds) Microbial interactions and communities. Academic Press, London New York, pp 287–322

Redente EF, Reeves FB (1981) Interactions between vesicular-arbuscular mycorrhiza and *Rhizobium* and their effect on sweetvetch growth. Soil Sci. 132, 410–415

Reid CPP (1974) Assimilation, distribution, and root exudation of ^{14}C by ponderosa pine seedlings under induced water stress. Plant Physiol 54:44–49

Reid CPP, Hurtt W (1970) Root exudation of herbicides by woody plants: allelopathic implications. Nature (London) 225:291

Rempe EKh (1972) Significance of vitamins synthesized by the root microflora. Fiziol Rast 19:663–668

Reyes AA (1979) Populations of the spinach wilt pathogen, *Fusarium oxysporum* f. sp. *spinaciae* in the root tissues, rhizosphere, and soil in the field. Can J Microbiol. 25:227–229

Reyes VG, Schmidt EL (1981) Populations of *Rhizobium japonicum* associated with the surface of soil-grown roots. Plant Soil 61:71–80

Rice EL (1974) Allelopathy. Academic Press, London New York

Rice EL (1979) Allelopathy – an update. Bot Rev 45:15–109

Richter M, Wilms W, Scheffer F (1968) Determination of root exudates in a sterile continuous flow culture. I. The culture method. Plant Physiol 43:1741–1746

Ridge EH, Rovira AD (1971) Phosphatase activity of intact young wheat roots under sterile and non-sterile conditions. New Phytol 70:1017–1026

Riley D, Barber SA (1969) Bicarbonate accumulation and pH changes at the soybean *Glycine max* (L.) Merr. root–soil interface. Soil Sci Soc Am Proc 33:905–908

Rishbeth, J (1963) Stump protection against *Fomes annosus*. III. Inoculation with *Peniophora gigantea*. Ann Appl Biol 52:63–77

Rittenhouse RL, Hale MG (1971) Loss of organic compounds from roots. II. Effect of O_2 and CO_2 tension on release of sugars from peanut roots under axenic conditions. Plant Soil 35:311–321

Rivière J (1960) Étude de la rhizosphère du blé. Ann Agron 11:397–440

Rivière J (1963) Rhizosphère et croissance du blé. Ann Agron 14:619–653

Robards AW, Jackson SM, Clarkson DT, Sanderson J (1973) The structure of barley roots in relation to the transport of ions into the stele. Protoplasma 77:291–311

Robinson RA (1976) Plant pathosystems. Springer, Berlin Heidelberg New York

Rodríguez-Kábana R (1982) The effects of crop rotation and fertilization on soil xylanase activity in a soil of the southeastern United States. Plant Soil 64:237–247

Rodríguez-Kábana R, Curl EA (1980) Nontarget effects of pesticides on soilborne pathogens and disease. Annu Rev Phytopathol 18:311–332

Rodríguez-Kábana R, Truelove B (1970) The determination of soil catalase activity. Enzymologia 31:217–236

Rodríguez-Kábana R, Truelove B (1982) The effect of crop rotation and fertilization on catalase activity in a soil of the southeastern United States. Plant Soil 69:97–104

Rodríguez-Kábana R, Backman PA, Curl EA (1977) Control of seed and soilborne plant diseases. In: Siegal MR, Sisler HD (eds). Antifungal compounds, Vol I. Marcel Dekker, New York, pp 11–161

Rogers CH, Watkins GM (1938) Strand formation in *Phymatotrichum omnivorum*. Am J Bot 25:244–246
Rogers HT, Pearson RW, Pierre WH (1942) The source and phosphate acitvity of exoenzyme systems of corn and tomato roots. Soil Sci 54:353–366
Rohde RA (1960) The influence of carbon dioxide on respiration of certain plant-parasitic nematodes. Proc Helm Soc Wash 27:160–164
Rorem ES (1955) Uptake of rubidium and phosphate ions by polysaccharide producing bacteria. J Bacteriol 70:691–701
Rosen H (1957) A modified ninhydrin colorimetric analysis for amino acids. Arch Biochem 67:10–15
Ross JP, Daniels BA (1982) Hyperparasitism of endomycorrhizal fungi. In: Schenck NC (ed). Methods and principles of mycorrhizal research. Am Phytopathol Soc, St Paul, Minn, pp 55–58
Rouatt JW, Katznelson H (1960) Influence of light on bacterial flora of roots. Nature (London) 186:659–660
Rouatt JW, Katznelson H (1961) A study of the bacteria on the root surface and in the rhizosphere soil of crop plants. J Appl Bacteriol 24:164–171
Rouatt JW, Lechevalier M, Waksman SA (1951) Distribution of antagonistic properties among actinomycetes isolated from different soils. Antibiot Chemother 1:185–192
Rouatt JW, Katznelson H, Payne TMB (1960) Statistical evaluation of the rhizosphere effect. Soil Sci Soc Am Proc 24:271–273
Rouatt JW, Peterson EA, Katznelson H (1963) Microorganisms in the root zone in relation to temperature. Can J Microbiol 9:227–236
Rouse DI, Baker R (1978) Modeling and quantiative analysis of biological control mechanisms. Phytopathology 68:1297–1302
Rovira AD (1956) Plant root excretions in relation to the rhizosphere effect. I. The nature of root exudate from oats and peas. Plant soil 7:178–194
Rovira AD (1959) Root excretions in relation to the rhizosphere effect. IV. Influence of plant species, age of plant, light, temperature, and calcium nutrition on exudation. Plant Soil 11:53–64
Rovira AD (1960) Rhizobium numbers in the rhizospheres of red clover and paspalum in relation to soil treatment and the numbers of bacteria and fungi. Aust J Agric Res 12:77–83
Rovira AD (1965a) Plant root exudates and their influence upon soil microorganisms. In: Baker KF, Snyder WC (eds) Ecology of soil-borne plant pathogens. Univ Calif Press, Berkeley, pp 170–186
Rovira AD (1965b) Interactions between plant roots and soil microorganisms. Annu Rev Microbiol 19:241–266
Rovira AD (1969) Plant root exudates. Bot Rev 35:35–57
Rovira AD (1979) Biology of the soil-root interface. In: Harley JL, Russell RS (eds) The soil-root interface. Academic Press, London New York, pp 145–160
Rovira AD, Bowen GD (1966) The effects of microorganisms upon plant growth. II. Detoxication of heat-sterilized soils by fungi and bacteria. Plant Soil 25:129–142
Rovira AD, Davey CB (1974) Biology of the rhizosphere. In: Carson EW (ed) The plant root and its environment. Univ Virginia Press, Charlottesville, pp 153–204
Rovira AD, Harris JR (1961) Plant root excretions in relation to the rhizosphere effect. V. The exudation of B-group vitamins. Plant Soil 14:199–214
Rovira AD, McDougall BM (1967) Microbiological and biochemical aspects of the rhizosphere. In: McLaren AD, Peterson GH (eds) Soil Biochemistry. Marcel Dekker, New York, pp 417–463
Rovira AD, Ridge EH (1973) Exudation of ^{14}C-labelled compounds from wheat roots: Influence of nutrients, micro-organisms and added organic compounds. New Phytol 72:1081–1087
Rovira AD, Newman EI, Bowen HJ, Campbell R (1974) Quantitative assessment of the rhizoplane microflora by direct microscopy. Soil Biol Biochem 6:211–216
Rovira AD, Foster RC, Martin JK (1979) Note on terminology: Origin, nature and nomenclature of the organic materials in the rhizosphere. In: Harley JL, Russell RS (eds) The soil-root interface. Academic Press, London New York, pp 1–4
Rush CM, Upchurch DR, Gerik TJ (1984) In situ observations of *Phymatotrichum omnivorum* with a borescope mini-rhizotron system. Phytopathology 74:104–105
Russell EJ, Hutchinson HB (1909) The effect of partial sterilization of soil on the production of plant food. J Agric Sci 3:111–145
Russell EW (1973) Soil conditions and plant growth, 10th edn. Longmans, London
Russell RS (1977) Plant root systems: Their function and interaction with the soil. McGraw Hill, London

Samstevich SA (1968) Gel-like excretions of plant roots and their influence upon soil and rhizosphere microflora. In: Ghilarov MS, Kovda VA, Novichkova-Ivanova LN, Rodin LE, Sveshnikova VM (eds) Methods of productivity studies in root systems and rhizosphere organisms. USSR Acad Sci, Nauka, Leningrad, pp 200–204

Samtsevich, SA, Borisova VN (1961) Effect of fertilizers on root microflora of winter wheat. Mikrobiologiya 30:1033–1041

Sands DC, Rovira AD (1970) Isolation of fluorescent pseudomonads with selective medium. Appl. Microbiol 20:513–514

Scharen AL (1960) Germination of oospores of*Aphanomyces euteiches* embedded in plant tissue. Phytopathology 50:274–277

Schäufele WR, Winner C (1979) Effects of crop rotation on parasitic oomycete damage ot feeding roots of sugar beet. In: Schippers B, Gams W (eds) Soil-borne plant pathogens. Academic Press, London New York, pp 343–349

Schenck NC (1981) Can mycorrhizae control root disease? Plant Dis 65:231–234

Schenck NC (ed) (1982) Methods and principles of mycorrhizal research. Am Phytopathol Soc, St Paul, Minn

Schenck NC, Kellam MK (1978) The influence of vesicular-arbuscular mycorrhizae on disease development. Fla Agric Exp Stn Tech Bull 797:1–16

Schippers B, Gams W (eds) (1979) Soil-borne plant pathogens. Academic Press, London New York

Schlub RL, Schmitthenner AF (1978) Effects of soybean seed coat cracks on seed exudation and seedling quality in soil infested with *Pythium ultimum*. Phytopathology 68:1186–1191

Schmidt EL (1979) Initiation of plant root-microbe interactions. Annu Rev Microbiol 33:355–376

Schmidt HW, Schonherr J (1982) Fine structure of isolated and non-insolated potato tuber periderm. Planta 154:76–80

Schönbeck F (1979) Endomycorrhiza in relation to plant diseases. In: Schippers B, Gams W (eds) Soil-borne plant pathogens. Academic Press, London New York, pp 271–280

Schreiber LR, Green RJ Jr (1963) Effect of root exudates on germination of conidia and microsclerotia of *Verticillium albo-atrum* inhibited by the soil fungistatic principle. Phytopathology 53:260–264

Schroth MN, Cook RJ (1964) Seed exudation and its influence on pre-emergence damping-off of bean. Phytopathology 54:670–673

Schroth MN, Hancock JG (1982) Disease-suppressive soil and root-colonizing bacteria. Science 216:1376–1381

Schroth MN, Hendrix FF Jr (1962) Influence of nonsusceptible plants on the survival of *Fusarium solani* f. *phaseoli* in soil. Phytopathology 52:906–909

Schroth MN, Hildebrand DC (1964) Influence of plant exudates on root-infecting fungi. Annu Rev Phytopathol 2:101–132

Schroth MN, Snyder WC (1961) Effect of host exudates on chlamydospore germination of the bean root rot fungus, *Fusarium solani* f. *phaseoli*. Phytopathology 51:389–393

Schroth MN, Weinhold AR, Hayman DS (1966) The effect of temperature on quantitative differences in exudates from germinating seeds of bean, pea, and cotton. Can J Bot 44:1429–1432

Scott FM (1965) The anatomy of plant roots. In: Baker KF, Snyder WC (eds) Ecology of soil-borne plant pathogens. Univ Calif Press, Berkeley, pp 145–153

Shamiyeh NB, Johnson LF (1973) Effect of heptachlor on numbers of bacteria, actinomycetes and fungi in soil. Soil Biol Biochem 5:309–314

Shaw S, Wilkins MB (1973) The source and lateral transport of growth inhibitors in geotropically stimulated roots of *Zea mays* and *Pisum sativum*. Planta 109:11–26

Shay FJ, Hale MG (1973) Effect of low levels of calcium on exudation of sugars and sugar derivatives from intact peanut roots under axenic conditions. Plant Physiol 51:1061–1063

Shen-Miller J, McNitt RE, Wojciechowski M (1978) Regions of differential cell elongation and mitosis, and root meristem morphology in different tissues of geotropically stimulated maize root apices. Plant Physiol 61:7–12

Shepherd AM (1970) The influence of root exudates on the activity of some plant-parasitic nematodes. In: Toussoun TA, Bega RV, Nelson PE (eds) Root diseases and soil-borne pathogens. Univ Calif Press, Berkeley pp 134–137

Sherwood JE, Klein DA (1981) Antibiotic-resistant *Arthrobactor* sp. and *Pseudomonas* sp. responses in the rhizosphere of blue grama after herbage removal. Plant Soil 62:91–96

Sherwood RT (1970) Physiology of *Rhizoctonia solani*. In: Parmeter JR Jr (ed) Biology and pathology of *Rhizoctonia solani*. Univ Calif Press, Berkeley, pp 69–92

Shields LM (1982) Algae. In: Page AL, Miller RH, Keeney DR (eds) Methods of soil analysis (2nd ed). Part 2. Chemical and microbiological properties. Am Soc Agron, Madison, Wisc, pp 1093–1101

Siegle H (1961) Über Mischinfertionen mit *Ophiobolus graminis* und *Didymella exitialis*. Phytopathol Z 42:305–348

Simon EW (1974) Phospholipids and plant membrane permeability. New Phytol 73:377–420

Simon EW (1981) Leakage from seeds and roots in relation to membrane permeability. In: Biological and chemical interactions in the rhizosphere. Symp Proc Ecol Res Comm. Swed Nat Sci Res Counc, Stockholm, pp 73–87

Singh BN (1946) A method of estimating the numbers of soil protozoa, especially amoebae, based on their differential feeding on bacteria. Ann Appl Biol 33:112–119

Singh BN (1955) Culturing soil protozoa and estimating their numbers in soil. In: Kevan DK McE (ed) Soil zoology. Academic Press, London, pp 403–411

Slankis V (1973) Hormonal relationships in mycorrhizal development. In: Marks GC, Kozlowski TT (eds) Ectomycorrhizae, their ecology and physiology. Academic Press, London New York, pp 231–298

Slankis V (1974) Soil factors influencing formation of mycorrhizae. Annu Rev Phytopathol 12:437–457

Smiley RW (1974) Take-all of wheat as influenced by organic amendments and nitrogen fertilizer. Phytopathology 64:822–825

Smiley RW (1975) Forms of nitrogen and the pH in the root zone and their importance to root infections. In: Bruehl GW (ed) Biology and control of soil-borne plant pathogens. Am Phytopathol Soc, St Paul, Minn, pp 55–62

Smiley RW (1979) Wheat-rhizoplane pseudomonads as antagonists of *Gaeumannomyces graminis*. Soil Biol Biochem 11:371–376

Smiley RW, Cook RJ (1973) Relationship between take-all of wheat and rhizosphere pH in soils fertilized with ammonium vs. nitrate-nitrogen. Phytopathology 63:882–890

Smit AJ, Woldendorp JW (1981) Nitrate production in the rhizosphere of *Plantago* species. Plant Soil 61:43–52

Smith AM (1976) Ethylene in soil biology. Annu Rev Phytopathol 14:53–73

Smith MS, Tiedje JM (1979) The effect of roots on soil denitrification. Soil Sci Soc Am J 43:951–955

Smith WH (1969 a) Germination of *Macrophomina phaseoli* sclerotia as affected by *Pinus lambertiana* root exudate. Can J Microbiol 15:1387–1391

Smith WH (1969 b) Release of organic materials from the roots of tree seedlings. For Sci 15:138–143

Smith WH (1970) Root exudates of seedlings and mature sugar maple. Phytopathology 60:701–703

Snyder WC, Schroth MN, Christou T (1959) Effect of plant residues on root rot of bean. Phytopathology 49:755–756

Southey JF (ed) (1970) Laboratory methods for work with plant and soil nematodes. H M Stn Off, London

Spanswick RM (1976) Symplastic transport in tissues. In: Lüttge E, Pitman MG (eds) Transport in plants, part B. Encyclopedia of plant physiology, new series, vol II. Springer, Berlin Heidelberg New York, pp 35–53

Sperber JI (1958 a) The incidence of apatite-solubilizing organisms in the rhizosphere and soil. Aust J Agric Res 9:778–781

Sperber JI (1958 b) Solution of apatite by soil microorganisms producing organic acids. Aust J Agric Res 9:782–787

Sprent JI (1975) Adherence of sand particles to soybean roots under water stress. New Phytol 74:461–463

Srivastava VB (1971) Investigations into rhizosphere microflora. VIII. Light and dark treatments in relation to root-region microflora. Plant Soil 35:463–470

Stanghellini ME, Stowell LJ, Kronland WC, von Bretzel P (1983) Distribution of *Pythium aphanidermatum* in rhizosphere soil and factors affecting expression of the absolute inoculum potential. Phytopathology 73:1463–1466

Starkey RL (1929) Some influences of the development of higher plants upon the mciroorganisms in the soil. I. Historical and introductory. Soil Sci 27:319–334

Starkey RL (1958) Interrelations between microorganisms and plant roots in the rhizosphere. Bacteriol Rev 22:154–172

Stirling GR, (1984) Biological control of *Meloidogyne javancia* with *Bacillus penetrans*. Phytopathology 74:55–60

Stoker R, Weatherley PE (1971) The influence of the root system on the relationship between the rate of transpiration and depression of leaf water potential. New Phytol 70:547–554
Stolp H (1973) The bdellovibrios: bacterial parasites of bacteria. Annu Rev Phytopathol 11:53–76
Stolp H, Petzold H (1962) Untersuchungen über einen obligat parasitischen Mikroorganismus mit lytischer Aktivität für *Pseudomonas*-Bakterien. Phytopathol Z 45:364–390
Stotzky G, Schenck S (1976) Observations on organic volatiles from germinating seeds and seedlings. Am J Bot 63:798–805
Stotzky G, Culbreth GW, Mish LB (1962) Apparatus for growing plants with aseptic roots for collection of root exudates and CO_2. Plant Physiol 37:332–341
Stout JD, Bamforth SS, Lousier JD (1982) Protozoa. In: Page AL, Miller RH, Keeney DR (eds) Methods of soil analysis, 2nd ed. Part 2. Chemical and microbiological properties. Am Soc Agron, Madison, Wisc, pp 1103–1120
Stover RH (1959) Growth and survival of root-disease fungi in soil. In: Holton CS, Fischer GW, Fulton RW, Hart H, McCallan SEA (eds) Plant pathology – problems and progress. Univ Wisc Press, Madison, Wisc, pp 339–355
Stover RH, Waite BH (1953)An improved method of isolating *Fusarium* spp. from plant tissue. Phytopathology 43:700–701
Strzelczyk E (1961) Studies on the interaction of plants and free-living nitrogen-fixing microorganisms. II. Development of antagonists of *Azotobacter* in the rhizosphere of plants at different stages of growth in two soils. Can J Microbiol 7:507–513
Subba-Rao NS, Bidwell RGS, Bailey DL (1961) The effect of rhizosphere fungi on the uptake and metabolism of nutrients by tomato plants. Can J Bot 39:1759–1764
Sulochana CB (1962a) Amino acids in root exudates of cotton. Plant Soil 16, 312–326
Sulochana, CB (1962b) B-vitamins in root exudates of cotton. Plant Soil 16, 327–334
Suslow TV, Schroth MN (1982a) Role of deleterious rhizobacteria as minor pathogens in reducing crop growth. Phytopathology 72:111–115
Suslow TV, Schroth MN (1982b) Rhizobacteria of sugar beets: Effects of seed application and root colonization on yield. Phytopathology 72:199–206
Sutcliffe J (1968) Plants and water. Inst Biol Stud Biol No 14. Arnold, London, p 23
Suzuki T, Kondo N, Fujii T (1979) Distribution of growth regulators in relation to the light-induced geotropic responsiveness in *Zea* roots. Planta 145:323–329
Tan EL, Loutit MW (1976) Concentration of molybdenum by extracellular material produced by rhizosphere bacteria. Soil Biol Biochem 8:461–464
Tan KH, Nopamornbodi O (1981) Electron microbeam analysis and scanning electron microscopy of soil-root interfaces. Soil Sci 131:100–106
Tang CS, Takenaka T (1983) Quantitation of a bioactive metabolite in the undisturbed rhizosphere – benzyl isothiocyanate from *Carica papaya* L. J Chem Ecol 9:1247–1253
Tang CS, Young CC (1982) Collection and identification of allelopathic compounds from the undisturbed root system of bigalta limpograss (*Hemarthria altissima*) Plant Physiol 69:155–160
Thirumalachar MJ, O'Brien MJ (1977) Suppression of charcoal rot in potato with a bacterial antagonist. Plant Dis Rep 61:543–546
Thom C, Humfeld H (1932) Notes on the association of microorganisms and roots. Soil Sci 34:29–36
Thompson LK, Hale MG (1983) Effects of kinetin in the rooting medium on root exudation of free fatty acids and sterols from roots of *Arachis hypogaea* L. "Argentine" under axenic conditions. Soil Biol Biochem 15:125–126
Thrower LB (1954) The rhizosphere effect shown by some Victorian heathland plants. Aust J Bot 2:246–267
Timonin MI (1940) The interactions of higher plants and soil microorganisms. III. Effect of by-products of plant growth on activity of fungi and actinomycetes. Soil Sci 52:395–413
Timonin MI (1946) Microflora of the rhizosphere in relation to the manganese deficiency disease of oats. Soil Sci Soc Am Proc 11:284–292
Tolmsoff WJ (1970) Metabolism of *Rhizoctonia solani*. In: Parmeter JR Jr (ed) Biology and pathology of *Rhizoctonia solani*. Univ Calif Press, Berkeley, pp 93–109
Toussoun TA, Patrick ZA (1963) Effect of phytotoxic substances from decomposing plant residues on root rot of bean. Phytopathology 53:265–270
Tribe HT (1957) On the parasitism of *Sclerotinia trifoliorum* by *Coniothyrium minitans*. Trans Br Mycol Soc 40:489–499
Tribe HT (1977) Pathology of cyst nematodes. Biol Rev 52:477–507

Tribe HT (1979) Extent of disease in populations of *Heterodera* with special reference to *H. schachtii*. Ann Appl Biol 92:61–72

Trolldenier G, Marckwordt U (1962) Untersuchungen über den Einfluß der Bodenmikroorganismen auf die Rubidium und Calciumaufnahme in Nährlösung wachsender Pflanzen. Arch Mikrobiol 43:148–151

Truelove B, Rodríguez-Kábana R, King PS (1977) Seed treatment as a means of preventing nematode damage to crop plants. J Nematol 9:326–330

Tsao PH (1970) Selective media for isolation of pathogenic fungi. Annu Rev Phytopathol 8:157–186

Tsutsumi M (1976) Conditions for collecting the potato root-diffusate and the influence on the hatching of the potato cyst nematode. Jpn J Nematol 6:10–13

Tuite J (1969) Plant pathologcial methods. Burgess, Minneapolis, Minn

Tukey HB Jr (1969) Implications of allelopathy in agricultural plant science. Bot Rev 35:1–16

Tullgren A (1917) En enkel apparat för automatiskt vittjande av sallgods. Entomol Tidskr 38:97–100

Turner DR, Chapman RA (1972) Infection of seedlings of alfalfa and red clover by concomitant populations of *Meloidogyne incognita* and *Pratylenchus penetrans*. J Nematol 4:280–286

Tzean SS, Estey RH (1978) Nematode-trapping fungi as mycopathogens. Phytopthology 68:1266–1270

Vančura V (1964) Root exudates of plants. I. Analysis of root exudates of barley and wheat in their initial phases of growth. Plant Soil 21:231–248

Vančura V (1967) Root exudates of plants III. Effect of temperature and "cold shock" on the exudation of various compounds from seeds and seedlings of maize and cucumber. Plant Soil 27:319–328

Vančura V (1980) Fluorescent pseudomonads in the rhizosphere of plants and their relation to root exudates. Folia Microbiol 25:168–173

Vančura V, Hanzlíková A (1972) Root exudates of plants. IV. Differences in chemical composition of seed and seedlings exudates. Plant Soil 36:271–282

Vančura V, Hovadík A (1965) Root exudates in plants. II. Composition of root exudates of some vegetables. Plant Soil 22:21–32

Vančura V, Staněk M (1975) Root exudates of plants. V. Kinetics of exudates from bean roots as related to the presence of reserve compounds in cotyledons. Plant Soil 43:547–559

Vančura V, Stotzky G (1976) Gaseous and volatile exudates from germinating seeds and seedlings. Can J Bot 54:518–532

Van de Bund CF (1972) Enkele waarnemingen aan *Lasioseius fimetorum* Karg. 1971 in een gezelschap van mijten, springstaarten en nematoden tussen de wortels van witte klaver. Entomol Berl 32:6–12

Vanderplank JE (1978) Genetic and molecular basis of plant pathogenesis. Springer, Berlin Heidelberg New York

Van Gundy SD (1982) Nematodes in: Page AL, Miller RH, Keeney DR (eds). Methods of soil analysis 2nd ed. Part 2. Chemical and microbiological properties. Am Soc Agron, Madison, Wisc, pp 1121–1130

Van Vuurde JWL, De Lange A (1978) The rhizosphere microflora of wheat grown under controlled conditions. II. Influence of the stage of growth of the plant, soil fertility and leaf treatment with urea on the rhizosphere soil microflora. Plant Soil 50:461–472

Van Vuurde JWL, Elenbaas PFM (1978) Use of fluorochromes for direct observation of microorganisms associated with wheat roots. Can J Microbiol 24:1272–1275

Van Vuurde JWL, Kruyswÿk CJ, Schippers B (1979) Bacterial colonization of wheat roots in root–soil model system. In: Schippers B, Gams W (eds) Soil-borne plant pathogens. Academic Press, London New York, pp 229–234

Venkata-Ram CS (1960) Foliar application of nutrients and rhizosphere microflora of *Camellia sinensis*. Nature (London) 187:621–622

Vermeer J, McCully ME (1982) The rhizosphere in *Zea*: new insights into its structure and development. Planta 156:45–61

Vidaver AK (1983) Bacteriocins: The lure and the reality. Plant Dis 67:471–474

Viglierchio DR (1961) Attraction of parasitic nematodes by plant root emanations. Phytopathology 51:136–142

Vincent JM (1965) Environmental factors in the fixation of nitrogen by the legume. In: Bartholomew WV, Clark FE (eds) Soil nitrogen. Am Soc Agron, Madison, Wisc, pp 384–435

Virtanen AI, Laine T, Hausen SV (1936) Excretion of amino acids from the root nodules of leguminous plants. Nature (London) 137:277

Vose PB, Ruschel AP (1979) Associative N_2 fixation, vols I, II. CRC Press, Boca Raton, Fla
Vrany J, Vančura V, Macura J (1962) The effects of foliar applications of some readily metabolized substances, growth regulators and antibiotics on rhizosphere microflora. Folia Microbiol 7:61–70
Waidyanatha UP De S, Yogaratnam N, Ariyaratne WA (1979) Mycorrhizal infection on growth and nitrogen fixation of *Pueraria* and *Stylosanthes* and uptake of phosphorus from two rock phosphates. New Phytol 82:147–152
Waksman SA (1947) Microbial antagonists and antibiotic substances. Commonwealth Fund, New York
Waksman SA (1959) The actinomycetes, vol I. Nature, occurrence, and acitivities. Williams & Wilkins, Baltimore
Wallace DH, Wilkinson RE (1975) Breeding for resistance in dicotyledonous plants to root rot fungi. In: Bruehl GW (ed) Biology and control of soil-borne plant pathogens. Am Phytopathol Soc, St Paul, Minn, pp 177–184
Wallace HR (1968) The dynamics of nematode movement. Annu Rev Phytopathol 6:91–114
Wallace HR (1978) Dispersal in time and space: Soil pathogens. In: Horsfall JG, Cowling EB (eds) Plant disease. An advanced treatise, vol II. Academic Press, London New York, pp 181–202
Wallwork JA (1970) Ecology of soil animals. Mc Graw-Hill, New York
Wallwork JA (1976) The distribution and diversity of soil fauna. Academic Press, London New York
Wang ELH, Bergeson GB (1974) Biochemical changes in root exudate and xylem sap of tomato plants infected with *Meloidogyne incognita*. J Nematol 6:194–202
Warcup JH (1950) The soil-plate method for isolation of fungi from soil. Nature (London) 166:117–118
Wareing PF, Phillips IDJ (1981) Growth and differentiation in plants, 3rd edn. Pergamon Press, Oxford, pp 194–198
Warnock AJ, Fitter AH, Usher MB (1982) The influence of a springtail *Folsomia candida* (Insecta, Collembola) on the mycorrhizal association of leek *Allium porrum* and the vesicular-arbuscular mycorrhizal endophyte *Glomus fasciculatus*. New Phytol 90:285–292
Watson AG, Ford EJ (1972) Soil fungistasis – a reapprasial. Annu Rev Phytopathol 10:327–348
Weatherley PE (1970) Some aspects of water relations. Adv Bot Res 3:171–206
Weinhold AR (1970) Significance of populations of major plant pathogens in soil: bacteria including Streptomycetes. In: Toussoun TA, Bega RV, Nelson PE (eds) Root diseases and soil-borne pathogens. Univ Calif Press, Berkeley, pp 22–24
Weinhold AR, Bowman T, Dodman RL (1969) Virulence of *Rhizoctonia solani* as affected by nutrition of the pathogen. Phytopathology 59:1601–1605
Wells HD, Bell DK, Jaworski CA (1972) Efficacy of *Trichoderma harzianum* as a biocontrol for *Sclerotium rolfsii*. Phytopathology 62:442–447
Went FW (1957) The experimental control of plant growth. Chronica Botanica, Waltham, Mass
Went FW, Thimann KV (1937) Phytohormones. Macmillan, New York
West PM (1939) Excretion of thiamin and biotin by the roots of higher plants. Nature (London) 144:1050–1051
Whalley WM, Taylor GS (1973) Influence of pea-root exudates on germination of conidia and chlamydospores of physiologic races of *Fusarium oxysporum* f. *pisi*. Ann Appl Biol 73:269–276
Wheeler H (1975) Plant pathogenesis. Springer, Berlin Heidelberg New York
White PR (1938) "Root-pressure" – an unappreciated force in sap movement. Am J Bot 25:223–227
Whiteman PC, Koller D (1964) Saturation deficit of the mesophyll evaporating surfaces in a desert halophyte. Science 146:1320–1321
Wiersum LK (1961) Utilization of soil by the plant root system. Plant Soil 15:189–192
Wieser W (1955) The attractiveness of plants to larvae of root knot nematodes. I. The effect of tomato seedlings and excised roots on *Meloidogyne hapla* Chitwood. Proc Helminthol Soc Wash DC 22:106–112
Wieser W (1956) The attractiveness of plants to larvae of root knot nematodes. II. The effect of excised bean, eggplant and soybean roots on *Meloidogyne hapla* Chitwood. Proc Helminthol Soc Wash DC 23:59–64
Wiggins EA, Curl EA (1979) Interactions of Collembola and microflora of cotton rhizosphere. Phytopathology 69:244–249
Wiggins EA, Curl EA, Harper JD (1979) Effects of soil fertility and cotton rhizosphere on populations of Collembola. Pedobiologia 19:75–82

Wilkins MB (1979) Growth control mechanisms in gravitropism. In: Haupt W, Feinleib ME (eds) Encyclopedia of plant physiology, new series vol VII. Physiology of movement. Springer, Berlin Heidelberg New York, pp 601–626

Wilkins MB (1984) Gravitropism. In: Wilkins MB (ed) Advanced plant physiology, Pitman, London, pp 163–185

Williams ST, Mayfield CI (1971) Studies on the ecology of actinomycetes in soil. III. The behavior of neutrophilic streptomycetes in acid soil. Soil Biol Biochem 3:197–208

Williamson FA, Wyn Jones RG (1973) The influence of soil microorganisms on growth of cereal seedlings and on potassium uptake. Soil Biol Biochem 5:569–575

Windels CE (1981) Growth of *Penicillium oxalicum* as a biological seed treatment on pea seed in soil. Phytopathology 71:929–933

Windels CE, Kommedahl T (1982) Pea cultivar effect on seed treatment with *Penicillium oxalicum* in the field. Phytopathology 72:541–543

Winogradsky S (1893) Sur L'assimilation de l'azote gazeux de l'atmosphère par les microbes, C R Acad Sci 116:1385–1388

Woldendorp JW (1963) The influence of living plants on denitrification. Meded Landbouwhogesch Wageningen 63:1–100

Woltz SS (1978) Nonparasitic plant pathogens, Annu Rev Phytopathol 16:403–430

Woolley TA (1982) Mites and other microarthropods. In: Page Al, Miller RH, Keeney DR (eds) Methods of soil analysis, 2nd ed part 2. Chemical and microbiological properties. Am Soc Agron, Madison, Wisc pp 1131–1142

Wright K, Northcote DH (1974) The relationship of root cap slimes to pectin. Biochem J 139:525–534

Wullstein LH, Pratt SA (1981) Scanning electron microscopy of rhizosheaths of *Oryzopsis hymenoides*. Am J Bot 68:408–419

Wullstein LH, Bruening ML, Bollen WB (1979) Nitrogen fixation associated with sand grain root sheaths (rhizosheaths) of certain xeric grasses. Physiol Plant 46:1–4

Wyllie TD (1962) Effect of metabolic by-products of *Rhizoctonia solani* on the roots of Chippewa soybean seedlings. Phytopathology 52:202–206

Yeates GW (1981) Nematode populations in relation to soil environmental factors: a review. Pedobiologia 22:312–338

Yemm EW, Cocking EC (1955) The determination of amino acids with ninhydrin. Analyst 80:209–213

Youssef YA, Mankarios AT (1975) Production of plant growth substances by rhizosphere mycoflora of broad bean and cotton. Biol Plant (Praha) 17:175–181

Zak B (1964) Role of mycorrhizzae in root disease. Annu Rev Phytopathol 2:377–392

Zentmyer GA (1961) Chemotaxis of zoospores for root exudates. Science 133:1595–1596

Zentmyer GA (1970) Tactic responses of zoospores of *Phytophthora*. In: Toussoun TA, Bega RV, Nelson PE (eds) Root diseases and soil-borne pathogens. Univ Calif Press, Berkeley, pp 109–111

Zentmyer GA (1980) *Phytophthora cinnamomi* and the diseases it causes. Monogr No 10, Am Phytopathol Soc, St Paul, Minn

Zentmyer GA, Erwin DC (1970) Development and reproduction of *Phytophthora*. Phytopathology 60:1120–1127

Zimmerman MH, Brown CL (1971) Trees: Structure and function. Springer, Berlin Heidelberg New York

Zuckerman BM, Jansson, Has-Börje (1984) Nematode chemotaxis and possible mechanisms of host prey recognition. Annu Rev Phytopathol 22:95–113

Zwarun AA (1972) Microbial competition for glucose in excised root experiments. Soil Sci Soc Am Proc 36:968–969

Subject Index

Volume 9

J. Palti

Cultural Practices and Infectious Crop Diseases

1981. 43 figures. XVI, 243 pages
ISBN 3-540-11047-X

Cultural Practices and Infectious Crop Diseases is the first monograph to deal comprehensively with the effects of agricultural practices on plant health. Following the effects of background factors (climate, soil, stress, and crop age), the author describes the impact of the major farming operations on the development of crop diseases, among them crop planning and alternation, multiple cropping, tillage, fertilitzation, moisture management, sowing, harvesting, and sanitation. The author concludes with a consideration of integrated disease control combining agricultural practices, resistance breeding and the use of chemical agents.
This volume will prove an invaluable aid to agriultural advisors, growers, researchers, students and teachers in appreciating the importance of approbriate practices management in the prevention of crop disease.

Volume 10

E. Bresler, B.L. McNeal, D.L. Carter

Saline and Sodic Soils

Principles - Dynamcis - Modeling

1982. 78 figures. X, 236 pages. ISBN 3-540-11120-4

"This book is an excellent compilation of the literature related to the management of saline and sodic soils; it also presents an in-depth mathematical analysis of the transportation and distribution of salts. For these subjects it is a valuable reference book for advanced courses in soil physics and chemistry and for research soil scientists. ..."
Soil Science

Volume 11

J.R. Parks

A Theory of Feeding and Growth of Animals

1982. 123 figures. XVI, 322 pages. ISBN 3-540-11122-0

The subject of this volume is the search for the deterministic elements in animal feeding and growth patterns that could form the basis of a testable theory, making this book a valuable guide to laying a scientific foundation for any undertaking in animal management and production technology.

Volume 12

J. Hagin, B. Tucker

Fertilization of Dryland and Irrigated Soils

1982. 64 figures. VII, 188 pages. ISBN 3-540-11121-2

This book is a treatment of the proper and judicious use of fertilizers with emphasis for semi-arid regions of the world. Enough background information is given on soil chemical processes and plant nutrition to provide the reader with an understanding of the rates of plant food needed and allows for the proper selection of fertilizer materials and methods of applications. The book is most complete on those subjects important to low or erratic rainfall areas. Conciseness has, however, been achieved by eliminating soil fertility considerations which are not pertinent to arid and semi-arid regions.

Volume 13

A. J. Koolen, H. Kuipers

Agricultural Soil Mechanics

1983. 128 figures. XI, 241 pages
ISBN 3-540-12257-5

Bringing together a vast array of information from all over the world on the effects of tilling and field traffic on soil, this unique work is the first systematized single-volume presentation of agricultural soil mechanics. The book opens with a discussion of soil mechanical properties: compactibility, deformability, resistance to breakage, and soil-material frictional properties. The second part covers the load bearing processes which may be induced by rollers and wheels, cones and plates, sleds and shear elements, tracks and cages, and the soil loosening processes caused by tines, blades and moldboards. The presentation is further broken down into examples of the occurrence of the various types of soil mechanical behavior, the physical principles underlying this behavior, and the practical and analytical use of such knowledge.
Profusely illustrated and complemented by carefully selected, up-to-date references, **Agricultural Soil Mechanics** is a source of vital information for applications ranging from the optimal design and choice of machinery and tire sizes to the continuing evaluation of agricultural work on production systems and soil conditions.

Volume 14

Energy and Agriculture

Editor: G. Stanhill

1984. 55 figures. XIII, 192 pages. ISBN 3-540-13476-X

Contents: Introduction to the Role of Energy in Agriculture. - Principles and Processes: Economic Impacts of Energy Prices on Agriculture. Energy Analysis of the Environmental Role in Agriculture. Genetic Engineering to Modify Energy Flow in Agriculture. - Energy Sources for Agriculture: Energy in Different Agricultural Systems: Renewable and Nonrenewable Sources. Agricultural Labour: From Energy Source to Sink. - Case Studies: Energy Use in the Food-Producing Sector of the European Economic Community. - Energy in Australian Agriculture: Inputs, Outputs and Policies. - Energy Use and Management in US Agriculture. - Subject Index.

Springer-Verlag
Berlin
Heidelberg
New York
Tokyo